Advances in Digital Image Correlation (DIC)

Advances in Digital Image Correlation (DIC)

Special Issue Editors

Jean-Noël Périé
Jean-Charles Passieux

MDPI • Basel • Beijing • Wuhan • Barcelona • Belgrade

Special Issue Editors

Jean-Noël Périé
Institut Clément Ader (ICA),
IUT GMP Toulouse
(Université de Toulouse)
France

Jean-Charles Passieux
Institut Clément Ader (ICA),
INSA Toulouse
(Université de Toulouse)
France

Editorial Office
MDPI
St. Alban-Anlage 66
4052 Basel, Switzerland

This is a reprint of articles from the Special Issue published online in the open access journal *Applied Sciences* (ISSN 2076-3417) from 2018 to 2020 (available at: https://www.mdpi.com/journal/applsci/special_issues/Advances_in_Digital_Image_Correlation).

For citation purposes, cite each article independently as indicated on the article page online and as indicated below:

LastName, A.A.; LastName, B.B.; LastName, C.C. Article Title. *Journal Name* **Year**, *Article Number*, Page Range.

ISBN 978-3-03928-514-3 (Pbk)
ISBN 978-3-03928-515-0 (PDF)

Contents

About the Special Issue Editors

Jean-Noël Périé Interests: identification of constitutive parameters; full field measurements; digital image correlation; experimental mechanics; composite materials.

Jean-Charles Passieux Interests: numerical methods in computational and experimental mechanics; global digital image correlation; identification of mechanical properties; high performance computing.

Editorial

Special Issue on Advances in Digital Image Correlation (DIC)

Jean-Noël Périé * and Jean-Charles Passieux *

Institut Clément Ader (ICA), Université de Toulouse, CNRS-INSA-UPS-Mines Albi-ISAE, 31400 Toulouse, France
* Correspondence: jean-noel.perie@iut-tlse3.fr (J.-N.P.) and passieux@insa-toulouse.fr (J.-C.P.)

Received: 20 January 2020; Accepted: 31 January 2020; Published: 24 February 2020

1. Introduction

Digital Image Correlation (DIC) has become the most popular full field measurement technique in experimental mechanics. It is a versatile and inexpensive measurement method that provides a large amount of experimental data. Because it can take advantage of a huge variety of image modalities, the technique allows covering a wide range of space and time scales. Stereo extends the scope of DIC to non-planar cases, which are more representative of industrial use cases. With the development of tomography, Digital Volume Correlation now gives access to volumetric data. It makes it possible to study the inner behavior of materials and structures.

However, the use of DIC data to quantitatively validate models or accurately identify a set of constitutive parameters is not yet straightforward. One of the reasons lies in the tricky compromises between measurement resolution and spatial resolution. Second, the question of the boundary conditions is still an open question. Another reason is that the measured displacements are not directly comparable with usual simulations. Finally, the use of full field data leads to new computational challenges.

2. Advances in DIC

In reviewing the 16 articles published in this special issue, it is interesting to see that they cover some of the current challenges and relevant topics facing the international Digital Image Correlation community. Applications of DIC to various scales of space and time or for the inspection of mechanical phenomena involving different types of materials (composite, metals, earth, biological tissues, etc.), in possibly complex environments. The question of large strains is also addressed. The papers address full-field measurements, their use for validation of mechanical models and for the identification of delicate mechanical properties. The coupling of DIC with other techniques is also an burning issue discussed in the special issue. Concerning the DIC variants, 2D DIC, stereo DIC and 3D Digital Volume Correlation (DVC) are also covered. Finally, the collection of articles also addresses algorithmic issues and questions related to efficient implementation.

More precisely, with regards to applications, in [1], transient kinematic measurements are performed with DIC for the inspection of the in-situ manufacturing of thermoplastic composite materials. The response of copper plates subjected to impulsive loading in complex fluid-structure environment, studied in [2] using high-speed stereo-DIC, illustrates the wide range of time scales that can be addressed by DIC. Along the same line, high-speed camera based DIC was used in [3] to observe ruptures during stick-slip motions of a simulated earthquakes. Still in the field of geomechanics, paper [4] measured earth surface dynamics and investigated the issue of application of DIC under severe environmental and lighting conditions and at very large space scales. The question, addressed in [5], of the thermal environment is also central for the (thermo-)mechanical analysis of materials, and it raises a whole set of experimental problems (texture, acquisition, filtering, etc.). Regarding an atypical application, balloons, the authors of [6] recall that the field of (very) large deformation is still wide open and depending on the use case, special specific experimental configurations may help. It is also a theme addressed by [7] where calibrated targets were used to evaluate measurement uncertainties in this large deformation regime. The possibility to bridge more intimately measurements and models is highlighted in [8] where experimental measurements are combined to a model to extract mechanical fields with a certain mechanical admissibility close to a shear crack at bi-material interface. A little further on in the coupling between models and measurements, [9] proposed an interesting methodology to quantitatively characterize mechanical (interlaminar) properties reputed to be difficult to identify using finite element model updating techniques. Among current topics, the coupling of DIC with other types of instrumentation techniques or more generally data fusion is discussed in Article [10]. A comparative analysis based on DIC and Accoustic Emission techniques is helpful to comprehend the characteristics of concrete fracture process zones. In addition to the classic 2D DIC, several variants are also illustrated in this special issue. For example, stereo-DIC is an ally of choice for the validation of models on complex or large poly-instrumented structures. The issue of calibrating several independent benches using valuable CAD information is discussed in [11]. Conversely, when non-planar tests are to be instrumented at small scales or in conditions of difficult access, stereo can be used with a single camera by adapting the mounting with, for example, prisms and mirrors [12]. Another variant of DIC, which is still in its infancy, relies on X-ray based digital volume imaging. Increasingly, the measurement of 3D fields in material bulk (DVC) is leading mechanical engineers to rethink the way they conduct tests, developing, in addition to new image correlation algorithms, special machines that allow in-situ testing. This trend is illustrated in article [13]. Volume measurement with X-ray tomography is not the only volume imaging method of interest in mechanics. For example, Optical Coherence Tomography (OCT) allows this kind of investigation to be carried out and is particularly interesting for biological materials. The results obtained, combined with identification techniques, make it possible to estimate some mechanical properties [14]. Last but not least, the last part concerns algorithmic issues. The choice of correlation metrics itself is still under investigation. For instance, in [15], the authors present different metrics based on the gradients of the image rather than on the grey level. The above-mentioned question of large deformations implies, in addition to experimental constraints, particular complexities from an algorithmic point of view. Initialization in the large deformations framework of the correlation algorithm is also a topical issue [16].

Funding: This research received no external funding.

Acknowledgments: This issue follows two quite successful special session related to Digital Image Correlation at the 18th International Conference on Experimental Mechanics (ICEM18) in Brussels. We would like to thank the European Society for Experimental Mechanics (EuraSEM) for inviting us to edit this special issue in MDPI Applied Sciences journal. We would also like to record our sincere gratefulness to all authors and reviewers who contributed to this special issue.

Conflicts of Interest: The authors declare no conflict of interest.

References

1. Shadmehri, F.; Hoa, S.V. Digital Image Correlation Applications in Composite Automated Manufacturing, Inspection, and Testing. *Appl. Sci.* **2019**, *9*, 2719. [CrossRef]
2. Dai, K.; Liu, H.; Chen, P.; Guo, B.; Xiang, D.; Rong, J. Dynamic Response of Copper Plates Subjected to Underwater Impulsive Loading. *Appl. Sci.* **2019**, *9*, '927. [CrossRef]
3. Zhuo, Y.Q.; Guo, Y.; Bornyakov, S.A. Laboratory Observations of Repeated Interactions between Ruptures and the Fault Bend Prior to the Overall Stick-Slip Instability Based on a Digital Image Correlation Method. *Appl. Sci.* **2019**, *9*, 933. [CrossRef]
4. Dematteis, N.; Giordan, D.; Allasia, P. Image Classification for Automated Image Cross-Correlation Applications in the Geosciences. *Appl. Sci.* **2019**, *9*, 2357. [CrossRef]
5. Du, Y.; Gou, Z.m. Application of the Non-Contact Video Gauge on the Mechanical Properties Test for Steel Cable at Elevated Temperature. *Appl. Sci.* **2019**, *9*, 1670. [CrossRef]

6. Koseki, K.; Matsuo, T.; Arikawa, S. Measurement of Super-Pressure Balloon Deformation with Simplified Digital Image Correlation. *Appl. Sci.* **2018**, *8*, 2009. [CrossRef]
7. Blenkinsopp, R.; Roberts, J.; Harland, A.; Sherratt, P.; Smith, P.; Lucas, T. A Method for Calibrating a Digital Image Correlation System for Full-Field Strain Measurements during Large Deformations. *Appl. Sci.* **2019**, *9*, 2828. [CrossRef]
8. Tal, Y.; Rubino, V.; Rosakis, A.J.; Lapusta, N. Enhanced Digital Image Correlation Analysis of Ruptures with Enforced Traction Continuity Conditions Across Interfaces. *Appl. Sci.* **2019**, *9*, 1625. [CrossRef]
9. Seon, G.; Makeev, A.; Schaefer, J.D.; Justusson, B. Measurement of Interlaminar Tensile Strength and Elastic Properties of Composites Using Open-Hole Compression Testing and Digital Image Correlation. *Appl. Sci.* **2019**, *9*, 2647. [CrossRef]
10. Dai, S.; Liu, X.; Nawnit, K. Experimental Study on the Fracture Process Zone Characteristics in Concrete Utilizing DIC and AE Methods. *Appl. Sci.* **2019**, *9*, 1346. [CrossRef]
11. Malowany, K.; Piekarczuk, A.; Malesa, M.; Kujawińska, M.; Więch, P. Application of 3D Digital Image Correlation for Development and Validation of FEM Model of Self-Supporting Arch Structures. *Appl. Sci.* **2019**, *9*, 1305. [CrossRef]
12. Dan, X.; Li, J.; Zhao, Q.; Sun, F.; Wang, Y.; Yang, L. A Cross-Dichroic-Prism-Based Multi-Perspective Digital Image Correlation System. *Appl. Sci.* **2019**, *9*, 673. [CrossRef]
13. Mao, L.; Liu, H.; Zhu, Y.; Zhu, Z.; Guo, R.; Chiang, F.p. 3D Strain Mapping of Opaque Materials Using an Improved Digital Volumetric Speckle Photography Technique with X-Ray Microtomography. *Appl. Sci.* **2019**, *9*, 1418. [CrossRef]
14. Meng, F.; Zhang, X.; Wang, J.; Li, C.; Chen, J.; Sun, C. 3D Strain and Elasticity Measurement of Layered Biomaterials by Optical Coherence Elastography based on Digital Volume Correlation and Virtual Fields Method. *Appl. Sci.* **2019**, *9*, 1349. [CrossRef]
15. Sjödahl, M. Gradient Correlation Functions in Digital Image Correlation. *Appl. Sci.* **2019**, *9*, 2127. [CrossRef]
16. Barranco-Gutiérrez, A.I.; Padilla-Medina, J.A.; Perez-Pinal, F.J.; Prado-Olivares, J.; Martínez-Díaz, S.; Gutiérrez-Frías, O.O. New Four Points Initialization for Digital Image Correlation in Metal-Sheet Strain Measurements. *Appl. Sci.* **2019**, *9*, 1619. [CrossRef]

 *applied sciences*

Article

Measurement of Super-Pressure Balloon Deformation with Simplified Digital Image Correlation

Kazuki Koseki [1,*], **Takuma Matsuo** [2,*] **and Shuichi Arikawa** [3,*]

1   Department of Mechanical Engineering, Graduate School of Science and Technology, Meiji University, Kawasaki City, Kanagawa Prefecture 214-8571, Japan
2   Department of Mechanical Engineering, School of Science and Technology, Meiji University, Kawasaki City, Kanagawa Prefecture 214-8571, Japan
3   Department of Mechanical Engineering Informatics, School of Science and Technology, Meiji University, Kawasaki City, Kanagawa Prefecture 214-8571, Japan
*   Correspondence: kkaz0526@gmail.com (K.K.); matsuo@meiji.ac.jp (T.M.); arikawa@meiji.ac.jp (S.A.); Tel.: +81-90-4392-9408 (K.K.)

Received: 28 August 2018; Accepted: 19 October 2018; Published: 22 October 2018

Abstract: A super pressure balloon (SPB) is an aerostatic balloon that can fly at a constant altitude for an extended period. Japan Aerospace Exploration Agency (JAXA) has been developing a light-weight, high strength balloon made of thin polyethylene films and diamond-shaped net with high tensile fibers. Previous investigations proved that strength requirements on SPB members are satisfied even though the net covering the SPB sometimes becomes damaged during the inflation test. This may be due to non-uniform expansion, which causes stress concentration, however, no method exists to confirm this hypothesis. In this study, we tested a new method called Simplified Digital Image Correlation method (SiDIC) to check if it can measure the displacement of the SPB by using a rubber balloon. After measuring the measurement accuracy of the Digital Image Correlation method (DIC) and SiDIC, we applied both DIC and SiDIC to a rubber balloon covered just with the net. Interestingly, SiDIC entailed a smaller amount of data but could measure the deformation more accurately than DIC. In addition, assuming the stress concentration, one part of the net was bonded to the balloon to restrict the deformation. SiDIC properly identified the undeformed region.

Keywords: super pressure balloon; stress concentration; strain; non-contact measurement; digital image correlation; large deformation

---

1. Introduction

A super-pressure balloon (SPB) is a vehicle that can fly at a constant altitude for an extended period to perform scientific observations at a fraction of the cost of using a satellite. The SPB maintains its internal gas at a pressurized state, which suppresses buoyancy fluctuation when the balloon volume changes due to atmospheric temperature variations between day and night [1,2]. JAXA has been developing a lightweight, high strength balloon made of thin polyethylene films and a diamond-shaped net with high strength tensile fibers. Previous research shows that the tensile strength of the net meets requirements on SPB member strength, though the nets covering the SPB sometimes become damaged during the inflation test [3–5]. This may be due to non-uniform expansion, which causes stress concentration, although no method exists to confirm this hypothesis [6–8]. Contact measuring devices like strain gauges are not suitable because the SPB is too large to monitor the whole balloon and because they can deform the balloon surface during the contact measurement. Conversely, non-contact measurement methods such as the Digital Image Correlation method (DIC) can be efficiently used for this application.

DIC—an optical method to measure changes in images—usually requires the use of patterns to be applied onto the specimen surface. This method is used not only for measuring the deformation of a test piece in a tensile test but also in fracture mechanics problems and bioengineering applications [9–11]. This method may be able to detect the stress concentration on the SPB [12]. However, it is not suitable to study the shape of SPBs, as ink spots on the thin film may affect its strength and weight properties. To measure the deformation as accurately as possible, it is necessary to spray the particles evenly and as finely as possible to a wide range. However, if we do this, a large amount of ink will be applied to the surface, not only will it weigh more but also the polyethylene film will not stretch uniformly due to curing of the ink.

To overcome this problem, a Simplified DIC (SiDIC) using intersection detection technology was developed, which allowed us to track the diamond-shaped weave of the net so that we could measure the deformation of the SPB during the pressurization process. In this study, we developed SiDIC and verified the measurement accuracy, using a rubber balloon and diamond-shaped plastic net. First, the measurement accuracy of DIC using a patterned rubber balloon was confirmed and the deformation size measured by DIC was consistent with the rough calculations. Next, the accuracy of SiDIC was tested using a rubber balloon with random spray patterns and covered by a diamond-shaped plastic net. The pictures taken before and after deformation were analyzed using DIC and SiDIC, and the results were compared. DIC and SiDIC measured very similar deformation fields. The two methods were then tested using a rubber balloon covered just with the net. It was found that SiDIC entailed a smaller amount of data although it measured the deformation more accurately than DIC. In addition, assuming the stress concentration, the net was bonded to the balloon to restrict the deformation. Remarkably, SiDIC could properly identify the undeformed region. In summary, SiDIC is a simple and efficient method for measuring the SPB's deformation field.

2. Simplified Digital Image Correlation Method

2.1. Digital Image Correlation

DIC is a non-contact method for measuring the amount of movement (displacement amount) on the specimen surface. A picture of the specimen surface is taken before and after deformation using a digital camera. An identical point on the specimen is determined in the images before and after deformation; the amount of movement of this point is used to obtain the amount of displacement undergone by the specimen surface. To obtain the amount of movement, the correlation of the light intensity value distribution, which defines the deformed position of the calculation region composed of a subset of pixels (see Figure 1), is expressed as Equation (1).

$$\Delta x_1 = \left(x_1' - x_1\right)C(x, y, x^*, y^*) = \frac{\sum F(x,y)G(x^*, y^*)}{\sqrt{\sum F(x, y)^2 G(x^*, y^*)^2}} \tag{1}$$

F: Light intensity value before deformation G: Light intensity value after deformation
(x, y): Coordinates before deformation (x*, y*): Coordinates after deformation

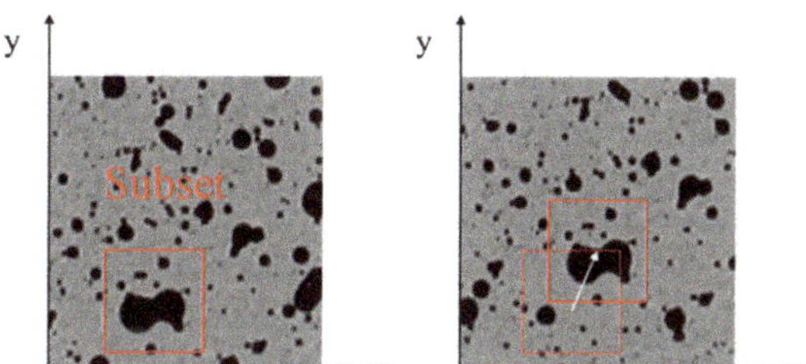

Figure 1. Process for calculation of deformation by DIC. (**a**) Reference image; (**b**) Deformed image.

2.2. The Simplified DIC and the Intersection Detection

SiDIC is a simple method to measure deformations. Instead of reading the light intensity values of the image, SiDIC recognizes the movement of the intersections to measure the deformation. Figure 2 shows how SiDIC measures the deformation. Basically, SiDIC reads the intersections of the net covering the balloon by using the intersection detection technology. SiDIC calculates the amount of deformation by reading the coordinates of the intersections. In this research, we detected the intersections by visual inspections. First, we read the coordinates of the intersections before and after deformation. Since actual intersections of net elements are represented by lines, there is no clear intersection in the image captured. In this experiment, the center of the line is defined as the intersection of the net. After reading the coordinates of the intersections, Equations (2) and (3) are used to calculate the amount of displacement. As a result, SiDIC measures the displacement from the amount of movement in the x and y directions before and after the deformation of each intersection.

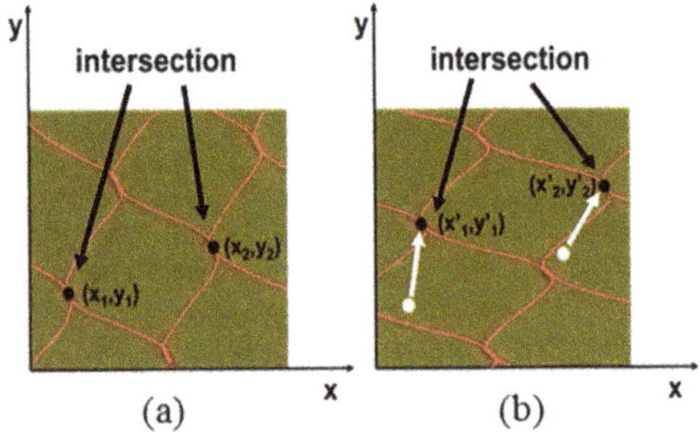

Figure 2. Coordinates of net intersections. (**a**) Reference image; (**b**) Deformed image.

$$\Delta x_1 = \left(x'_1 - x_1\right). \tag{2}$$

$$\Delta y_1 = \left(y'_1 - y_1\right) \tag{3}$$

3. DIC and SiDIC Experiments on a Rubber Balloon

3.1. Experimental Setups

To verify the measurement accuracy of SiDIC, first, the measurement accuracy of DIC was determined using a random spray patterned rubber balloon as shown in Figure 3a. We used a rubber balloon with a maximum diameter of 20 cm and the size of the particles covering the balloon surface was 1~5 [mm] (1.5~12 [pixel]). Next, to verify the measurement accuracy of SiDIC, a balloon which not only has random spray patterns but is also covered by a net was used as shown in Figure 3b. This makes it possible to use both methods. The thickness of the net is 1 mm. We took pictures after the balloon surface dug into the net holes. In the SPB, it is known that the net and film deforms together in this state. Therefore, we assume it will deform together as well. To test whether DIC and SiDIC could be applied to the actual SPB, a rubber balloon covered only by a net, as shown in Figure 3c, was used.

Generally, the shape of three-dimensionally deformed objects is measured using the three-dimensional DIC (3D-DIC). Since the planar limitation comes from the two-dimensional nature of the images shot by the camera, the solution is to use more than one camera. From images taken from two different angles of the same object, it is possible to estimate its 3D shape [13]. In this method, it is assumed that two-dimensional deformation measurement using DIC of each image taken from different angles is performed correctly. Therefore, in this research, we confirm whether the 2D deformation measurement of each image is done correctly. In this experiment, a digital camera is used to take an image of the balloon before and after deformation. Therefore, not the actual deformation of the surface of the balloon but the two-dimensional deformation amount on the image is measured. The unit deformation measured from the image corresponds to a pixel.

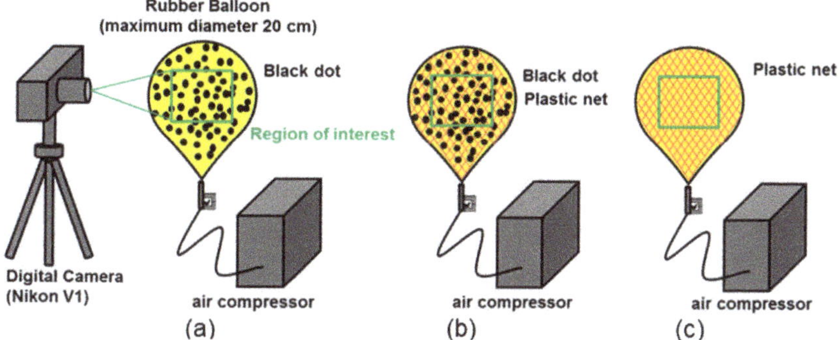

Figure 3. Experimental setups for the displacement measurement. (**a**) Random spray pattern; (**b**) Net + spray pattern; (**c**) Plastic net.

3.2. The Measurement Accuracy of DIC Applied to a Rubber Balloon

We liken the SPB to a rubber balloon. We sprayed a black pattern onto the rubber balloon and analyzed the deformation using DIC. Figure 4 shows the displacement distribution in the x-direction obtained from the DIC analysis. It shows how much control point coordinates moved after deformation using a color scale.

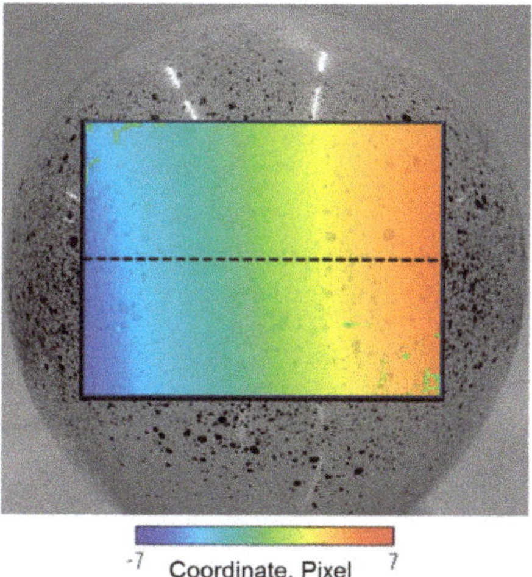

Figure 4. Black dotted balloon deformation.

From Figure 4, the displacement is clearly readable. Next, we verified the measurement accuracy of DIC. In this study, it is assumed that the balloon expands uniformly in the circumferential direction. The balloon is assumed to behave as a sphere and strain values are determined from Equations (4)–(7) by comparing the maximum radius value of the balloon before and after the deformation. To calculate the rough theoretical value, we measured the maximum radius from the pictures taken before and after deformation. After measuring the radius, we used Equation (4) to calculate the theoretical value of the strain. The radius before deformation (R) was 321 pixels, and the radius after deformation (R′) was 327 pixels. The experimental value was obtained using Equations (5) through (7). Notations "x" and "x'" indicate the x-direction deformation results by DIC. Computed strains are shown in Figure 5. From Figure 5, the experimental value showed a similar value compared to the theoretical value. Therefore, we found that the displacement and strain could be measured using classical DIC. The reason why the error occurred was that the balloon was assumed to be a sphere and to inflate uniformly.

$$\varepsilon_T = \frac{\Delta R(pixel)}{R(pixel)} \tag{4}$$

$$\alpha = R \times \sin^{-1}\frac{x}{R} \tag{5}$$

$$\beta = R' \times \sin^{-1}\frac{x + \Delta x}{R'} \tag{6}$$

$$\varepsilon'_E = \frac{\beta - \alpha}{\alpha} \tag{7}$$

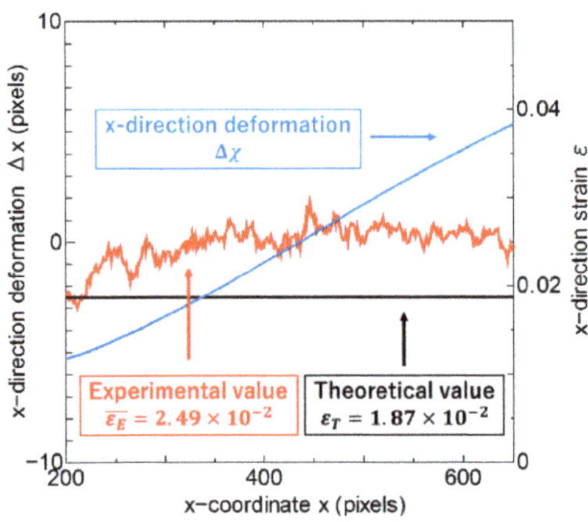

Figure 5. X-direction deformation and deviation between the distribution map for the region of interest theoretical and experimentally measured values of strain.

3.3. Comparison of DIC and SiDIC Measurements Accuracy

Next, the measurement accuracy of SiDIC was verified by using balloon (b) in Figure 3. Figure 6 shows the displacement distribution in the x-direction obtained from DIC. We analyzed the deformation at the dotted line control path using SiDIC and compared the corresponding results of DIC. Figure 7 shows the compared results. The continuous line represents DIC results while points denote SiDIC results. It can be seen that the two techniques give almost the same results, thus confirming the validity of SiDIC. In addition, similar results were obtained in the y-direction deformation. The reason for the measurement error of SIDIC is because the intersection is visually detected.

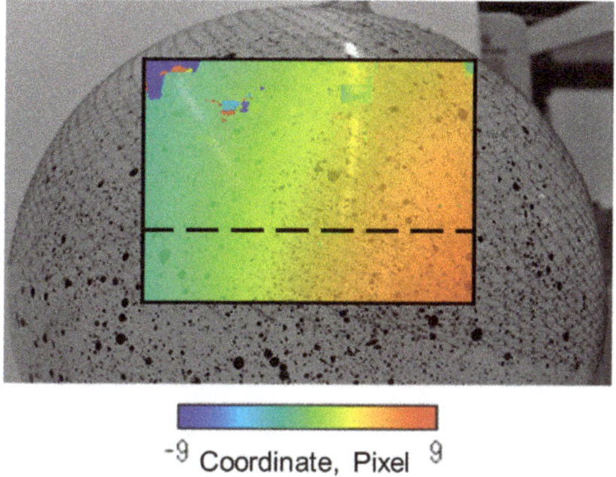

Figure 6. Horizontal displacement determined by DIC superimposed onto the photograph of the balloon.

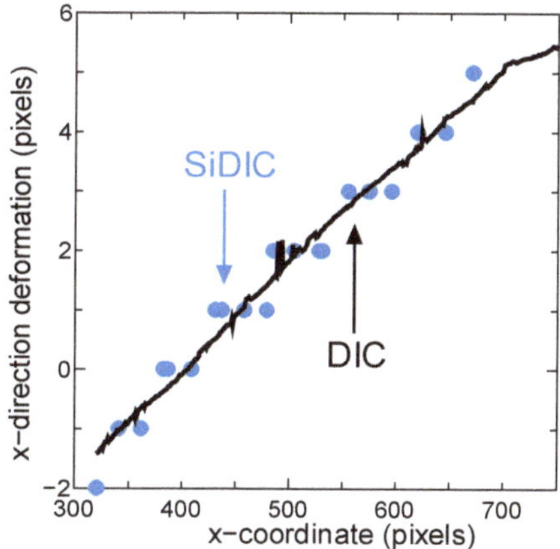

Figure 7. Comparison of horizontal displacement distributions obtained by DIC and SiDIC for the control path highlighted in Figure 6.

3.4. DIC and SiDIC Measurements of Net-Covered Balloon Displacements

Next, DIC was applied to a balloon covered just with a net, as shown in Figure 3c. The x-direction deformation map shown in Figure 8a was obtained. Also, SiDIC was applied, and Figure 8b shows the corresponding results. Figure 8a shows that most parts of the net were well read, though some were not measured properly such as the areas limited by the red circles in the figure. In addition, similar results were obtained in the y-direction deformation. Since DIC reads light intensity values, errors arise when this quantity cannot be read properly.

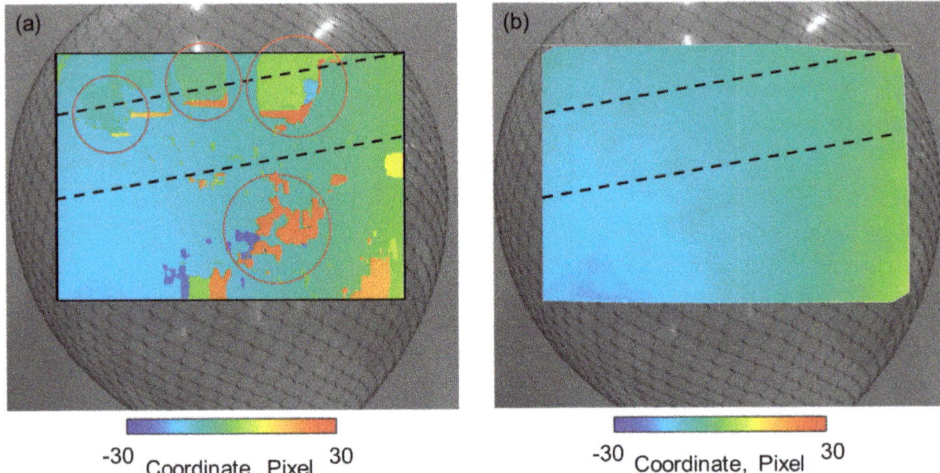

Figure 8. Horizontal displacement distribution by (**a**) DIC; (**b**) SiDIC, on the photograph of net covered balloon.

We measured the deformation field along the middle and upper dotted control lines and compared the results of DIC and SiDIC. Figure 9a presents results for the center dotted line. Figure 9a shows that DIC recovered fairly well on the deformation field although with some localized errors. SiDIC results were more stable yet overall consistent with those obtained by DIC. The same conclusion can be drawn from Figure 9b for the upper control path where stronger oscillations in displacement values are present. The observed behavior occurred because SiDIC works on a smaller amount of data than DIC. Furthermore, DIC may misrecognize displacements of net intersections and it is sensitive to light reflection on the balloon surface. Figure 10 shows an example of the misrecognition algorithm. From Figure 10, the intersection actually moved to point (2) after deformation. DIC errors occurred when it misrecognizes the intersections after the deformation as (1) or (3). On the other hand, the method read the coordinates of the intersections to prevent errors, thus measuring the deformation correctly.

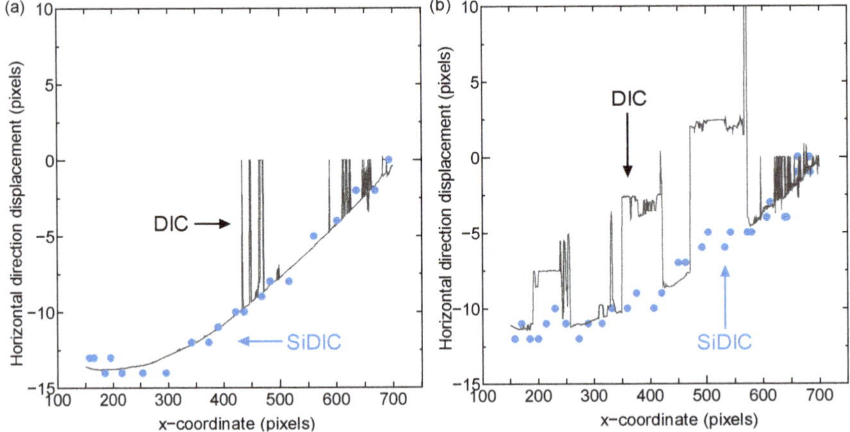

Figure 9. Comparison of horizontal displacement distributions obtained by DIC and SiDIC for the control paths highlighted in Figure 8: (**a**) Central dotted line; (**b**) Upper dotted line.

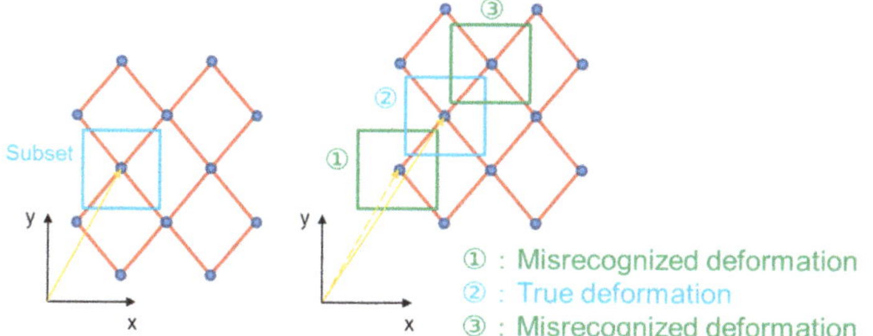

Figure 10. Misrecognition of the deformation of the net by DIC.

3.5. Measurements of Undeformed Regions Using SiDIC

In addition, assuming stress concentration to be an important issue for the SPB design, the net was bonded to the balloon using a strong instant adhesive to restrict the deformation (Figure 11). In this experiment, a randomly sprayed rubber balloon covered with a plastic net was used and one part of the net was boned. Figure 12 shows the x-direction displacement map obtained by SiDIC. From Figure 12, the bonded area shows "0" deformation. This experiment confirmed that SiDIC could properly identify undeformed regions. Hence, SiDIC can detect anomalies and asymmetry of the deformation field.

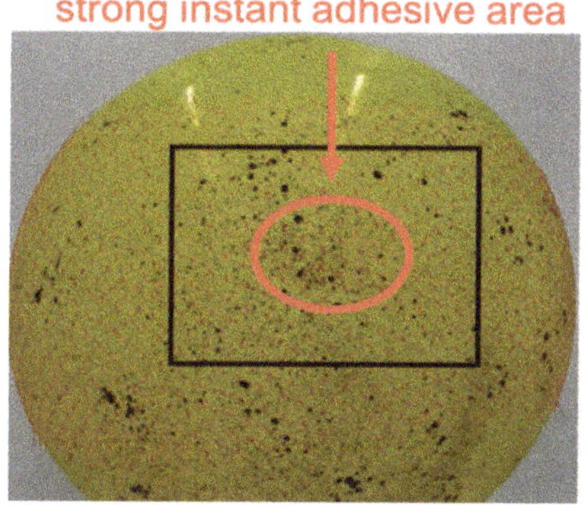

Figure 11. Measurement around no deformation.

Figure 12. Displacement distribution region performed by SiDIC in the x-direction.

4. Conclusions

In this research, we tested a new method called "Simplified Digital Image Correlation method (SiDIC)" for detecting non-uniform deformation of the super pressure balloon. In order to confirm the measurement accuracy of the developed SiDIC, first, we assessed the measurement accuracy of DIC using a rubber balloon covered with random spray patterns. Next, we used a rubber balloon covered with both spray patterns and a net to analyze it with DIC and SiDIC. Results of SiDIC and DIC were found to be in good agreement. Next, we applied both DIC and SiDIC to a rubber balloon covered just with a net supposing as an SPB. As a result, DIC recognized the net as a pattern, although it could not measure the whole deformation accurately. On the other hand, SiDIC measured the deformation clearly. Furthermore, SiDIC was able to identify undeformed regions when balloon deformation was restricted by bonding the net to the rubber shell. Therefore, it can be used as a simple deformation measurement method for the balloon. In this research, we used a two-dimensional DIC to find out whether SiDIC could successfully measure the deformation. However, 2D-DIC cannot measure the deformation in the depth direction, which means we have not measured the real deformation of the balloon. In addition, we used visual detection to detect the intersections and the experiment we used to identify the measurement accuracy was rough. From the results, we will continuously upgrade the measurement accuracy and develop an intersection detecting program at the same time; we are attempting to measure the balloon deformation more accurately by evolving SiDIC to "3D-SiDIC". We are planning to discuss how to fuse SiDIC and 3D-DIC in the future.

Author Contributions: K.K. conceived, designed, and performed the experiments, analyzed the data and wrote the paper. T.M. and S.A. provided the laboratory support and improved the manuscript.

Funding: This work was supported by JSPS KAKENHI Grant Number 17H01352.

Acknowledgments: The authors are grateful to Yoshitaka Saito and Ken Goto for suggesting the topic treated in this paper. We also thank them for sharing images and data of the Super pressure balloon with us.

Conflicts of Interest: The authors declare that there is no conflict of interest.

References

1. Henry, C.; David, P. Development of the NASA ultra-long duration balloon. In Proceedings of the NASA Science Technology Conference (NSTC2007), East Adelphi, MD, USA, 19–21 June 2007.
2. Cathey, H.M., Jr. The NASA super pressure balloon—A path to flight, Advances in Space Research. *Adv. Space Res.* **2007**, *44*, 23–38. [CrossRef]
3. Saito, Y.; Shoji, Y.; Matsuzaka, Y.; Mizuta, E.; Matsushima, K.; Tanaka, S. *A Super-Pressure Balloon with a Diamond-Shaped Net*; JAXA-RR-010-003; JAXA Research and Development Report: Tokyo, Japan, 2011; pp. 21–40.
4. Saito, Y.; Shoji, Y.; Matsuzaka, Y.; Matsushima, K.; Tanaka, S.; Kajiwara, K.; Shimadu, S. Development of a Super-pressure balloon with a diamond-shaped net. *Adv. Space Res.* **2014**, *54*, 1525–1529. [CrossRef]
5. Tanaka, R.; Matsuo, T.; Saito, Y.; Akita, D.; Nakashino, K.; Goto, K. Evaluation of tensile strength of net for Supper pressure balloon. In Proceedings of the Mechanical Engineering Congress (MECJ-16), Fukuoka, Japan, 11–14 September 2016.
6. Akita, D.; Saito, Y.; Goto, K.; Nakashino, K.; Matsuo, T.; Matsushima, K.; Hashimoto, H.; Shimadu, S. Development of a super-pressure balloon with a diamond-shaped net for high-altitude long-duration flights. In Proceedings of the Spatial Structures in the 21st Century IASS Annual Symposium, Tokyo, Japan, 26–30 September 2016.
7. Nakashino, K.; Saito, Y.; Goto, K.; Akita, D.; Matsuo, T.; Matsushima, K.; Hashimoto, H.; Shimadu, S. Development of Super pressure balloon with diamond-shaped net and numerical study of its structural characteristics. In Proceedings of the 4th AIAA Spacecraft Structures Conference (AIAA SCITECH 2017), Grapevine, TX, USA, 9–13 January 2017.
8. Nakamura, S.; Nakashino, K.; Saito, Y.; Goto, K.; Akita, D.; Matsuo, T. Deployment characteristics of Super pressure balloon with a diamond-shaped net. In Proceedings of the 25th Space Engineering Conference (SEC16), Yamaguchi, Japan, 21–22 December 2016.

9.	Grytten, F.; Daiyan, H.; Polanco-Loria, M.; Dumoulin, S. Use of digital image correlation to measure large strain tensile properties of ductile thermoplastics. *Polym. Test.* **2009**, *28*, 653–660.

10.	Jorge, A.; John, L. Investigation of crack growth in functionally graded materials using digital image correlation. *Eng. Fract. Mech.* **2002**, *69*, 1695–1711.

11.	Shao, X.; Dai, X.; Chen, Z.; He, X. Real-time 3D digital image correlation method and its application in human pulse monitoring. *Appl. Opt.* **2016**, *55*, 696–704. [CrossRef] [PubMed]

12.	Joseph, W.; Shirong, W.; Joseph, B.; Kiley, M. Super pressure balloon non-linear structural analysis and correlation using photogrammetric measurements. In Proceedings of the AIAA 5th Aviation, Technology, Integration, and Operations Conference (ATIO), Arlington, VA, USA, 28 September 2005.

13.	Pramod, R.; Erwin, H. *Optical Methods for Solid Mechanics*; Wiley-VCH: Weinheim, Germany, 2012; pp. 183–228.

Article

A Cross-Dichroic-Prism-Based Multi-Perspective Digital Image Correlation System

Xizuo Dan [1,2], Junrui Li [2], Qihan Zhao [1], Fangyuan Sun [1], Yonghong Wang [1,*] and Lianxiang Yang [2]

[1] School of Instrument Science and Opto-electronics Engineering, Hefei University of Technology, Hefei 230009, China; danxizuo@163.com (X.D.); zhaoqihan@mail.hfut.edu.cn (Q.Z.); sunfangyuan@mail.hfut.edu.cn (F.S.)

[2] Department Mechanical Engineering, Oakland University, Rochester, MI 48309, USA; jli23456@oakland.edu (J.L.); yang2@oakland.edu (L.Y.)

[*] Correspondence: yhwang@hfut.edu.cn; Tel.: +86-139-5519-8216

Received: 31 December 2018; Accepted: 4 February 2019; Published: 16 February 2019

Featured Application: A 3D-DIC system based on a single 3CCD color camera and a cross dichroic prism is proposed in this paper, this system can be applied to situations where three cameras are required for DIC measurement.

Abstract: A robust three-perspective digital image correlation (DIC) system based on a cross dichroic prism and single three charge-coupled device (3CCD) color cameras is proposed in this study. Images from three different perspectives are captured by a 3CCD camera using the cross dichroic prism and two planar mirrors. These images are then separated by different CCD channels to perform correlation calculation with an existing multi-camera DIC algorithm. The proposed system is considerably more compact than the conventional multi-camera DIC system. In addition, the proposed system has no loss of spatial resolution compared with the traditional single-camera DIC system. The principle and experimental setup of the proposed system is described in detail, and a series of tests is performed to validate the system. Experimental results show that the proposed system performs well in displacement, morphology, and strain measurement.

Keywords: digital image correlation; multi-perspective; single camera; cross dichroic prism

1. Introduction

Digital image correlation (DIC) technology was proposed in the 1980s [1,2]. As a robust, noncontact, full-field, and high-precision measurement method, this technology is not sensitive to the measurement environment. Thus, this method has been successfully applied to measure displacement and strain in most cases. The DIC method, especially 3D-DIC, has been extended to numerous research fields by researchers [3–6]. Therefore, DIC has become an important method in the field of experimental mechanics [7–10]. Traditional 3D-DIC technology obtains images by using two black and white charge-coupled devices (CCDs) or complementary metal oxide semiconductor (CMOS) cameras. A dual-camera system meets the measurement requirements in most cases. However, due to limitations in the field of view, it is difficult to obtain satisfactory results by using a dual-camera system if a multi-angle analysis is required. Therefore, researchers use multiple cameras to perform DIC measurements. The multi-camera system provides an obvious advantage over the traditional dual-camera system, in that it can cover a large area of measurement and thereby expand the measurement range DIC offers. In addition, for measurement areas with complex profiles, the multi-camera system can effectively reduce errors. However, DIC images are always obtained using expensive industrial cameras. For the dual- and multi-camera system, a trigger device must be added in the system in order to meet the required image acquisition synchronization. However, all these factors increase the measurement cost.

In recent years, researchers have proposed different methods for 3D-DIC measurement using a single camera. These methods can be mainly divided into two categories. The first divides the camera CCD into two parts, wherein the images of two different angles of the object are presented in two parts of the CCD with the aid of the designed optical path. Pankow proposed a four-mirror adapter-assisted single-camera 3D-DIC system to measure full-field out-of-plane displacement histories at high frame rates [11]. Genovese used a compact system of a biprism and a camera to perform stereo-DIC measurement [12]. Barone used two planar mirrors and a low-frame-rate camera to measure 3D vibration [13], which is also an interesting application of 3D-DIC measurement using a single camera. This method can also be used with the aid of transmission diffraction grating [14,15]. The second category uses a color camera, which allows different color channels to record images from different perspectives. Li and Yu used a 3CCD color camera to perform 3D-DIC measurement [16,17]. The approaches presented in these two papers are very similar; the only difference is that Yu's system has one more mirror than Junrui's system to avoid image flipping. Yu proposed using a single high-speed color CMOS camera to perform high-speed 3D-DIC measurements without an additional triggering device [18]. Zhong used a dichroic filter to replace the beam splitter and color filters [19]. This method has considerable advantages, such as simple optical paths, a system requirement of only one color camera, two mirrors, and a cube prism. Another remarkable advantage of this method is that it does not reduce image resolution, which brings the accuracy of the measurements closer to that offered by the traditional dual camera DIC system. Wang used two beam splitters and three mirrors to perform multi-perspective DIC measurement [20], thereby effectively using the three channels of the 3CCD camera despite its complexity.

In this study, we propose a compact setup by using a cross dichroic prism to perform multi-perspective DIC measurements. The images from three different view angles are obtained by the three channels of the 3CCD color camera through two mirrors and the cross dichroic prism. Bandpass filters are unnecessary due to the high performance of the cross dichroic prism, and the three view angles can simultaneously occupy the entire CCD without reducing the resolution and consequently maintain DIC measurement accuracy. At the same time, the system has been simplified compared with our previous system [19]. There is no longer a need for bandpass filters in the proposed system; only two planar mirrors are required as opposed to the three Wang needed, and the two beam splitters are replaced by a cross dichroic prism, which reduces costs. However, the images acquired by the system proposed in this paper were not flipped and rotated, which reduced the computational complexity of DIC. Details of the method will be described in the following text. After introducing the experimental procedure, three typical experiments were performed to evaluate the metrological performances of the proposed method. The 3D displacement, 3D shapes, and strain can be determined using the developed system.

2. Methodology

The simplified DIC system is based on a 3CDD color camera. This camera type is equipped with a refraction prism, which divides the light into R, G, and B channels and simultaneously records via three independent CCD chips. The resolution of each CCD is the same as that of the entire color camera. The captured color image of this camera is a 24-bit bitmap, which consists of three 8-bit bitmaps from different channels. The main advantage this camera type offers is that almost no color aliasing exists among the three channels; thus, it can be used for three-perspective DIC measurements. Another important advantage of this system is that it does not require a triggering device for simultaneous acquisition, as the images of the three channels are acquired simultaneously using the 3CCD camera. The 3CCD color camera used in this work is an HW-F202 with 1624×1236 resolution, provided by Hitachi in Beijing, China. Figure 1 shows the camera's spectrum. If the illumination is based on the spectrum range in this figure, then no color aliasing will occur. Figure 2 presents the schematic of the cross dichroic prism, which is a combination of four triangular prisms that combine the three color beams R, G, and B to form the color image. Hence, images of different viewing angles can pass through the cross dichroic prism from three different directions and transmit into the camera lens from the same direction.

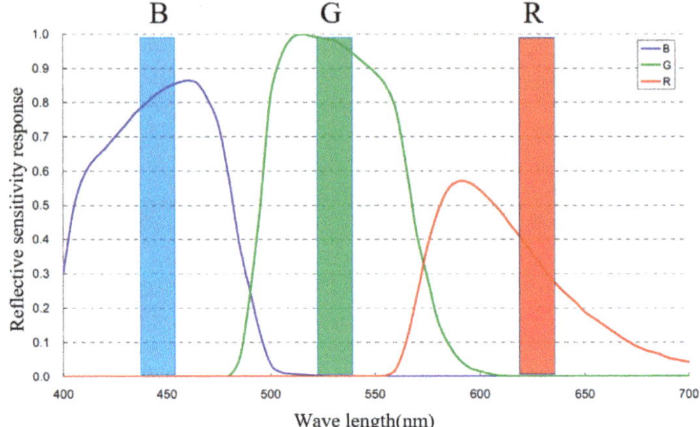

Figure 1. Quantum efficiency of the color camera.

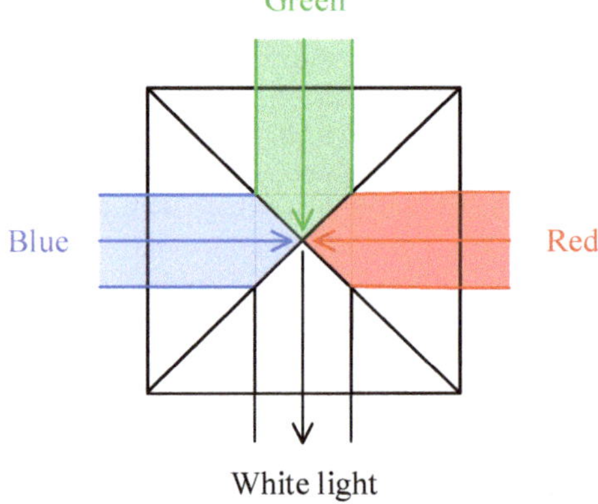

Figure 2. Schematic of the cross dichroic prism.

Figure 3 shows the optical arrangement of the proposed measurement system. M1 and M2 are planar mirrors.

In theory, white light illumination can be used for this system as long as the cross dichroic prism has the right bandwidth. In this work, as the camera had different reflective sensitivity responses for various light spectrums, three LEDs corresponding to the camera spectrum were used for illumination to obtain the images under similar brightness. Spectral sensitivity of the color camera showed that the sensitivity of the red channel was lower than those of the green and blue channels, thus requiring the use of a brighter red light source. The red LED we used in the experiment was adjustable in brightness. A cross dichroic prism was fixed in front of the lens and 3CCD camera, and its filter bandwidth was designed similarly to the color bands in Figure 1. The cross dichroic prism used in this system can be seen as a combination of two planar mirrors and three bandpass filters, theoretically ensuring that no distortion will be introduced. As this system was equivalent to three cameras installed at different angles, its calibration and image correlation can refer to the ordinary multi-camera DIC system. Chen's solution is suitable for the proposed system [21].

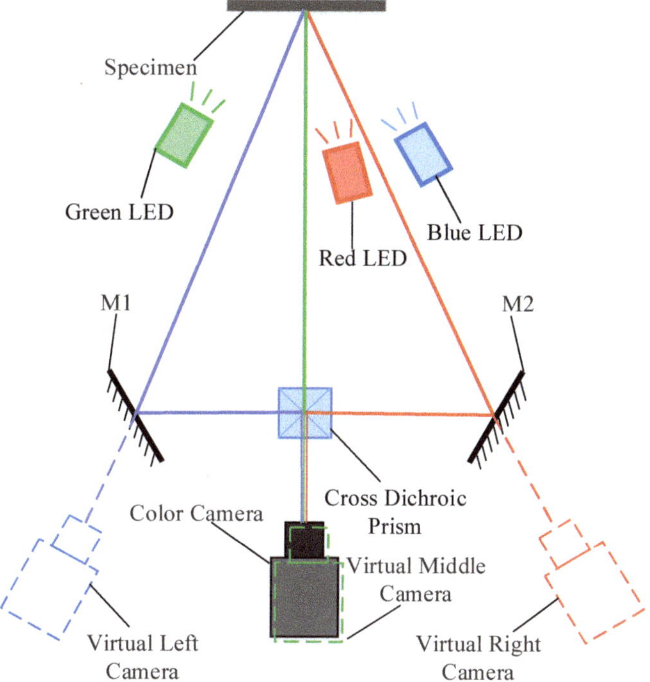

Figure 3. Scheme of the improved multi-perspective 3D-DIC system.

3. Experiments and Results

Three different types of experiments were performed to verify the feasibility of the proposed system in various applications. First, a morphology test was conducted to measure the shape of a specimen with a curved surface. Second, rigid body displacement measurement was performed, whereby in-plane and out-of-plane displacement was conducted using a piezoelectric drive micro displacement platform and a flat plate. Finally, a tensile experiment was conducted to verify the accuracy of strain measurement. The standard 3D-DIC algorithm can be utilized directly to perform the evaluation, and the Istra4D provided by Dantec Dynamics was used for the evaluation because of its good performance in multi-camera calibration and DIC calculations.

Figure 4 shows the experimental system presented in this study. In this system, the optical path of the red light is shorter than those of the blue and green lights; thus, the three channels may not be simultaneously focused. The aperture of the lens should be adjusted to small to make all channel images as clear as possible. The focal length of the lens used in the experiments was 35 mm. Adjusting the location and angle of the mirror can change the relative angle between the images of different channels. The cross dichroic prism was fixed at 8 mm in front of the camera lens. Careful adjustment was required to ensure the high quality of images of the three channels and similarity of brightness and size of the specimen being measured in the three images.

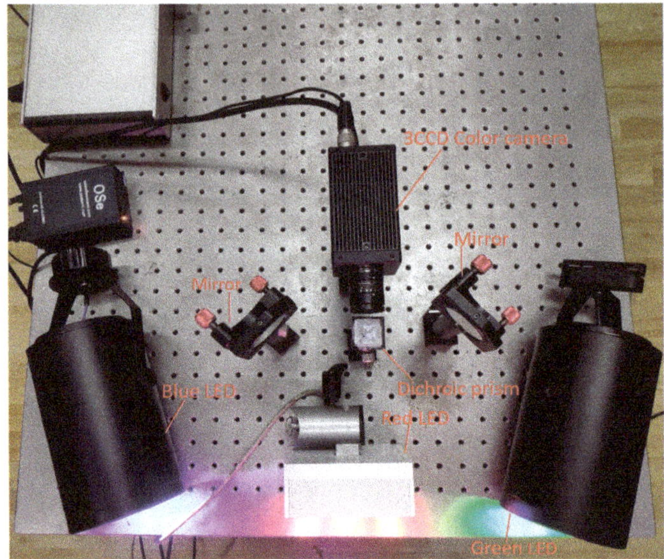

Figure 4. Experimental setup of the improved multi-perspective 3D-DIC system.

Figure 5 shows that an 8 × 8 chessboard was used to calibrate the system. The size of the small square of the calibration board was 11 mm × 11 mm. The three circles on the calibration board were used to mark the direction. Eight 24-bit images of the calibration board in different positions and directions were captured, and each 24-bit color image was converted into three 8-bit grayscale images by the R, B, and G channels. The R, B, and G channels data of the 24-bit color image were converted into three grayscale images, respectively; every image can be seen as captured by a virtual camera. The intrinsic and extrinsic parameters of the three virtual cameras can be calculated by Zhang's calibration algorithm. The reprojection errors are shown in Figure 6.

Figure 5. Calibration images (including the original and channel images).

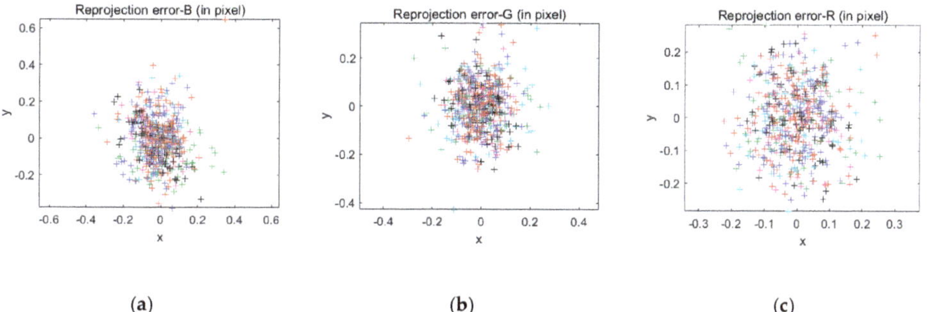

Figure 6. The reprojection errors: (**a**) blue channel, (**b**) green channel, and (**c**) red channel.

3.1. Rigid Body Displacement Experiment

One of 3D-DIC's major advantages is its robustness and high precision in spatial displacement measurements. In-plane and out-of-plane rigid-body movement tests were performed to validate the feasibility and accuracy of the proposed system in displacement measurement. A 100 mm × 100 mm flat plate with speckles sprayed on the surface was fixed on a piezoelectric-drive micro-displacement platform provided by Winner Optics with a resolution of 13 nm. The flat plate was displaced from 0 mm to 0.2 mm in 0.02 mm intervals in the in-plane (X) and out-of-plane (Z) directions. After setting up the experiment system, illumination intensity was adjusted to ensure similar brightness of the gray image from each channel.

While calibrating this experiment, the Z direction of the established coordinate system was parallel to the Z direction of the displacement platform. A subset size of 31 × 31 pixels and a grid step of 5 pixels were adopted in the correlation calculation. The displacement results was expressed by the average displacement of 5 × 5 subsets at the image center. Figure 7 shows the measured displacements and errors in the X and Z directions. As ensuring that the X and Z directions of the movement were exactly the same with the directions of the selected calculation coordinate was difficult, the synthesis values of X and Z directions displacements were selected to be the measured value, which were approximately coincident with the real values. Errors from the measurement were less than 0.005 mm; standard deviations of the X and Z direction synthetic displacement were 0.0025 and 0.0033 mm, respectively. Results show that the proposed system has high accuracy in displacement measurement. The displacement error map of the final step in the X displacement test is shown in Figure 8.

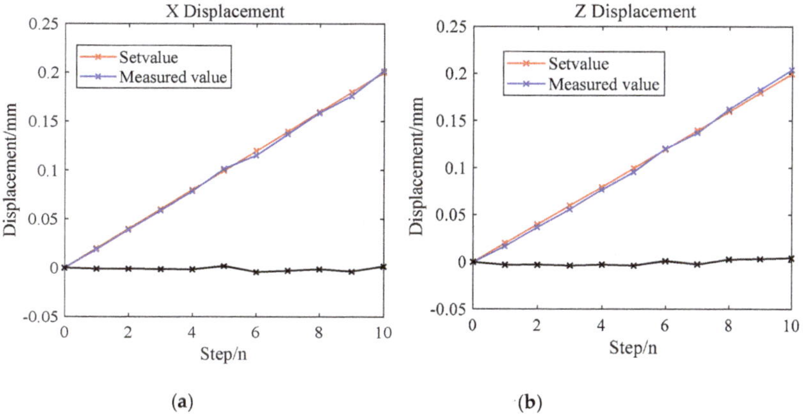

Figure 7. The displacement results: (**a**) X direction and (**b**) Z direction.

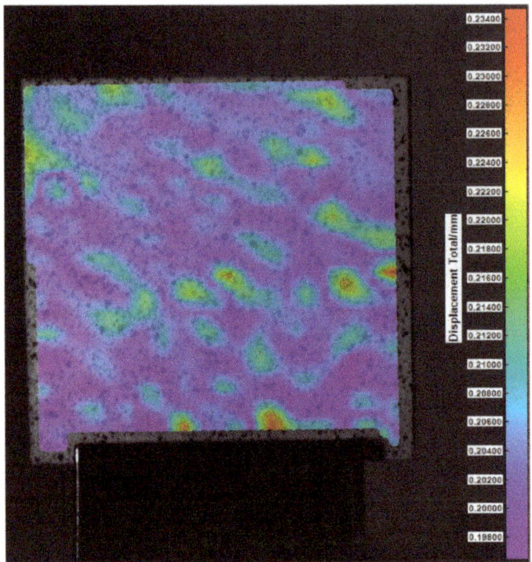

Figure 8. X direction displacement error map (the final step).

3.2. Morphology Measurement Experiment

The dual-camera DIC system can be used to measure regular morphology. However, for objects with complex surfaces, good results are difficult to obtain due to field of view limitations. Similarly, for big objects, the field of view of the two cameras cannot completely cover the area to be measured. The presented system involves three perspectives, and the left and right views can be adjusted independently to satisfy the requirements of complex topography measurements or of big objects, thus enabling it to achieve better results than the two-perspective DIC. The specimen used for this test was a flat aluminum plate with a cylindrical bulge in the center, and the presented system was used to measure the surface profile of the specimen. The results of three-perspective DIC calculation were compared with those of two perspectives. The left and right parts of Figure 9 show the results of DIC calculation using three and two perspectives, and the middle green channel was not included in the calculation of two perspectives. Similar parameters were used in both calculations. Figure 9 illustrates that the morphology result under three perspectives is better than that under two perspectives. Clear defects exist at the boundary between the cylindrical protrusion and plane in the morphological cloud map calculated by two channels. The reason is that the boundary line has a large error in the DIC calculations, whereas the middle channel can provide redundant information to eliminate errors. In our previous work, the DIC calculation of three fields of view had considerable advantages in measuring complex morphologies compared with two fields of view. In the calculation of double views, remarkable concave and convex regions cannot be obtained accurately through correlation calculation, whereas the three-perspective system can obtain complete information. The system proposed in this study also has this advantage. The statistical error maps in Figure 10 show that the topographical errors calculated from the three perspectives (left) are considerably smaller than the results of the two-perspective calculations (right), which represents the uncertainty of the 3D coordinates of each point.

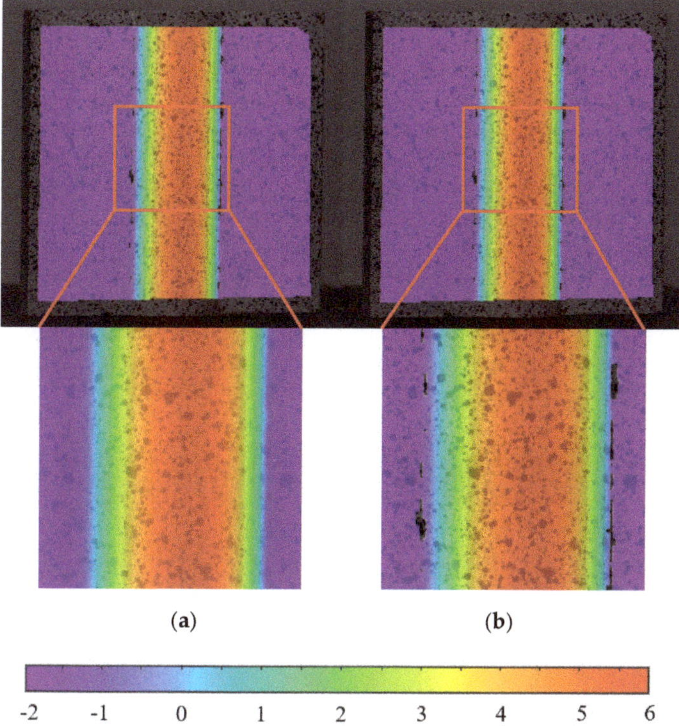

Figure 9. Cloud map of the measured contour/mm. (**a**), morphology result calculated under three perspectives; (**b**), morphology result calculated under two perspectives (green channel was not included).

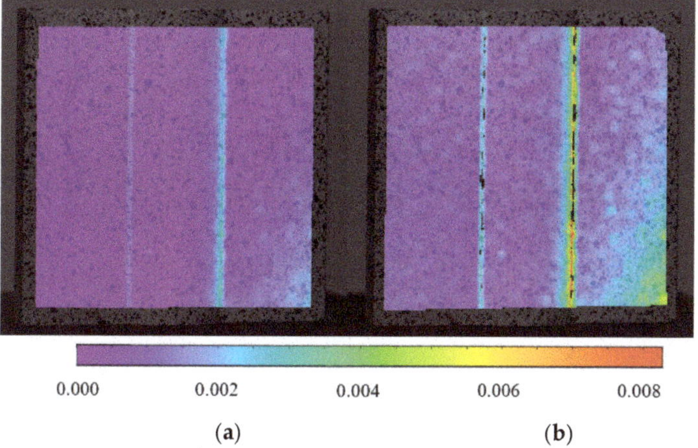

Figure 10. Contour statistical error map/mm. (**a**) contour statistical error calculated under three perspectives; (**b**) under two perspectives (green channel was not included).

3.3. Tensile Test for Strain Measurement

DIC strain measurement is currently the most important noncontact strain measurement method. The system presented in this study is also applicable to strain measurement. A tensile test was performed to verify the capability of the proposed system for strain measurement along with a commercial 3D-DIC system, as shown in Figure 11. A steel sheet was stretched until fracture occurred. Stretching speed is set to 2 mm/min. The proposed single camera and commercial 3D-DIC systems (Q-400 provided by Dantec Dynamics) were used to obtain the images and calculate the strain. The two systems were triggered by the same trigger source for image acquisition, and the acquisition frequency is 1 frame/s. Figure 12 shows the strain cloud before fracture, as calculated by the two systems. The left and right parts were obtained by the proposed system and the Q-400, respectively. Figure 13 displays the engineering strain curves measured by the two systems. The strain data represented by the curves are the mean values of the strain on the short straight line in Figure 12. Figure 13 presents two strain curves that are consistent. The curve calculated by the proposed system is smooth because the DIC calculation under three perspectives has the advantage of error reduction.

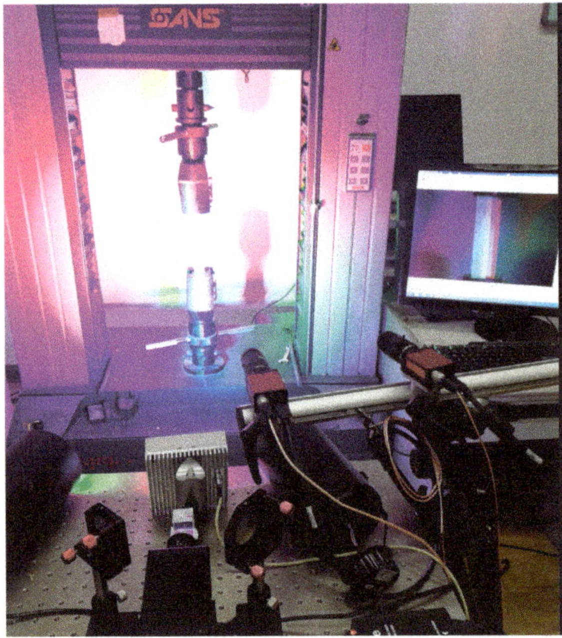

Figure 11. The setup for the strain test.

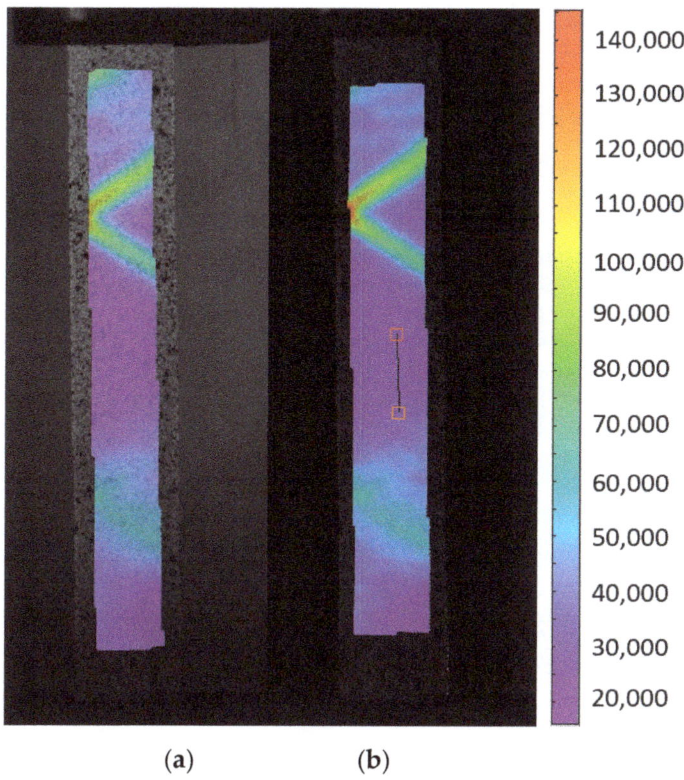

(a)   (b)

Figure 12. Cloud map of strain before fracture/strain. (**a**) proposed system; (**b**) Q-400 DIC system.

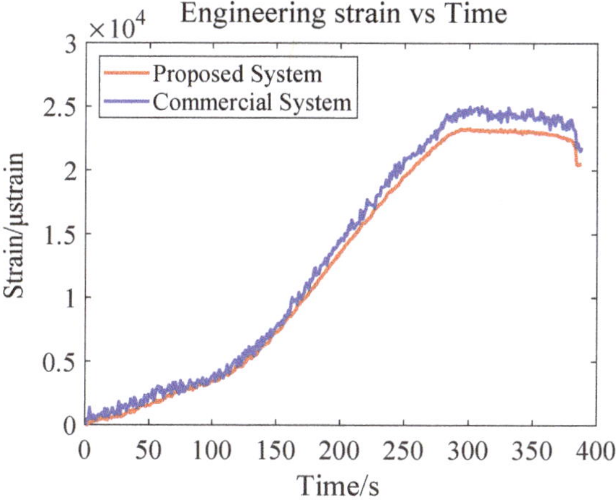

Figure 13. Engineering strain curve.

4. Conclusions

In this study, a novel 3D-DIC system based on a single 3CCD color camera and a cross dichroic prism is proposed. Images from three different perspectives were captured using a 3CCD camera through the cross dichroic prism and two planar mirrors, and those images were used to perform

DIC calculations. Three different types of experiments were performed to verify the feasibility and accuracy of the proposed system. Results showed that this multi-perspective pseudo-vision system performed well in all three experiments.

This system has the advantages of using a single color camera and not requiring any synchronous triggering device to ensure synchronous image acquisition, resulting in a compact structure. In comparison with the existing CCD segmentation single camera 3D-DIC system, the proposed system adopts a 3CCD color camera to record images from different perspectives by using R, G, and B channels. As a result, each view can occupy an entire CCD without reducing the spatial resolution, and the need for an additional lens distortion calibration process is eliminated. In addition, this system has one additional view compared with the conventional dual-camera DIC system. Thus, the calculation results are more accurate due to the three perspectives of area coverage. This system requires only one cross dichroic prism and two planar mirrors compared with our previous system of two prisms, three mirrors, and three bandpass filters.

The proposed system still has shortcomings in some respects. The cross dichroic prism should be carefully selected to match the three channels of the color camera. Otherwise, images of the different channels will be aliased. The proposed system also requires high monochromaticity of each channel of the cross dichroic prism. The illumination brightness of the proposed system must be adjustable to maintain the similarity of the image brightness of each channel. Additionally, the optical path requires a stable environment, which limits its scope of application.

Author Contributions: X.D. prepared the whole manuscript, including writing the original draft, explaining techniques/technologies, and analyzing experimental results based on the instructions of L.Y and Y.W. J.L. provided the initial idea of this paper. The other co-authors provided some specific information and helped in writing and editing the original draft.

Funding: This work was partially supported by the National Key Research and Development Program of China (Grant number: 2016YFF0101803) and the National Natural Science Foundation of China (Grant number: 11672045).

Acknowledgments: The authors would like to express their sincere thanks to Siyuan Bao from Hefei University of Technology for his very important job for the experiments in this article.

Conflicts of Interest: The authors declare no conflict of interest.

References

1. Yamaguchi, I. Speckle displacement and deformation in the diffraction and image fields for small object deformation. *Opt. Acta* **1981**, *28*, 1359–1376. [CrossRef]
2. Peters, W.H.; Ranson, W.F. Digital imaging techniques in experimental stress analysis. *Opt Eng.* **1982**, *21*, 427–431. [CrossRef]
3. Shao, X.; Dai, X.; Chen, Z.; He, X. Real-time 3D digital image correlation method and its application in human pulse monitoring. *Appl. Opt.* **2016**, *55*, 696–704. [CrossRef] [PubMed]
4. Sutton, M.A.; Ke, X.; Lessner, S.M.; Goldbach, M.; Yost, M.; Zao, F.; Schreier, H.W. Strain field measurements on mouse carotid arteries using microscopic three-dimensional digital image correlation. *J. Biomed. Mater. Res. Part A* **2008**, *84*, 178–190. [CrossRef] [PubMed]
5. Dai, X.; Yun, H.; Pu, Q. Measuring thickness change of transparent plate by electronic speckle pattern interferometry and digital image correlation. *Opt. Commun.* **2010**, *283*, 3481–3486. [CrossRef]
6. Pan, B.; Qian, K.; Xie, H.; Asundi, A. Two-dimensional digital image correlation for in-plane displacement and strain measurement: A review. *Meas. Sci. Technol.* **2009**, *20*, 062001. [CrossRef]
7. Chu, T.C.; Ranson, W.F.; Sutton, M.A. Applications of digital-image-correlation techniques to experimental mechanics. *Exp. Mech.* **1985**, *25*, 232–244. [CrossRef]
8. Bornert, M.; Brémand, F.; Doumalin, P.; Dupre, J.-C. Addendum to: Assessment of digital image correlation measurement errors: Methodology and results. *Exp. Mech.* **2015**, *49*, 1.
9. Xiao, Z. Three-dimensional digital image correlation system for deformation measurement in experimental mechanics. *Opt. Eng.* **2010**, *49*, 103601.
10. Li, J.; Xin, X.; Yang, G.; Zhang, G.; Siebert, T.; Yang, L. Whole-field thickness strain measurement using multiple camera digital image correlation system. *Opt. Lasers Eng.* **2017**, *90*, 19–25. [CrossRef]

11. Pankow, M.; Justusson, B.; Waas, A.M. Three-dimensional digital image correlation technique using single high-speed camera for measuring large out-of-plane displacements at high framing rates. *Appl. Opt.* **2010**, *49*, 3418. [CrossRef] [PubMed]
12. Genovese, K.; Casaletto, L.; Rayas, J.A.; Flores, V.; Martinez, A. Stereo-Digital Image Correlation (DIC) measurements with a single camera using a biprism. *Opt. Lasers Eng.* **2013**, *51*, 278–285. [CrossRef]
13. Barone, S.; Paolo, N.; Alessandro, P.; Armando, V.R. 3D vibration measurements by a virtual-stereo-camera system based on a single low frame rate camera. *Procedia Struct. Integr.* **2018**, *12*, 122–129. [CrossRef]
14. Pan, B.; Wang, Q. Single-camera microscopic stereo digital image correlation using a diffraction grating. *Opt. Express* **2013**, *21*, 25056–25068. [CrossRef]
15. Xia, S.; Gdoutou, A.; Ravichandran, G. Diffraction assisted image correlation: A novel method for measuring three-dimensional deformation using two-dimensional digital image correlation. *Exp. Mech.* **2013**, *53*, 755–765. [CrossRef]
16. Junrui, L.; Xizuo, D.; Wan, X.; Wang, Y.; Yang, G.; Yang, L. 3D digital image correlation using single color camera pseudo-stereo system. *Opt. Laser Technol.* **2017**, *95*, 1–7.
17. Yu, L.; Pan, B. Color stereo-digital image correlation method using a single 3CCD color camera. *Exp. Mech.* **2017**, *57*, 649–657. [CrossRef]
18. Yu, L.; Pan, B. High-speed stereo-digital image correlation using a single color high-speed camera. *Appl. Opt.* **2018**, *57*, 9257–9269. [CrossRef]
19. Zhong, F.Q.; Shao, X.X.; Quan, C. 3D digital image correlation using a single 3CCD colour camera and dichroic filter. *Meas. Sci. Technol.* **2018**, *29*. [CrossRef]
20. Wang, Y.; Dan, X.; Li, J.; Wu, S.; Yang, L. Multi-perspective digital image correlation method using a single color camera. *Sci. China Technol. Sci.* **2018**, *61*, 61–67. [CrossRef]
21. Chen, F.; Chen, X.; Xie, X.; Feng, X.; Yang, L. Full-field 3D measurement using multi-camera digital image correlation system. *Opt. Lasers Eng.* **2013**, *51*, 1044–1052. [CrossRef]

applied
sciences

Article

Laboratory Observations of Repeated Interactions between Ruptures and the Fault Bend Prior to the Overall Stick-Slip Instability Based on a Digital Image Correlation Method

Yan-Qun Zhuo [1,*], Yanshuang Guo [1] and Sergei Alexandrovich Bornyakov [2]

[1]   State Key Laboratory of Earthquake Dynamics, Institute of Geology, China Earthquake Administration, Beijing 100029, China; guoysh@ies.ac.cn

[2]   Institute of the Earth's Crust, Siberian Branch, Russian Academy of Sciences, Irkutsk 664033, Russia; bornyak@crust.irk.ru

*   Correspondence: zhuoyq@ies.ac.cn; Tel.: +86-1062-009-010

Received: 4 February 2019; Accepted: 26 February 2019; Published: 5 March 2019

Abstract: Fault geometry plays important roles in the evolution of earthquake ruptures. Experimental studies on the spatiotemporal evolution of the ruptures of a fault with geometric bands are important for understanding the effects of the fault bend on the seismogenic process. However, the spatial sampling of the traditional point contact type sensors is quite low, which is unable to observe the detailed spatiotemporal evolution of ruptures. In this study, we use a high-speed camera combined with a digital image correlation (DIC) method to observe ruptures during stick-slip motions of a simulated bent fault. Meanwhile, strain gages were also used to test the results of the DIC method. Multiple cycles of the alternative propagation of ruptures between the two fault segments on the both sides of the fault bend were observed prior to the overall failure of the fault. Moreover, the slip velocity and rupture speed were observed getting higher during this process. These results indicate the repeated interactions between the ruptures and the fault bend prior to the overall instability of the fault, which distinguishes the effect of the fault bend from the effect of asperities in straight faults on the evolution of ruptures. In addition, improvement in the temporal sampling rate of the DIC measurement system may further help to unveil the rupture evolution during the overall instability in future.

Keywords: earthquake rupture; fault geometry; spatiotemporal evolution; strain gage; spatial sampling rate; rupture speed; slip velocity; high-speed camera

---

1. Introduction

The experimental study of the evolution of earthquake ruptures is of great significance for understanding the underlying physical process of earthquake preparation and occurrence. Geological surveys and field observation data showed that fault geometry plays important roles in the initiation and propagation of earthquake ruptures [1–8]. Numerical simulations analyzed the influences of the fault bend on the rupture process and the fault slip distribution, which revealed that the angle of the fault bend, the normal stress, and the loading mode play important roles in the initiation and propagation of the ruptures [9–14]. Specifically, the rupture zone and overall slip distribution on the fault are controlled by the angle of the fault bend, while rupture velocity and detailed slip distribution around the bend are influenced by time-dependent normal stress changes caused by the rupture [12]. The fault bend will serve as an initiation and/or a termination point for the rupture via reducing normal stress on the dilatational segment and increasing normal stress on the compressive segment of the bent fault during dynamic ruptures [11]. The angle of the fault bend and the sliding direction of a dip-slip bent fault control the time and location of the rupture nucleation [13]. In addition, the rupture process of the Chi-Chi earthquake [10,15], Landers earthquake [16], and İzmit earthquake [17] were reconstructed via numerical simulations to further reveal the mechanism of the influence of fault geometry on the earthquake rupture process using seismic and geodetic data combined with bent fault models.

Physical experiments have been carried out to study the rupture process and the evolution of the relevant physical fields of bent faults. It was observed that the two segments on each side of the fault bend became active alternatively during sliding, which implied that the fault bend plays a role of a valve during sliding [18]; a two-step rupture propagation process was observed prior to the overall instability of the fault. Namely, the dynamic rupture started on a fault segment is stopped near the fault bend, which is restarted near the bend on the other fault segment after a certain delay time and leads to the overall slip of the entire fault without being arrested by the presence of the fault bend [19]. The influence of fault bends on the growth of sub-Rayleigh and intersonic dynamic shear ruptures was also studied in the laboratory [20]. In addition, the changes and characteristics of the deformation fields before and after the instability of the bent faults were also studied in the laboratory [21–24].

From the above results based on experiments conducted in rock materials [18,19,21–24], alternative activities were observed between the two fault segments on both sides of the bend before the overall instability of the fault. Since these experimental results were obtained via point contact type observation methods, such as strain gages, the spatial sampling was quite sparse. As a result, the detailed spatiotemporal evolution of ruptures cannot be observed. Accordingly, at least two problems remain unclear: (1) Is there only one alternative propagation of ruptures between the two fault segments on both sides of the bend prior to the overall instability of the fault as proposed in [19]? (2) What is the characteristic of the ruptures during their alternative propagation between the two fault segments? The solution to the two problems depends on the observation of the detailed spatiotemporal evolution of the ruptures. Therefore, it is necessary to further study the rupture evolution of the bent fault using high spatiotemporal sampling observation methods.

An experimental study depends strongly on observations. Quantitative measurement of fault slip, slip velocity, and strain via intensive sampling in time and space is very important to reveal the detailed spatiotemporal evolution of fault ruptures. The observation methods for fault slip and deformation can be divided into contact type (such as resistance strain gages or displacement gages) and non-contact type (e.g., the digital image correlation method). The contact type observation requires the sensors to be in contact with the sample, while the non-contact type observation allows the sensor to be separated from the sample. The contact type sensor occupies a certain area, which limits the number of sensors used in measurement. For instance, only dozens of strain gages can be used to cover a fault of tens of centimeters long, which were usually used in previous experimental studies [18,19,21–24]. However, via the digital image correlation (DIC) method, thousands of pixels can be used by camera photography to observe the slip and deformation along a fault of equivalent length [25–30]. Therefore, compared with the contact type observation, the spatial sampling rate of the DIC method is dramatically higher. On the other hand, the measurement precision of the contact type sensor is usually higher than that of the non-contact type sensor. For example, the measurement precision of the strain gage can easily reach 1 $\mu\varepsilon$ (micro-strain) [23], which is difficult to achieve by the DIC method. Therefore, it will be effective and economical for measurements to comprehensively utilize the two types of observations via making their respective advantages and verifying the results of each other.

Thus, the DIC method combined with strain gages are used in this study to observe the detailed spatiotemporal evolution of ruptures along a bent fault.

2. Materials and Methods

2.1. Sample and Loading Conditions

Most of the devastating earthquakes are located in the upper crust, which has a granodioritic bulk composition [31]. Therefore, the use of the granodiorite as a sample to simulate the earthquake rupture process is representative. A granodiorite sample with size of $300 \times 300 \times 50$ mm was cut through to form a bent fault. The fault surface was ground with a diamond wheel with a particle size of 150#. The roughness of the fault surface was ~100 μm before loading. The elastic modulus, Poisson's ratio, and shear modulus of the sample were 60 GPa, 0.27, and 24 GPa, respectively, which were tested via a uniaxial press machine. As shown in Figure 1, the bend point divided the fault into two segments of equal length. The segment with a small angle (42.5°) from the direction of σ_1 was referred to as segment S_I, while the other segment with a larger angle (47.5°) from the direction of σ_1 was referred to as segment S_{II}. The angle between the two segments was 5°. The bend point was located on one diagonal of the sample and was offset 6.549 mm from the geometric center of the sample. Axis D coincides with the fault trace. The common origins of axes D, X, and Y were located at the bend point of the fault. The coordinates of segments S_I and S_{II} on axis D were negative and positive, respectively. During the experiment, the sample was placed in a horizontal biaxial hydraulic servo control loading apparatus for loading. The maximum load in each axis of the loading apparatus was 1000 kN. The range of the displacement rate of the piston in each axis of the loading apparatus was from 0.01 to 100 μm/s. In order to ensure that the loading system was stable and the loading process was not interrupted even in the case of stick-slip motion of the sample, and to produce the suitable stick-slip cycle durations for observation, we used the following loading mode. The loads along the X-direction and Y-direction were synchronously increased from 0 to 4.63 MPa, then the load along the X-direction was held constant at 4.63 MPa, while the Y-direction was transferred to a displacement rate control of 0.5, 0.1, and 0.05 μm/s successively to make the sample generate dextral stick-slip motions. The sample used in this experiment was a repetition of previous studies [23,24] and the loading procedure was also similar to previous studies [23,24]. The variations of the differential stress (σ_1-σ_2) applied to the end of the sample by the apparatus and the displacement of the piston of the loading end along the Y-direction (dy) with time in the experiment are shown in Figure 2. Each stress drop in Figure 2 corresponds to a stick-slip instability event of the fault. See our previous paper [25] for details on the loading system.

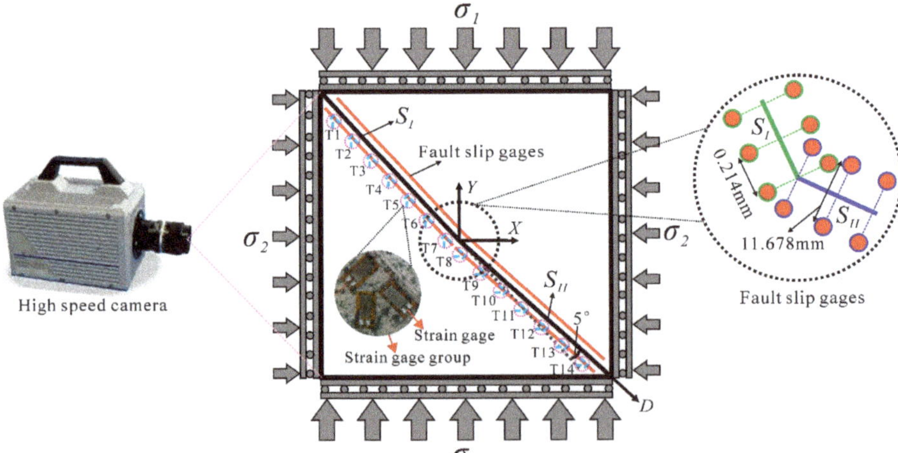

Figure 1. Experimental design. The field of view of the high-speed camera covers the entire sample surface. The D-axis coincides with the fault trace. The common origins of the D-, X-, and Y-axis coincide with the bend point of the fault. The fault is divided into segments S_I and S_{II}. The red lines on both sides of the fault are each composed of 1700 measuring points (pixels), which are symmetric with the fault and offset 5.839 mm from the fault. T1–T14 are the numbers of 14 strain gage groups mounted along the fault on the bottom sample surface. The inset pointing to the T5 strain gage group shows details of the

arrangement of the three strain gages forming the strain gage group on the sample surface. The illustration on the right is an enlarged view of the fault slip gages in the dashed circle around the fault bend. The red solid circles are the measuring points (pixels) with a 5.839 mm offset from the fault. The green and blue dashed lines connecting the measuring points (corresponding to segments S_I and S_{II}, respectively) are auxiliary lines, indicating that the two connected measuring points are symmetrically distributed with respect to the fault and form a fault slip gage. The spacing of the fault slip gages in the same fault segment is 0.214 mm.

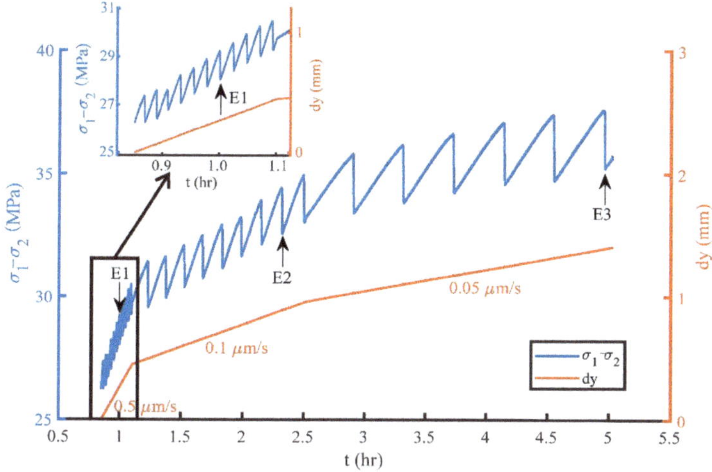

Figure 2. Variations of the differential stress (σ_1-σ_2) and piston displacement along the Y-axis (*dy*) with time (*t*). The inset is a magnified view of the black rectangular zone showing details of the stress drops during a loading rate of 0.5 µm/s. E1, E2, and E3 are the numbers of the stick-slip events indicated by the arrows, which are also the events observed by the high-speed camera.

2.2. Digital Image Correlation Method to Observe Fault Slip

To improve the spatial sampling rate of the fault slip measurement, a high-speed camera (Photron Fastcam SA2, Japan) was used to capture images of the upper sample surface during three stick-slip events (E1, E2, and E3 in Figure 2). The recording duration of each event was 7.127 s. The sampling rate was 1000 frames per second. The resolution of each image was 1792 × 1792 pixels. The actual size of each pixel corresponding to the sample surface was 157.8 × 157.8 µm². Since the images needed to be exported from the camera buffer to the computer (that takes 1 h or more depending on the data transfer rate of the equipment) after each recording to ensure the next acquisition could be performed, not all of the stick-slip events could be recorded. As a result, only three stick-slip events were recorded in the experiment. Since stick-slip events of similar recurrent periods occurred repeatedly in each loading rate as shown in Figure 2, the three recorded stick-slip events were representative for the events at the same loading rate.

The DIC method, which is an object recognition method based on pattern matching via a correlation algorithm in computer graphics [29,30,32–34], was used to process the images and calculate the displacement field of the upper sample surface. The region of interest (ROI), which is a rectangle zone covering the whole fault, was chosen to calculate the displacement field. The determination of the size of the subregion is as follows. The change of the correlation coefficient (*CC*) with the side length (*R*) of the square subregion used to calculate *CC* in the DIC method was tested and shown in Figure 3. *CC* decreased and remained unchanged before and after *R* reached 25 pixels, respectively. However, the standard deviations of *CC* decreased as *R* increased. Based on the principle of selecting the minimum *R* under the condition that the *CC* was sufficiently high via comprehensive consideration of the mean and standard deviation values of *CC* [26,28], *R* = 25 pixels was used to calculate the *CC* in the DIC method. The subregion was moved pixel by pixel in the calculated image when calculation was performed. All of the images were calculated with respect to the first image in each stick-slip event and, accordingly, the cumulative displacement field of the ROI was obtained. The threshold for determining whether a calculation result was reliable depended on the *CC*. Namely, the calculated displacement on a location was reliable when the *CC* in the location is larger than 0.99973, which was twice the standard deviation lower than the average value of *CC* at *R* = 25 shown in Figure 3.

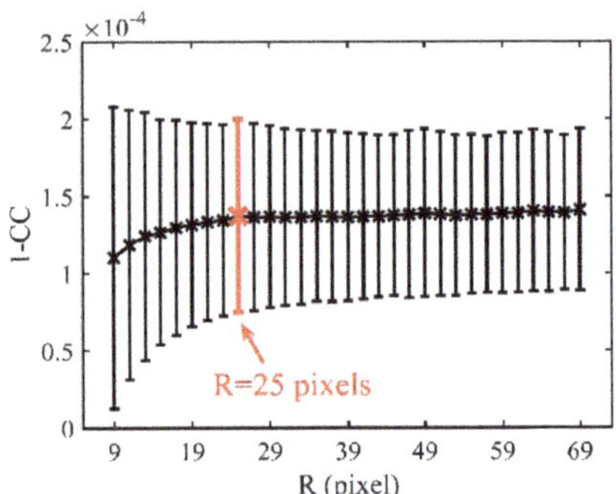

Figure 3. Influence of the side length of the subregion (*R*) on the correlation coefficient (*CC*) during the calculation of the DIC method. Crosses denote the mean value of 1–*CC*, error bars indicate the standard deviations, and the cross and error bar in red correspond to the optimal subregion side length.

Two bent lines on both sides of the fault (the red symmetric lines about the fault in Figure 1) were selected to calculate the fault slip. Each line was offset 5.839 mm from the fault, which made the fault not intersect the subregion with the center point located at the line and ensured the accuracy of the fault slip measurement [26,28]. The two bent lines each contained 1700 measuring points (pixels), of which 810 were in segment S_I and 890 were in segment S_{II}. The spacing between the measuring points was 0.214 mm. The two lines were symmetrically distributed with the fault. Each pair of symmetric measuring points from the two lines formed a fault slip gage. As a result, a total of 1700 fault slip gages with spacing of 0.214 mm were used to observe the detailed spatial distribution of the fault slip. The displacement error of the DIC method was ±5 μm, which was obtained under the condition that the sample was static without loading. Segmentation smoothing of the time series of the fault slip was performed to improve the measurement precision. See our previous papers [26,30] for the DIC method and data processing method used in this paper.

2.3. Strain Gage to Observe Shear Strain along the Fault

An array of 42 resistance strain gages were mounted along the fault on the bottom sample surface, which formed 14 strain gage groups, as shown in Figure 1. The strain gage groups were used to measure the shear strain along the fault and test the results derived from the DIC method. Each strain gage had a resistance grid of 3 × 5 mm with a resistance value of 120.1 ± 0.1 Ω. The sensitivity coefficient of the strain gage was 2.10 ± 1%. Data were acquired by a 96-channel strain acquisition system [21]. The analog-to-digital conversion, sampling rate, and observation error of the device was 16 bit, 100 Hz, and ±1.5 με, respectively. The angles between the three strain gages in each strain gage group and the measured fault segment were 0°, 45°, and 90°. Correspondingly, the offsets of the centers of the three gages from the measured fault segment were 5.0, 12.2, and 6.1 mm, respectively. The plane strain tensor could be obtained by each strain gage group (the specific method was described in previous studies [21,24]), and subsequently, the shear strain of the fault could be calculated.

3. Results

The results obtained from the DIC method show that the rupture process of the bent fault can be divided into two stages: an alternative propagation stage followed by an overall instability stage. The two stages were observed in all three recorded stick-slip events in the experiment.

3.1. The Alternative Propagation Stage

The alternative propagation of the rupture between the two fault segments prior to the overall instability of the fault can occur in multiple cycles. During the process, the rupture speed increases from several tens of mm/s to several tens of m/s. Meanwhile, the slip velocity within the rupture also grows from several μm/s to several mm/s. Although the rupture at this stage can propagate between the two fault segments, the rupture speed usually has a jump when the rupture propagates across the fault bend. This indicates the influence of the fault bend on the propagation of the rupture, especially when the slip velocity within the rupture is high as shown in Figure 4b, Figure 5c, and Figure 6b.

Two alternative propagation cycles of the rupture were observed in event E1, as shown in Figure 4. The first cycle began in segment S_I and propagated to segment S_{II}, as shown in Figure 4a. During this process, the rupture speed accelerated from 16 mm/s in segment S_I to 151 mm/s in segment S_{II}, and the slip velocity accelerated up to 3 μm/s. The rupture in the second cycle propagated in the opposite direction with respect to that in the first cycle, during which the rupture speed increased from 0.645 m/s in segment S_{II} to 15 m/s in segment S_I, and the slip velocity increased up to 200 μm/s, as shown in Figure 4b.

Three alternative propagation cycles of the rupture were observed in event E2, as shown in Figure 5. The rupture propagated from segment S_I to segment S_{II} in the first and the last cycles, as shown in Figure 5a,c, but propagated from segment S_{II} to segment S_I in the second cycle, as shown in Figure 5b. In the first cycle, only the propagation in the segment S_{II} was observed. The rupture speed increased from 25 to 90 mm/s and the slip velocity increased up to 1.2 μm/s. During the second cycle, the rupture propagated from segment S_{II} at a speed of 0.642 m/s to segment S_I at a speed of 2.7 m/s. Meanwhile, the slip velocity increased up to 50 μm/s. In the last cycle, the rupture propagated from segment S_I at a speed of 6.5 m/s to segment S_{II} at a speed of 30 m/s and the slip velocity increased up to 1200 μm/s.

In event E3, there were also two alternative propagation cycles of the rupture, as shown in Figure 6, which initiated in segment S_{II} and propagated to segment S_I in the first cycle, as shown in Figure 6a, and propagated in the opposite direction in the second cycle, as shown in Figure 6b. During the first cycle, the rupture speed increased from 142 mm/s in segment S_{II} to 270 mm/s in segment S_I and the slip velocity accelerated up to 2.5 μm/s. While in the second cycle, the rupture speed increased from 0.128 m/s in segment S_I to 13 m/s in segment S_{II} and the slip velocity accelerated up to 500 μm/s.

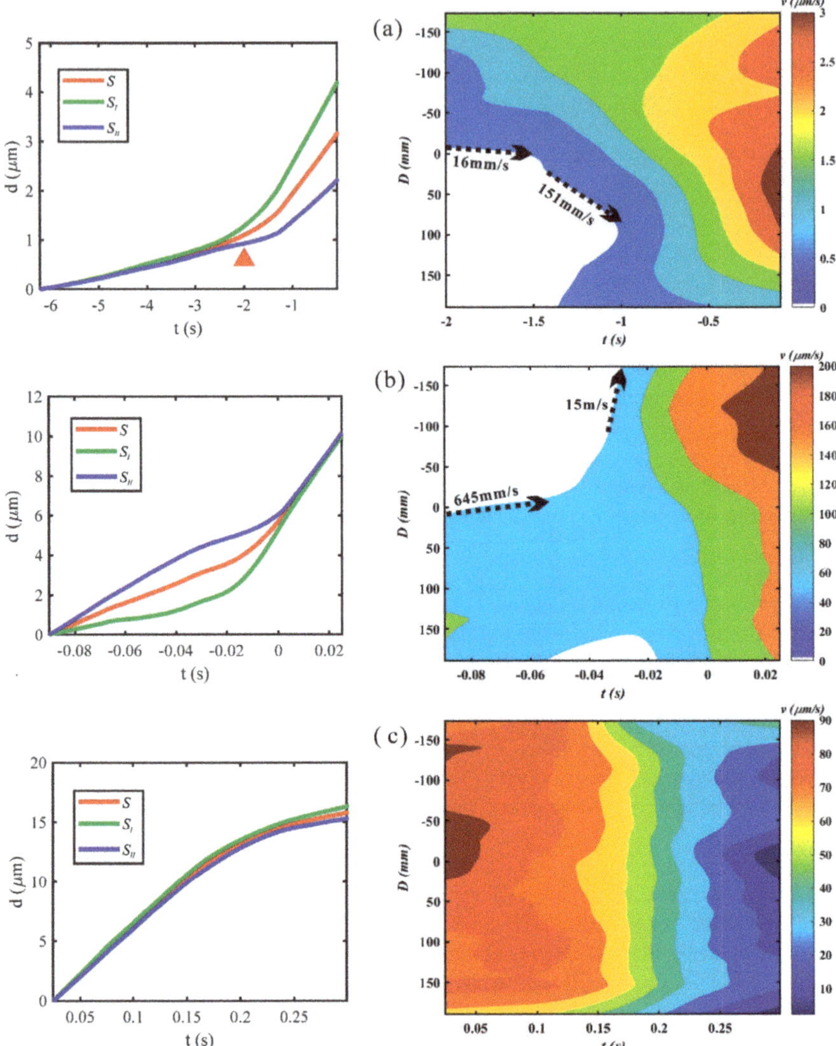

Figure 4. Spatiotemporal distribution of fault slips and slip velocity before and after the overall instability of event E1. The subgraphs on the left side show the change of the average slip of the whole fault (S), the segment S_I (S_I), and the segment S_{II} (S_{II}) with time. $D = 0$, $D < 0$, and $D > 0$ denote the fault bend point, the segment S_I, and the segment S_{II}, respectively. The subgraphs on the right side show the spatiotemporal distribution of the fault slip velocity (v). The black dotted arrows and numbers represent the directions and propagation speeds of ruptures, respectively. Note that the time span of the left subgraph is larger than that of the right subgraph in (**a**). The red triangle in the left subgraph in (**a**) denotes the beginning of the obvious slip velocity that can be observed; spatiotemporal evolution of slip velocity after which is shown in the right subgraph in (**a**). The time spans of the left and right subgraphs in (**b**,**c**) are consistent. The time axes from (**a**) to (**c**) are continuous. The common timing point of the data from the DIC method and strain gages is set zero for each stick-slip event. Accordingly, time before and after the common timing point is negative and positive, respectively.

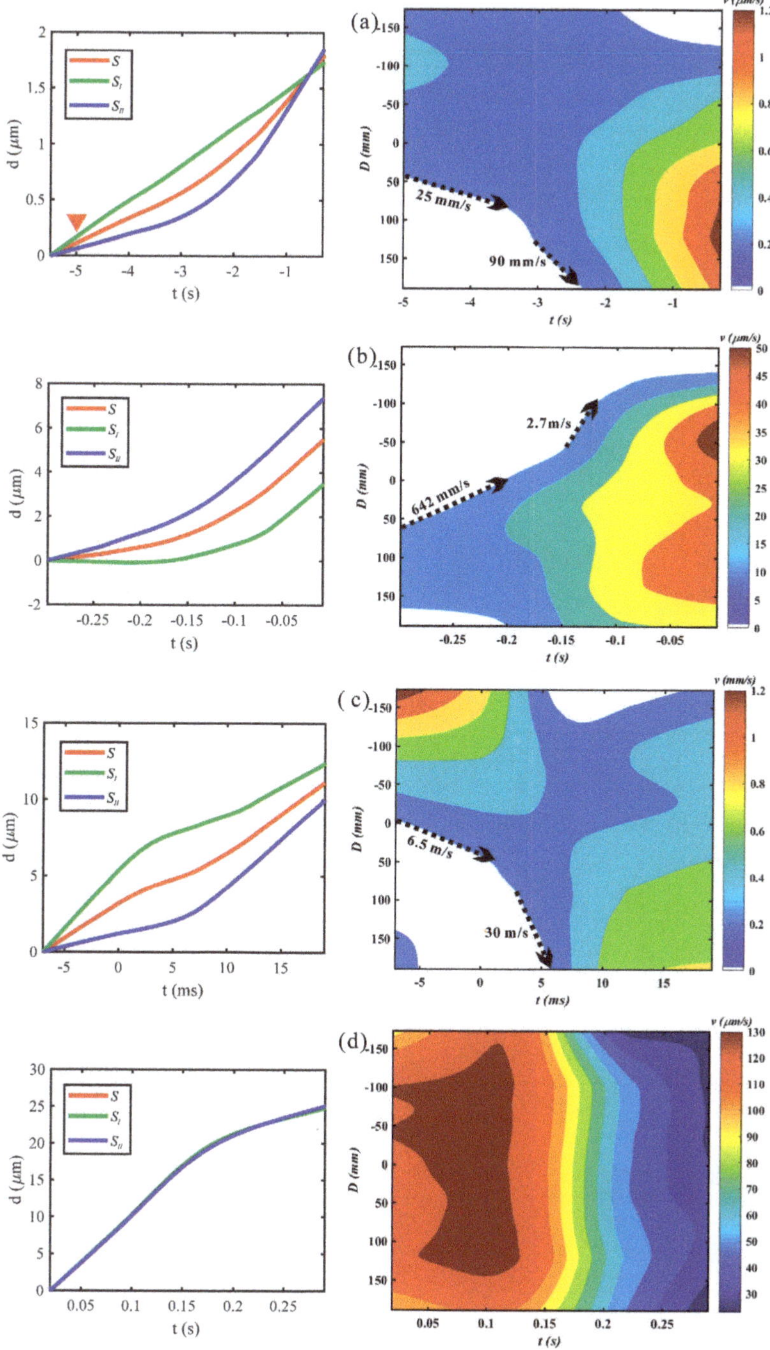

Figure 5. Spatiotemporal distribution of fault slips and slip velocity before and after the overall instability of event E2. The time spans of the left and right subgraphs in (**b**–**d**) are consistent. The time axes from (**a**) to (**d**) are continuous. Other labels are the same as shown in Figure 4.

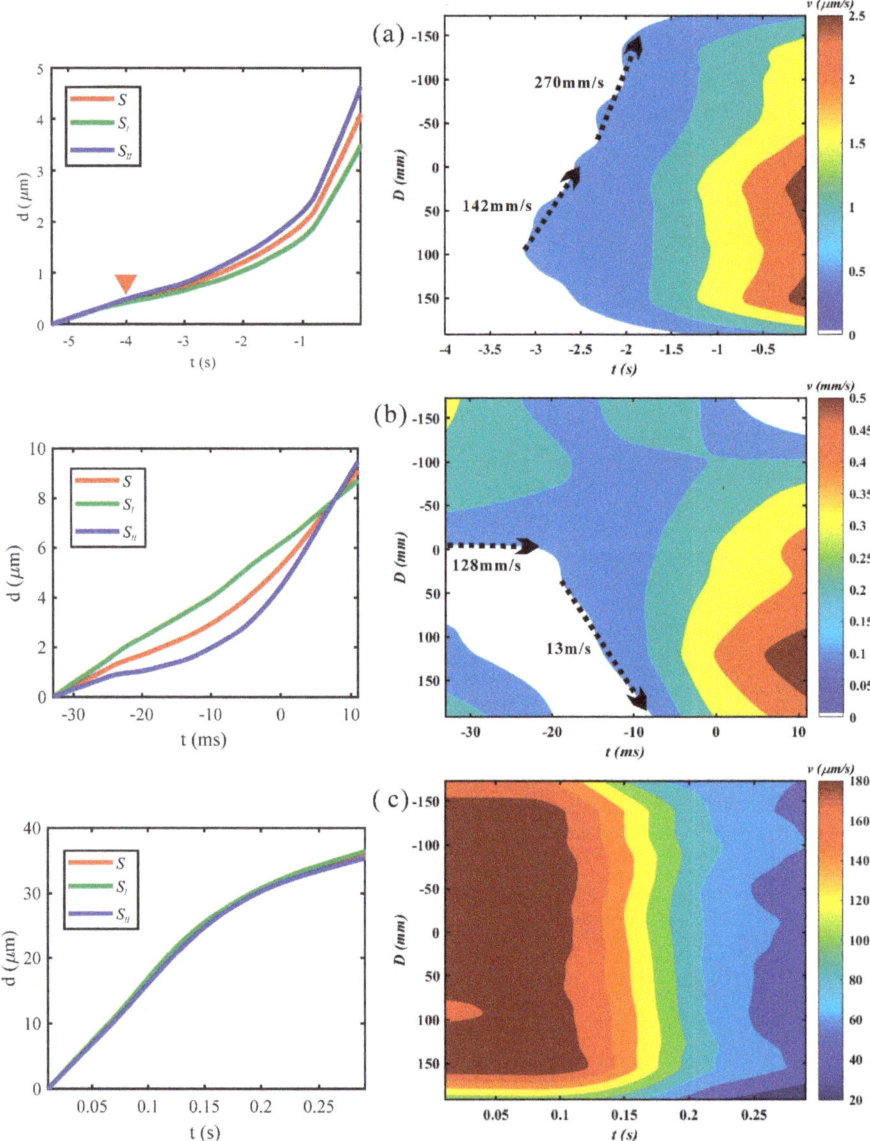

Figure 6. Spatiotemporal distribution of fault slips and slip velocity before and after the overall instability of event E3. The time axes from (**a**) to (**c**) are continuous. Labels are the same as shown in Figure 4.

3.2. The Overall Instability Stage

The overall instability stage immediately followed the alternative propagation stage, which was characterized by the almost uniform distribution of slip velocity along the whole fault, as shown in Figure 4c, Figure 5d, and Figure 6c. This indicates that the influence of the fault bend on the rupture propagation was weak in this stage. There were two distinct features at this stage compared with the alternative propagation stage: (1) the slip velocity at this stage was in the order of tens to hundreds of µm/s, as shown in Figure 4c, Figure 5d, and Figure 6c, which was lower than that in the last cycle of the alternative propagation stage, as shown in Figure 4b, Figure 5c, and Figure 6b. (2) The rupture speed in this stage was too high and beyond the observation capability of the high-speed camera due to the sampling rate of 1000 frames per second, and as a result, slip velocity contours were almost perpendicular to the time axis in Figure 4c, Figure 5d, and Figure 6c.

4. Discussion

4.1. Reliability Test of the DIC Method via Strain Gage Measurement

The above fault slip velocity measured by the DIC method spanned a wide range of at least 3 orders of magnitude, as shown in Figure 4, Figure 5, and Figure 6. Although the high slip velocity was easily observed by the DIC method with a high signal-noise ratio, the reliability of the observed low slip velocity needs to be tested, especially in the first cycle of the alternative propagation stage where the slip velocity can be as low as below 1 µm/s, as shown in Figure 4a, Figure 5a, and Figure 6a. Such a low slip velocity leads to a problem of whether the observed velocity was a fault slip or a strain buildup. To make it clear, a high precision measurement of the shear strain along the fault is necessary because local shear strain will drop as local instability occurs in the fault [35,36]. Therefore, we used the evolution of shear strain at the 14 locations along the fault to test whether the observed slip velocity in the first cycle of the alternative propagation stage, as shown in Figure 4a, Figure 5a, and Figure 6a, was a fault slip or a strain buildup. Since the sampling rate of the strain acquisition system was only 100 Hz, the shear strain measurement was unable to test the rupture in other cycles of the alternative propagation stage.

Figure 7 shows the evolution of the shear strain in a duration of 7 s covering the first cycle of the alternative propagation stage of the three stick-slip events. The subgraphs (a) and (b), (c) and (d), (e) and (f) in Figure 7 correspond to events E1, E2, and E3, respectively. The changes of the average shear strain in the whole fault, the segment S_I, and the segment S_{II} with time are shown on the left side of Figure 7. The changes of the shear strain of each strain gage group with time are shown on the right side of Figure 7. The red triangles in Figure 7a,c,e mark the turning points in the average shear strain of the fault segments. Figure 7a shows that the average shear strain in segment S_I increased and decreased just before and after the red triangle ($t \sim -2.8$ s), respectively, while the average shear strain continuously increased in segment S_{II} at the red triangle. These indicate that segment S_I began to slip at $t \sim -2.8$ s. Figure 7b shows that the rupture initiated near group T1, of which the shear strain dropped first. These are consistent with the results observed in Figure 4a.

The two red triangles in Figure 7c show that the average shear strain in segment S_{II} changed from ascending to descending at $t \sim -2.8$ s, while the average shear strain in segment S_I changed from decreasing to rising at $t \sim -2.4$ s. This indicates that the rupture gradually shifted from segment S_I to segment S_{II} between $t \sim -2.8$ s and $t \sim -2.4$ s, which is consistent with the evolution of the rupture observed in Figure 5a.

The red triangle in Figure 7e shows that the average shear strain began to rise slowly in segment S_I but decreased in segment S_{II} after $t \sim -3.1$s, indicating that the rupture initiated in segment S_{II} before it propagated to segment S_I during this period. Figure 7f shows that the shear strain of group T11 dropped first, which indicates that the rupture initiated near group T11. These are consistent with the observations in Figure 6a.

The consistency of the observations between the strain gages and the DIC method indicates that the rupture evolution shown in Figure 4a, Figure 5a, and Figure 6a are reliable.

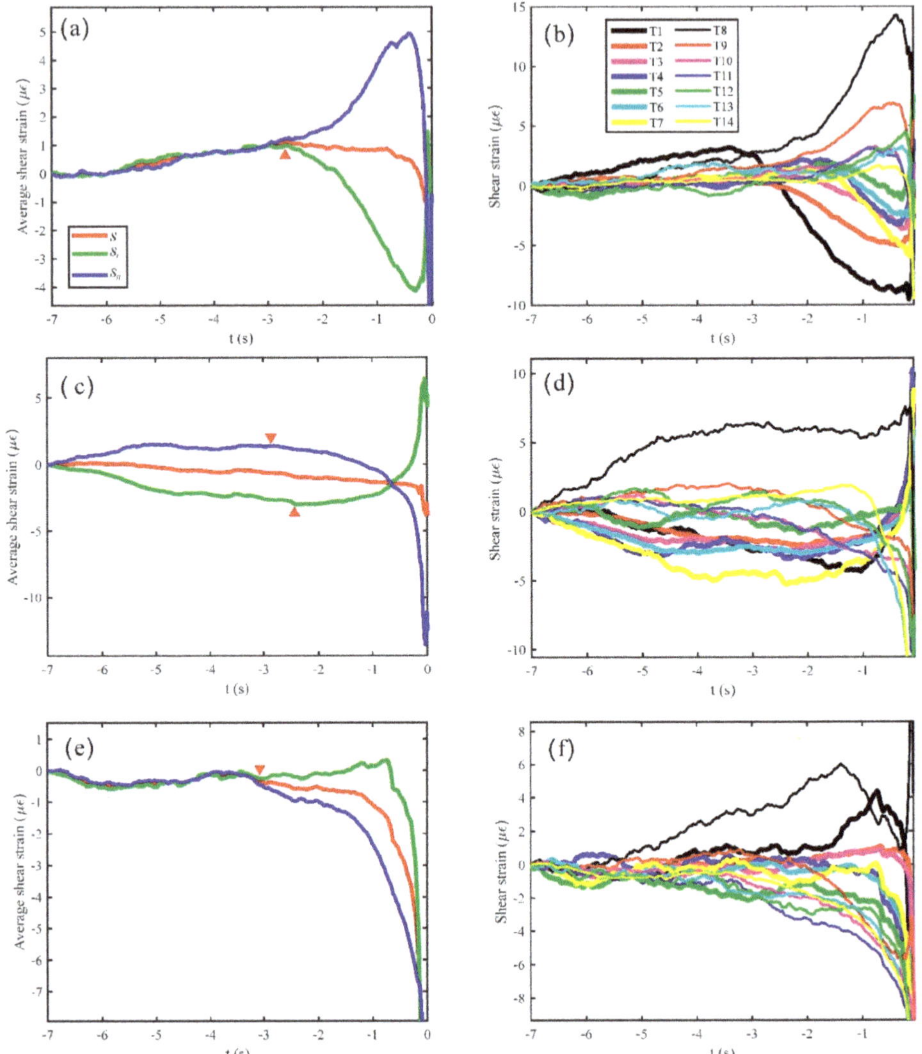

Figure 7. Variations of the shear strain with time. (**a–f**) correspond to events E1, E2, and E3, respectively. (**a,c,e**) are the average shear strains in the whole fault (S), the segment S_I (S_I), and the segment S_{II} (S_{II}), respectively. The red triangles indicate the turning points in the trend of the average shear strain. (**b,d,f**) are the shear strains in each strain gage group. Groups T1–T7, which are mounted in the segment S_I, are displayed with thick lines. Groups T8–T14, which are mounted in the segment S_{II}, are displayed with thin lines. The common timing point of the data from the DIC method and strain gages is set zero for each stick-slip event. Accordingly, the time before and after the common timing point is negative and positive, respectively.

4.2. The Influences of the Fault Bend on the Rupture Propagation and Further Research

The previous experiments have observed the process of the alternative propagation of the rupture between the two fault segments on both sides of the fault bend [18,19,24]. The process that the rupture initiated at one segment was stopped at the fault bend and restarted after a time delay near the fault bend at another segment prior to the overall instability of the fault [19], which was similar to the last cycle of the alternative propagation stage in our experiment. However, more details of the alternative propagation of the rupture were observed in our experiment, especially the multiple cycles of the alternative propagation process, during which the rupture speed and slip velocity were increasing. This is owing to the high spatial sampling observation of the DIC method.

The boundaries of the slip velocity contours are perpendicular to the time axis in Figure 4c, Figure 5d, and Figure 6c, indicating that the rupture had a rapid propagation speed during the overall instability, which was beyond the recording capability of the high-speed camera with a sampling rate of 1000 frames per second. As a result, the rupture process during the overall instability of the fault could not be fully recorded. Thus, what we can discuss is the alternative propagation stage that was fully recorded. Previous studies have pointed out that the fault bend can serve as an initiation or termination for earthquake ruptures, e.g., [7]. The observed alternative propagation of the rupture in our experiment, as shown in Figures 4–6, was also essentially influenced by the fault bend. Namely, the rupture initiated in one fault segment was hindered by the fault bend and accelerated before and after it propagated across the fault bend to another segment, respectively, as shown in Figure 4a,b, Figure 5a–c, and Figure 6a,b. This effect indicates the interaction between the fault bend and the rupture, which is similar to the effect that the asperities hinder and promote the rupture before and after the failure of the asperities in straight faults, respectively [25,35]. On the other hand, unlike straight faults where the failed asperities have been deformed and do not interact with the rupture in the same seismogenic process [25,35], a fault bend is a geometric structure of the fault and is accordingly difficult to be deformed by the rupture. As a result, the fault bend can interact with the rupture in multiple cycles during the same seismogenic process, which was observed in the alternative propagation stage in our experiment.

The process leading to the overall instability of the fault observed in the straight fault is usually accompanied with acceleration in rupture propagation and fault slip, e.g., [26,30,35]. However, this process of the bent fault observed in our experiment is different. Although the rupture accelerated continuously in trend, the slip velocity during the overall instability stage of the bent fault was slower than that in the last cycle of the alternative propagation stage, as shown in Figure 4, Figure 5, and Figure 6. Study of the mechanism of this phenomenon will be very important for further understanding of the conditions for the overall instability of the bent fault. We think that the observation of the detailed spatiotemporal evolution of the rupture propagation during the overall instability of the bent fault may help solve the problem. This depends on the improvement of the temporal sampling rate of the DIC observation system.

5. Conclusions

In this experiment, we used a high-speed camera combined with a DIC method to observe the rupture process of a bent fault with a 5° bending angle between its two segments under biaxial loading, which was also supported by the observation of strain gages. The results show that the evolution of the ruptures was composed of two stages: the alternative propagation between the two fault segments on both sides of the fault bend followed by the overall instability of the fault. Thanks to the dense spatial sampling of the DIC method, multiple alternative propagation cycles of the ruptures between the two fault segments with accelerating rupture speed and slip velocity were found prior to the overall instability of the fault, which indicate the repeated interactions between the ruptures and the fault bend prior to the overall instability of the fault.

Author Contributions: Conceptualization, Y.Z. and Y.G.; Data curation, Y.Z. and Y.G.; Formal analysis, Y.Z.; Funding acquisition, Y.Z.; Investigation, Y.Z., Y.G., and S.A.B.; Methodology, Y.Z. and Y.G.; Software, Y.Z.; Validation, Y.Z.; Visualization, Y.Z.; Writing—original draft, Y.Z.; Writing—review & editing, Y.Z., Y.G., and S.A.B.

Funding: This research was funded by the National Natural Science Foundation of China, grant numbers 41702226 and 41572181.

Acknowledgments: Yuntao Ji and Qiang Li helped to conduct the experiments. Jin Ma and S. I. Sherman supplied constructive suggestions. We thank the four reviewers for their helpful suggestions and comments.

Conflicts of Interest: The authors declare no conflict of interest.

References

1. Sangha, S.; Peltzer, G.; Zhang, A.; Meng, L.; Liang, C.; Lundgren, P.; Fielding, E. Fault geometry of 2015, mw7.2 murghab, tajikistan earthquake controls rupture propagation: Insights from insar and seismological data. *Earth Planet. Sci. Lett.* **2017**, *462*, 132–141. [CrossRef]
2. Elliott, A.J.; Oskin, M.E.; Liu-zeng, J.; Shao, Y.X. Persistent rupture terminations at a restraining bend from slip rates on the eastern altyn tagh fault. *Tectonophysics* **2018**, *733*, 57–72. [CrossRef]
3. Acharya, H.K. Influence of fault bends on ruptures. *Bull. Seismol. Soc. Am.* **1997**, *87*, 1691–1696.
4. Barka, A.A.; Kadinsky-Cade, K. Strike-slip fault geometry in turkey and its influence on earthquake activity. *Tectonics* **1988**, *7*, 663–684. [CrossRef]
5. Ando, R.; Imanishi, K.; Panayotopoulos, Y.; Kobayashi, T. Dynamic rupture propagation on geometrically complex fault with along-strike variation of fault maturity: Insights from the 2014 northern nagano earthquake. *Earth Planets Space* **2017**, *69*, 130. [CrossRef]
6. Wang, W.-M.; He, Y.-M.; Yao, Z.-X. Complexity of the coseismic rupture for 1999 chi-chi earthquake (Taiwan) from inversion of gps observations. *Tectonophysics* **2004**, *382*, 151–172. [CrossRef]
7. King, G.; Nábělek, J. Role of fault bends in the initiation and termination of earthquake rupture. *Science* **1985**, *228*, 984–987. [CrossRef] [PubMed]
8. King, G.C.P. Speculations on the geometry of the initiation and termination processes of earthquake rupture and its relation to morphology and geological structure. *Pure Appl. Geophys.* **1986**, *124*, 567–585. [CrossRef]
9. Aochi, H.; Fukuyama, E.; Matsu'ura, M. Spontaneous rupture propagation on a non-planar fault in 3-d elastic medium. *Pure Appl. Geophys.* **2000**, *157*, 2003–2027. [CrossRef]
10. Oglesby, D.D.; Day, S.M. The effect of fault geometry on the 1999 chi-chi (Taiwan) earthquake. *Geophys. Res. Lett.* **2001**, *28*, 1831–1834. [CrossRef]
11. Duan, B.; Oglesby, D.D. Multicycle dynamics of nonplanar strike-slip faults. *J. Geophys. Res. Solid Earth* **2005**, *110*. [CrossRef]
12. Kase, Y.; Day, S.M. Spontaneous rupture processes on a bending fault. *Geophys. Res. Lett.* **2006**, *33*. [CrossRef]
13. Fang, Z.; Xu, G.; Oglesby, D.D. Geometric effects on earthquake nucleation on bent dip-slip faults. *Int. J. Appl. Mech.* **2011**, *3*, 99–117. [CrossRef]
14. Lozos, J.C.; Oglesby, D.D.; Duan, B.; Wesnousky, S.G. The effects of double fault bends on rupture propagation: A geometrical parameter study. *Bull. Seismol. Soc. Am.* **2011**, *101*, 385–398. [CrossRef]
15. Zeng, Y.; Chen, C.-H. Fault rupture process of the 20 september 1999 chi-chi, Taiwan, earthquake. *Bull. Seismol. Soc. Am.* **2001**, *91*, 1088–1098. [CrossRef]
16. Aochi, H.; Fukuyama, E. Three-dimensional nonplanar simulation of the 1992 landers earthquake. *J. Geophys. Res. Solid Earth* **2002**, *107*, ESE4-1–ESE4-12. [CrossRef]
17. Aochi, H.; Madariaga, R.L. The 1999 Izmit, Turkey, earthquake: Nonplanar fault structure, dynamic rupture process, and strong ground motion. *Bull. Seismol. Soc. Am.* **2003**, *93*, 1249–1266. [CrossRef]
18. Ma, J.; Ma, W.; Ma, S.; Deng, Z.; Liu, L.; Liu, T. Experimental study and numerical simulation on physical fields during the deformation of a 5° bend fault. *Seismol. Geol.* **1995**, *17*, 318–326.
19. Kato, N.; Satoh, T.; Lei, X.; Yamamoto, K.; Hirasawa, T. Effect of fault bend on the rupture propagation process of stick-slip. *Tectonophysics* **1999**, *310*, 81–99. [CrossRef]
20. Rousseau, C.-E.; Rosakis, A.J. On the influence of fault bends on the growth of sub-rayleigh and intersonic dynamic shear ruptures. *J. Geophys. Res. Solid Earth* **2003**, *108*. [CrossRef]
21. Liu, L.; Ma, J.; Ma, S. Characteristics and evolution of background strain field on typical structure models. *Seismol. Geol.* **1995**, *17*, 349–356.
22. Ma, J.; Liu, L.; Ma, S. Fault geometry and departure of precursors from the epicenter. *Earthq. Res. China* **1998**, *12*, 59–67.
23. Guo, Y.-S.; Ma, J.; Yun, L. Experimental study on stick-slip process of bending faults. *Seismol. Geol.* **2011**, *33*, 26–35.
24. Yun, L.; Guo, Y.-S.; Ma, J. An experimaental study of evolution of physical field and the alternative activities during stick-slip of 5° bend fault. *Seismol. Geol.* **2011**, *33*, 356–368.

25. Zhuo, Y.-Q.; Liu, P.; Chen, S.; Guo, Y.; Ma, J. Laboratory observations of tremor-like events generated during preslip. *Geophys. Res. Lett.* **2018**, *45*, 6926–6934. [CrossRef]
26. Zhuo, Y.-Q.; Guo, Y.; Chen, S.; Ji, Y.; Ma, J. Laboratory observations of linkage of preslip zones prior to stick-slip instability. *Entropy* **2018**, *20*, 629. [CrossRef]
27. Zhuo, Y.-Q.; Bornyakov, S.A.; Guo, Y.-S.; Ma, J.; Sherman, S.I. Influences of obliquity angle difference on the evolution of fen-wei rift: A study from segemented transtension clay model. *Seismol. Geol.* **2016**, *38*, 259–277.
28. Zhuo, Y.-Q.; Ma, J.; Guo, Y.-S.; Ji, Y.-T. Identification of the meta-instability stage via synergy of fault displacement: An experimental study based on the digital image correlation method. *Phys. Chem. Earth Parts A/B/C* **2015**, *85–86*, 216–224. [CrossRef]
29. Ji, Y.; Hall, S.A.; Baud, P.; Wong, T.F. Characterization of pore structure and strain localization in majella limestone by x-ray computed tomography and digital image correlation. *Geophys. J. Int.* **2015**, *200*, 699–717. [CrossRef]
30. Zhuo, Y.; Guo, Y.; Ji, Y.; Ma, J. Slip synergism of planar strike-slip fault during meta-instable state: Experimental research based on digital image correlation analysis. *Sci. China Earth Sci.* **2013**, *56*, 1881–1887. [CrossRef]
31. Rudnick, R.L.; Gao, S. 3.01—Composition of the continental crust. In *Treatise on Geochemistry*; Holland, H.D., Turekian, K.K., Eds.; Pergamon: Oxford, UK, 2003; pp. 1–64.
32. Sutton, M.A.; Wolters, W.J.; Peters, W.H.; Ranson, W.F.; McNeill, S.R. Determination of displacements using an improved digital correlation method. *Image Vis. Comput.* **1983**, *1*, 133–139. [CrossRef]
33. Peters, W.H.; Ranson, W.F. Digital imaging techniques in experimental stress analysis. *OPTICE* **1982**, *21*, 213427. [CrossRef]
34. Yamaguchi, I. A laser-speckle strain gauge. *J. Phys. E Sci. Instrum.* **1981**, *14*, 1270–1273. [CrossRef]
35. McLaskey, G.C.; Kilgore, B.D. Foreshocks during the nucleation of stick-slip instability. *J. Geophys. Res. Solid Earth* **2013**, *118*, 2982–2997. [CrossRef]
36. Dieterich, J.H. Preseismic fault slip and earthquake prediction. *J. Geophys. Res.* **1978**, *83*, 3940. [CrossRef]

 *applied sciences*

Article

Application of 3D Digital Image Correlation for Development and Validation of FEM Model of Self-Supporting Arch Structures

Krzysztof Malowany [1,*], Artur Piekarczuk [2], Marcin Malesa [1], Małgorzata Kujawińska [1] and Przemysław Więch [2]

[1]   Warsaw University of Technology, Institute of Micromechanics and Photonics, 8 Św. A. Boboli St., 02-525 Warsaw, Poland; m.malesa@mchtr.pw.edu.pl (M.M.); m.kujawinska@mchtr.pw.edu.pl (M.K.)

[2]   Instytut Techniki Budowlanej, 1 Filtrowa St., 00-611 Warsaw, Poland; a.piekarczuk@itb.pl (A.P.); p.wiech@itb.pl (P.W.)

*   Correspondence: k.malowany@mchtr.pw.edu.pl

Received: 15 January 2019; Accepted: 25 March 2019; Published: 28 March 2019

Abstract: Many building structures, due to a complex geometry and non-linear material properties, are cumbersome to analyze with finite element method (FEM). A good example is a self-supporting arch-shaped steel sheets. Considering the uncommon geometry and material profile of an arch (due to plastic deformations, cross section of a trough, a goffer pattern), the local loss of stability can occur in unexpected regions. Therefore, the hybrid experimental-numerical methodology of analysis and optimization of arch structures have been proposed. The methodology is based on three steps of development and validation of a FEM with utilization of a digital image correlation (DIC) method. The experiments are performed by means of 3D DIC systems adopted sequentially for each measurement step conditions from small size sections, through few segment constructions up to full scale in situ objects.

Keywords: experimental-numerical method; digital image correlation; finite element method; static analysis; arch structures

---

1. Introduction

The development of large-scale and complex engineering structures creates new challenges for designers and constructors, who need to meet the demands of increasing safety, extended component lifetime and simultaneously reduced investment and operation costs. To fulfill these requirements new materials (e.g., composites) and assembly technologies are being developed. An example of this type of construction is a self-supporting arch structure. Such structures had been initially built as temporary buildings used for military purposes. Adaptations of this technology for civil purposes, required extension of the designed lifetime and consideration of different environmental conditions, and therefore design problems, especially in terms of stability and load transfers occurred [1–5]. In order to ensure safety and proper operational parameters during their lifetime, hybrid experimental-numerical methods are being used during design and exploitation stages [6]. The common practice in experimental mechanics is to validate the numerical model using point wise sensors (e.g., strain gauges). They are attached to a tested structure in places in which the highest stress concentrations are expected (based on the analysis of a numerical model). This is a simple and low-cost approach, but a problem can occur if an inaccurate numerical model does not indicate all the places in which stress concentrations occur. This, in turn, may cause errors in the process of validation of the numerical model. Therefore, an advanced evaluation of numerical model is more often supported with the data obtained by means of full-field optical measurement methods, which determine displacements and deformations in critical areas of the objects [7,8]. In the case of investigations of large engineering structures, the most commonly used techniques are terrestrial laser scanning [9–11] and the 3D digital image correlation (3D DIC) method [1,12–17]. Terrestrial laser scanning systems enable 3D measurements of shape of object. By comparison of the acquired shapes in different load conditions, deformation of an object under the load can be calculated [10,11]. The method is simple to use, however, due to time required for measurement (from few second up to few minutes), its utilization is limited to static measurements. 3D DIC measurement systems combine digital image correlation and triangulation methods. 3D DIC provides directly 3D displacement vector **d** distribution (displacement maps (u,v,w) in x, y and z directions respectively) in a measured field of view [18]. Utilization of the 3D DIC method requires modification of a measured surface (applying a paint coat that provides a random texture) therefore it is more difficult to use compared to terrestrial laser scanning systems. Nonetheless the 3D DIC system enables measurements of an investigated object under varying in time conditions, and therefore this method have been used in the presented application. Numerical modeling of a self-supporting arch structure is cumbersome, considering the uncommon geometry and material characteristics (due to plastic deformations, the cross section of a trough, goffer pattern). Thus, the development and validation of a FEM model of a full-sized construction made of self-supporting arch sheets described in Section 2, was divided into three steps starting from small sized sections (Section 4), through few segment constructions (Section 5), to full scale in situ objects (Section 6). Each step required a tailored approach to the measurements with 3D DIC due to different: accuracies, sizes of fields of view, forms of outcome data, environmental conditions. Some aspects of this work had been presented in our previous papers. In papers [2,3] the investigation of 1 m long section of arch-shaped steel sheets was presented, the goals of this work was to simulate local loss of stability and to determine a geometric model of the surface shape for FEM analysis. In papers [1,13] the investigation of segments of arch-shaped steel sheets in laboratory conditions were presented; the goals of this work were to investigate the global stability and to determine the mechanical behavior of supports. The measurements presented in paper [13] were performed with utilization of multi-camera DIC system with overlapping areas of fields of view of cameras. In paper [16] the performed measurements of the full-scale construction made of arch-shaped self-supporting metal plates was presented, and in order to enable these measurements the multi-camera DIC system with distributed field of view (FOV) was developed. In this paper we summarize the development of 3D DIC systems and present the combined three-steps hybrid experimental-numerical methodology of analysis and optimization of arch structures.

2. Characteristic of the Investigated Object and General Procedure Supporting Development and Validation of a Finite Element Model

2.1. Specification of Arch-Shaped Steel Sheets Used as a Self-Supporting Arch Structures

A self-supporting arch-shaped covering of steel sheet sections is used in civil engineering [19]. Typical radii of the assembled arch coverings varies in the range of 6 m to 30 m (Figure 1a). Due to the simple design, quick installation and relatively low implementation costs, this type of covering gained significant popularity. Arch-shaped steel sheets are cold formed in two stages. At first, the flat metal plates (stored in a roller) are bended in order to receive plates with trapezoidal cross section, then the plates are goffered (local plastic deformation) in order to receive arches (Figure 1b). A full scale object is obtained from a number of single arches connected to each other with double lock standing seams. Considering the trapezoidal cross section, the differences in radius of the arch cause waiving in the surface between shelves.

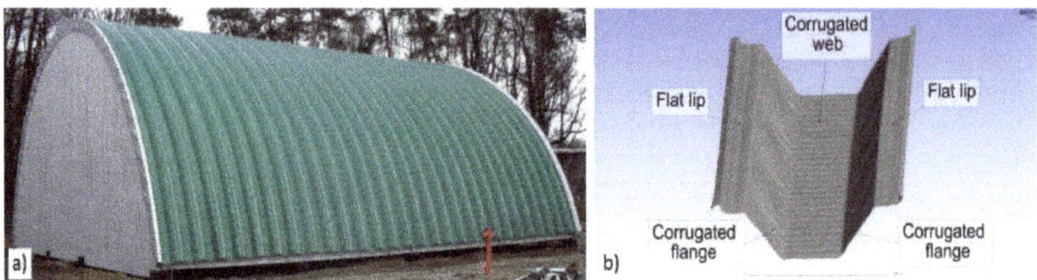

Figure 1. (**a**) Outside image of the measured hall as an example of typical self-supporting arch-shaped covering, (**b**) section surfaces.

2.2. Three-Step Development and Validation of a Numerical Model

Considering the uncommon geometry and material characteristic of an arch, development of a FEM model of such a structure required a specific approach and its validation at different stages of advancement. The development of a FEM model of the full-sized construction made of the self-supporting arch has been heavily supported by the experiments performed by means of 3D DIC within three steps. At first, the tests have been carried out on single sections of the arch in order to determine which geometric model of the surface shape (planar, corrugation or corrugation and wavy model) should be applied in the further steps for FEM analysis. The considered sections were 1 m long and 0.7 m wide. In the next step, a structure composed of four individual segments with the geometry selected at step 1 has been used in laboratory tests with controlled loading conditions in order to investigate the global stability and to determine the mechanical behavior of supports, in particular to define the rotational stiffness of supports used in a FEM model of arches. The measured object was of 12 m span and 2.8 m wide. The final step was based on the tests of a full-scale object: 8 m high covering the area of 18 × 18 m. The outcomes of this test is the validation of a complete FEM model in which the knowledge gained in steps 1 and 2 had been implemented. It has been proven that the model accurately simulates construction deformations under the environmental loads. The model can be scaled to larger constructions. The general procedure supporting the development and validation of a FEM model has been summarized and presented in Figure 2. This procedure is described in detail in Sections 5 and 6.

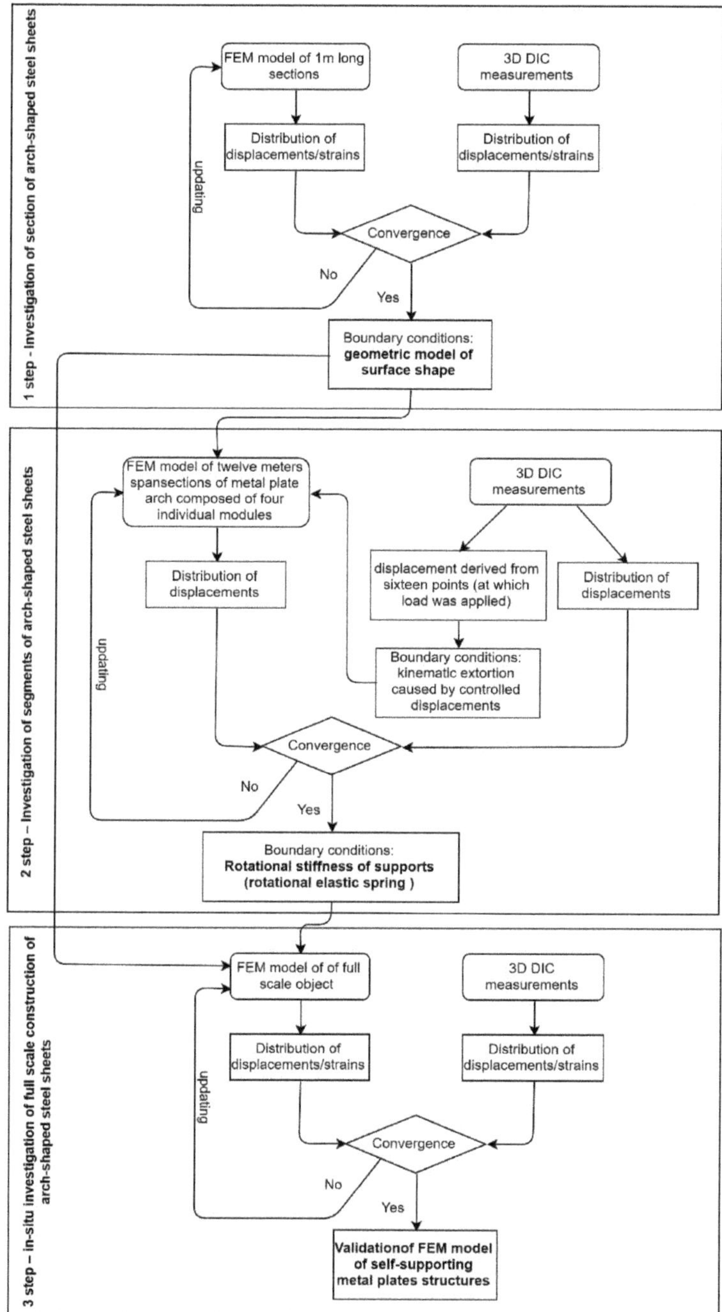

Figure 2. Flowchart of the 3-steps updating and validation of numerical model of self-supporting arch structures with utilization of 3D digital image correlation method.

3. Digital Image Correlation Method

At each stage of the procedure described in Section 2 full-field measurements of displacement vector **d** (u,v,w) of the investigated structures are required. The method which is best fitted to the measurement requirements is digital image correlation [18]. DIC is based on acquisition of a set of images of a tested object which is subjected to load. The surface of an object under investigation has to be covered with a random texture. The 2D DIC version uses a single camera. One of the acquired images is selected as a reference for the others. The reference image is divided into small regions (or subsets), a position of each subset is tracked in the remaining images, using the maximum zero-mean normalized sum of squared difference function as the criterion (or any other correlation metric). The image can be divided into hundreds or thousands of subsets, thus 2D DIC provides in-plane displacement maps over the selected area of interest (AOI). The 3D DIC is a technique that combines the 2D DIC with stereovision by using two cameras for observation of the same AOI. 3D DIC provides: the 3D shape of a surface, in-plane and out-of-plane displacement maps. The in-plane strain maps are calculated by differentiation of the in-plane displacement maps. According to [20], the minimum displacement measurement error can be less than 0.001 pixels, however, it must be noted that in real applications the accuracy of measurements strongly depends on factors such as image noise and stability of experimental conditions. Moreover, the out-of-plane measurement error is larger than the in-plane one, and strongly depends on a stereo angle between the cameras as set [20,21]. The accuracy of displacement measurements is scalable with the resolution of the cameras (larger camera's resolution and smaller FOV indicate higher displacement measurement accuracy). The displacements in the "x" direction are given as "u" in [pixels] and after scaling are expressed as "U" in [mm]; similarly, displacements in the "y" and "z" direction are given as "v [pixels] ", "V[mm] ", and "w[pixels] ", "W[mm] ", respectively.

4. Investigation of Section of Arch-Shaped Steel Sheets

At this stage, the test bench was designed in order to simulate local loss of stability in 1 m long sections [2,3]. The examined sample was fastened between two horizontal rigid plates, with defined degrees of freedom. The force was eccentrically applied in the direction of the axis that passes through the center of gravity of the cross sections. With this arrangement, the compression force on 1 m long interval of the arch (having a radius of 18 m) was mapped.

4.1. 3D Digital Image Correlation Setup

The 3D DIC system used in this measurements comprises two AVT Pike F-1600 (4872 × 3248 pixels) monochromatic cameras equipped with 28 mm lenses, set on an angle of 30°. The setup was mounted on an aluminum frame to enable easy geometric modifications. The surface of the examined specimen was illuminated with two 200 W light-emitting diode (LED) lamps (13,000 lumen) equipped with a light diffuser ("soft box"), in order to eliminate shadows on the surface. The FOV of the system was 1.5 m × 1 m, covering the entire area of the sample (Figure 3).

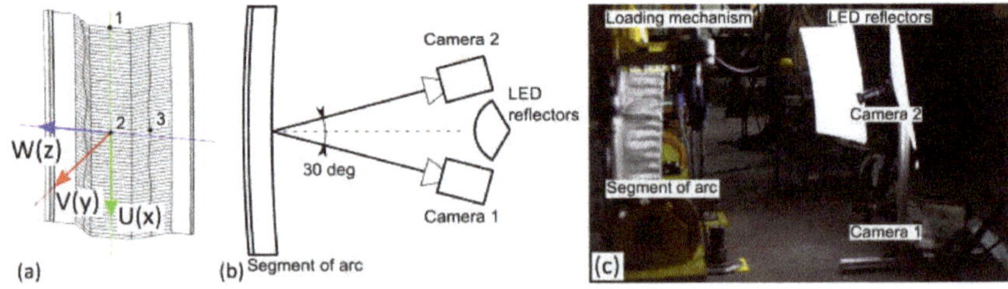

Figure 3. The experimental setup: (**a**) orientation of coordinate system and location of three points adopted for further analysis; (**b**) scheme and (**c**) photo of the measurement system based on 3D DIC [2].

4.2. FEM Model of the Section of Arch-Shaped Steel Sheets

Calculations covered 3 numerical models with different specificity of geometry mapping. The first model (A) (Figure 4a) was devoid of characteristic web corrugations and waviness, the second model (B) (Figure 4b) had corrugated surfaces mapped but with no waves, and the third one (C) (Figure 4c) had all the web corrugations and waves characteristic of such a profile. The models (Figure 4a–c) were developed in the ANSYS graphic module making use of the data from real element measurements. Furthermore, when describing the behaviour of the particular models, the aforementioned A, B and C signs will be used. The elastic-plastic steel model was used for the calculations, developed on the basis of the tests of steel samples [3]. The support and load conditions were accepted according to the assumptions included in the article [3]. Calculations and tests were performed in the form of axial compression of the sample.

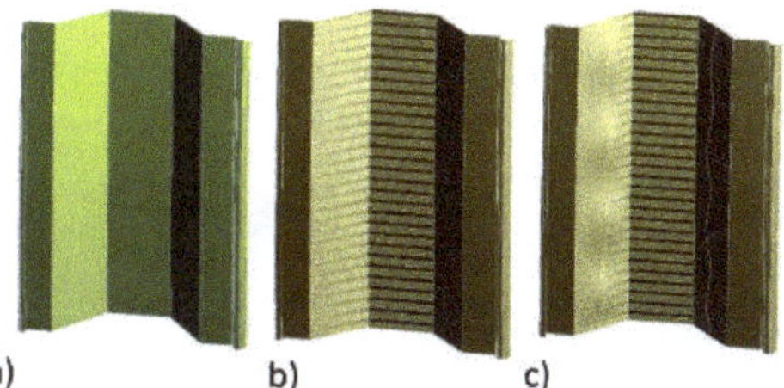

Figure 4. The geometry of numerical models: (**a**) planar model (A), (**b**) corrugation model (B) and (**c**) corrugation and wavy model (C).

4.3. Utilization of 3D DIC Measurements in the Process of Validation of FEM Model of the Section

The results of 3D DIC measurement were utilized in order to determine whether the simplification of the geometry of the model is allowed. At first, in order to perform quantitative analysis, the comparison between the displacements extracted from selected points of the structure from three models and experimental data has been performed. Points are distributed over the entire surface of the sample, in places corresponding to maximum displacements, according to FEM models. Locations of the points allowed for the assessment of representative movements for the entire sample. Exemplary data are presented in Figure 5, point 2 concerns the maximum V displacements of the sample. The best correlation between experimental and numerical results have been obtained with the FEM model comprising the most detailed geometry (C model). The results obtained from the two remaining models (of simplified geometry) differ significantly from the experimental results. Therefore, the simplifications of a FEM model have not been allowed in further analysis. Subsequently, the full-field qualitative comparison of the failure mode maps between DIC and numerical (C model) displacement maps has been carried out. Exemplary maps are presented in Figure 6. The displacement maps obtained with the numerical analysis show good agreement with the experimental data, considering its character and values. Some discrepancies that can be observed could be caused by unavoidable deviation of geometry of the sample and support conditions (e.g., inaccuracy in direction of applied force). The presented analysis validated the assumptions made in the C model, and has proven them to be useful in the analysis below.

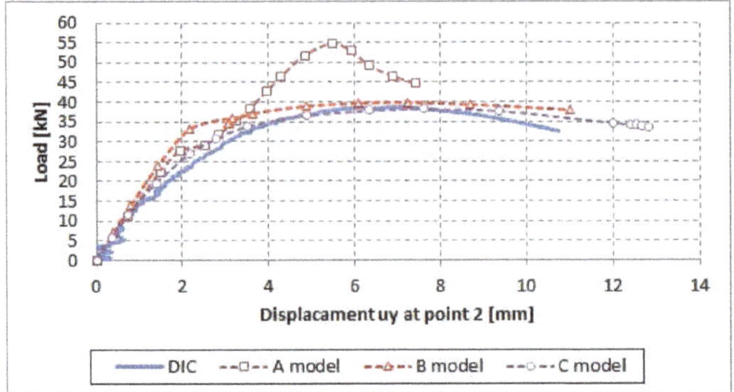

Figure 5. The comparison of V displacement functions obtained for three numerical models and experimental data (results for point 2) [2].

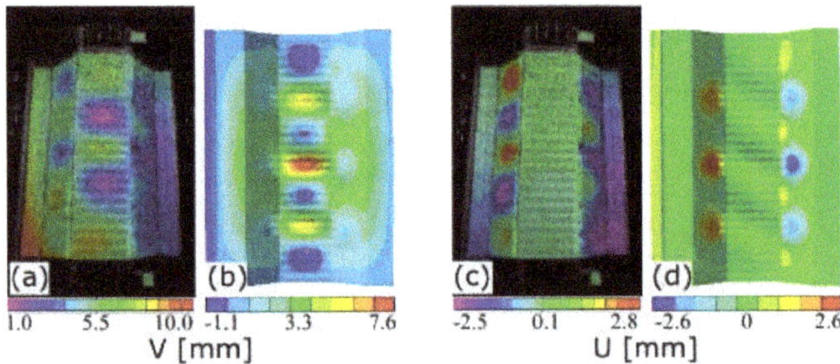

Figure 6. V displacement maps derived from (**a**) experiment and (**b**) the C model, and U displacement maps derived from (**c**) experiment and (**d**) the C model [2].

5. Investigation of Segments of Arch-Shaped Steel Sheets

The 12 m span sections of metal plate arch composed of four individual modules have been examined with the use of custom-made laboratory stands, which made it possible to apply force equivalent to a natural load caused by snow and wind [22,23]. The loading mechanism consisted of pulleys and beams which transferred point load (applied with hydraulic actuator) into 16 points (4 points for each module). The load was recorded by the actuator mounted onto the main beam, which reduced the force loss of the pulleys.

5.1. Multi-Camera DIC System with Overlapping Field of View

Considering the length of 12 m of arch segments, the measurements have been carried out with the use of multi-camera DIC system in which the field of view of neighboring 3D setups overlapped each other. As a result, the obtained, stitched FOV was 7 × 4 m (Figure 7a). For stitching we used the method described in the papers [13,24,25], while the general measurement procedure of the segments of arch-shaped steel sheets are presented in [13]. The multi-camera DIC system used in this measurements comprised eight 5 MPx (2448 × 2048) Pointgrey cameras equipped with 8mm focal length lenses. The cameras were connected to the control computer in order to synchronize the data acquisition procedure. The calibration procedure comprised two steps. In the first stage, 3D DIC setups have been calibrated separately with the checkerboard before the measurements. The quality of calibrations has been expressed as a reprojection error, which was smaller than 0.05 px for all 3D DIC setups. After the scaling (from pixels to mm), the accuracy of displacement measurements of each 3D DIC setup can be estimated at 0.05 mm. In the second stage, the transformation of individual coordinate systems of separate 3D DIC systems into a common coordinate system has been determined. The fields of view of neighboring 3D DIC setups overlapped each other (Figure 7) in order to make it possible to capture the images of the same calibration target (checkerboard in this case) with two systems simultaneously. Checkerboard corners (markers) viewed by each camera have been detected and their positions in two 3D DIC coordination systems were obtained. The knowledge of position of markers in two separate coordination systems was used to obtain geometrical transformation between these two systems. The common coordination system (CS) is associated with one of the two systems. Here the coordination system of DIC setup 2 has been selected as the global CS. The transformation between CSs of setup no. 1 and 3 into CS of setup no. 2 has been determined directly. The transformation of CS of setup no. 4 was obtained indirectly—at the beginning, the transformation into CS of setup no. 3 was obtained, and then transformation from CS of setup no. 3 into CS of setup no. 2 was performed. The transformation errors obtained were below 0.5 mm. In order to correlate CSs of the measurements with numerical simulations, the data from the global coordination system (associated with setup no. 2) was transferred to the coordinate system, in which the xy plane was parallel to the ground and z axis was perpendicular to the ground (Figure 7b). AOI covers the area of corrugated web surface of two middle arches and additionally eight points on two external arches, corresponding to the locations of force extortion. Data obtained from upper (flat lip according to Figure 1) and middle (corrugated flange according to Figure 1) surfaces of the middle arches have been thresholded due to a higher correlation error (because of the loss of depth of focus).

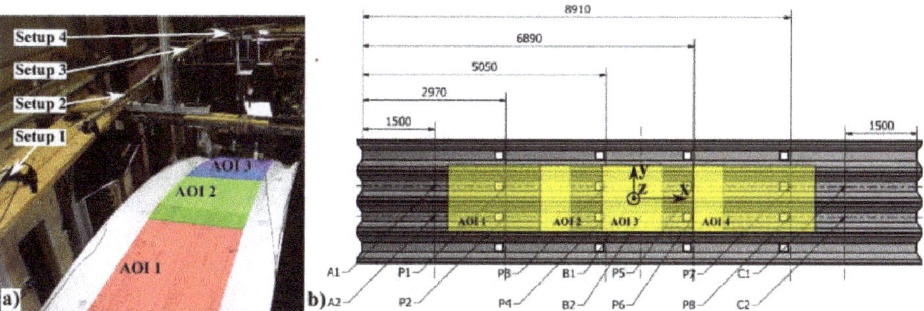

Figure 7. (**a**) Location of the cameras and field of views during the measurements; (**b**) scheme of the investigated segments of arch-shaped steel sheets with marked areas of interest (AOIs), characteristic points (for analysis) and orientation of coordinate system (top view) [4].

5.2. FEM Model of Segments of Arch-Shaped Steel Sheets

The geometry of the full-sized model has been adopted on the basis of the analysis performed in Section 4.2). The model with corrugation and waviness of the middle surfaces (model C according to the description from Section 4.2) was used for calculations. The full-sized numerical model is an image of the examined element in terms of dimensions and load mode.

In order to accurately simulate the load caused by snow cover, the kinematic extortion forced by controlled displacements of 16 points (at which load was applied) was used (Figure 7b, points A1, A2, B1, B2, C1, C2). Constraints utilized in FEM model (in particular rotational stiffness of supports) were updated through comparison with experimental data, and this process is described below.

5.3. Utilization of 3D DIC Measurements in Process of Validation of FEM Model of Segments of Section of Arch-Shaped Steel Sheets

Displacements of the chosen points (Figure 7b, points A1, A2, B1, B2, C1, C2) were derived on the basis of the multi-camera DIC analysis. In order to update the support conditions of the FEM model, a comparison between displacements' distribution obtained from the FEM model and multi-camera DIC system was made. In Figure 8, an exemplary comparison has been presented that shows half of maximum load. The FEM simulation comprises two supports on both ends of the arch. Both supports are modelled by a rotational elastic spring that allows the structural member to rotate (limited by rotational stiffness), but not to translate in any direction. As the first approximation, the rotational stiffness of supports is based on the simplified theoretical calculations. Subsequently, through comparison of experimental results and theoretical calculation, the value of rotational stiffness of elastic support was determined. The results of the comparison of experimental and numerical displacements after updating the numerical one at selected points is presented on a diagram of displacements towards 3 directions (U, V, W) in the function of force increment (Figure 9). The results of the comparison of DIC tests and FEM numerical analyses are presented as balance paths with reference to proper reference points. The directions of the displacements are referred to the coordinate system as in Figure 7b. The displacements are marked as follows: U(X), V(Y) and W(Z). The horizontal displacements U (X)—Figure 9a, i.e., in the arch plane are compatible in the reference points U (3, 4), U (5, 6) and U (7, 8), whereas balance path of the reference points U (1, 2) obtained in the tests diverge from those determined in calculations. The situation is similar in vertical displacements W (Z)—Figure 9c. This means that the test element was deformed asymmetrically. The diagram comparing horizontal displacements of the reference points in the arch plane of V(Y)—Figure 9b proves that. The test model tilts erratically in the range up to 3 mm whereas the calculations show slight deviation of the reference points within 0.5 mm. The asymmetrical deformation of the test element is probably caused by imprecise assembly and slight deviations in the load symmetry. The computational model does not include random events related to the assembly or load mode. The only disorders of the geometry of the computational model are related to the introduction of geometrical imperfections, which only slightly change the symmetry of the displacements of the reference points.

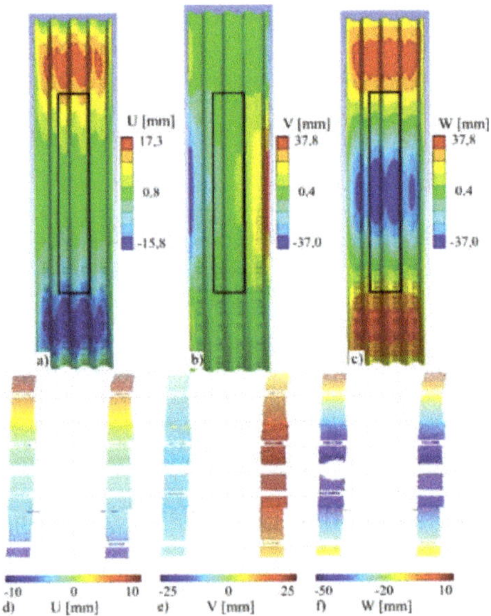

Figure 8. The comparison of displacements distribution: top view of displacements obtained from the finite element model (**a–c**), of the entire specimen, and experimental results (**d–f**) for the AOI covering the 7 m long area of lower surface (corrugated web according to Figure 1) of two middle arches. Rectangular areas marked with lines correspond to the areas on the experimental specimen that has the two middle arches that are being measured [4].

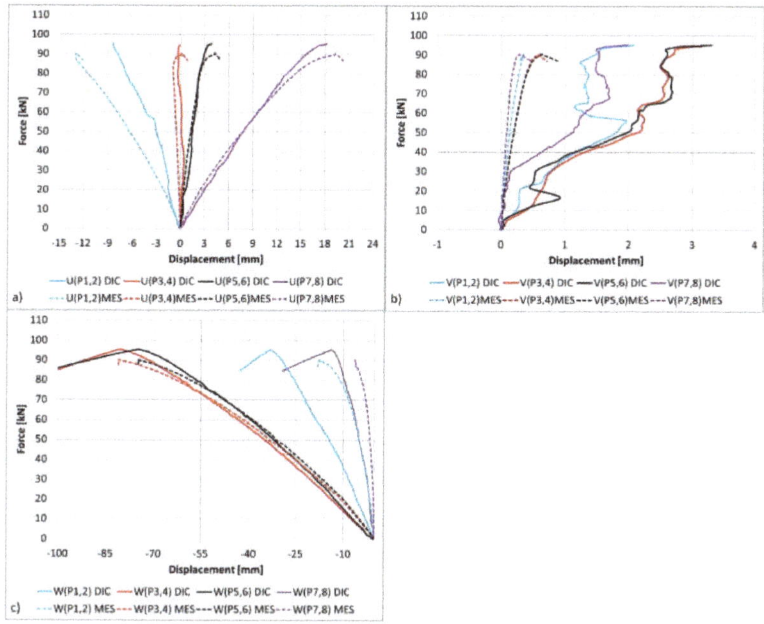

Figure 9. The comparison of load (**a**) U, (**b**) V, (**c**) W displacement functions obtained for numerical models and experimental data (results for referential points).

6. In Situ Investigation of Full-Scale Construction of Arch-Shaped Steel Sheets

Finally the full-scale construction made of arch-shaped self-supporting metal plates (Figure 1a) has been examined. The dimensions of the hall were ([span × length]/[height]) 18 × 18 m/7 m. Such a construction can be exposed to environmental loads caused by the presence of snow or wind [22,23] and changes of temperature.

6.1. Multi-Camera DIC System with Distributed Field of View

In order to cover the localization of points corresponding to the numerically predicted maximum displacements (caused by environmental loads), the measurements system consisted of three 3D DIC setups (six cameras in total). Each FOV of 3D DIC setup covered the area of 2.5 × 1.5 m and distance between neighboring AOIs was approximately 4 m. In order to perform the measurements in distributed FOV, the dedicated multi-camera DIC system was developed. This system and is described in detail in [16]. The multi-camera DIC system used in these measurements comprised six 5 MPx (2448 × 2048) Pointgrey cameras equipped with 8 mm focal length lenses. The cameras were connected to the control computer in order to synchronize the data acquisition procedure. The calibration procedure comprised two steps. In the first step, each 3D DIC system was calibrated with utilization of a standard chessboard calibration target. In the second calibration step, the transformation of individual coordinate systems of separate 3D DIC systems into a common coordinate system has been determined, with the additional support of a laser tracker [26]. Geometrical transformations were determined by using multiple positions of a fiducial marker, which was sequentially placed in the FOV of each 3D DIC setups. The 3D positions of the marker in local coordinate systems were determined with the use of three-dimensional computer vision algorithms [18]. Simultaneously, the positions of the marker in the global coordinate system were determined by a laser tracker. A global coordination system was related to the foundations of the building. Transformation parameters (rotation: Rx; Ry; Rz and translation: Tx; Ty; Tz) between local and global coordinate systems were obtained by using the singular value decomposition (SVD) method [27]. The obtained accuracy of displacement measurements in each 3D DIC setup was below 0.3 mm (detailed description of calculation of displacement error is presented in [16]).

A multi-camera DIC system was placed on the 6 m high and 10 m wide buildings' temporary scaffolding (Figure 10), and the measurements were carried out for a few months (from March to October 2016). Therefore, the temperature influence on displacements of measurement setup was noticeable. In order to neutralize this influence, the displacements of the 3D DIC systems as a function of temperature, caused by the thermal expansions of aluminum scaffolding were measured by means of a laser tracker. The correction obtained was included in the results of displacement measurements. The temperature inside the hall was measured using a weather station.

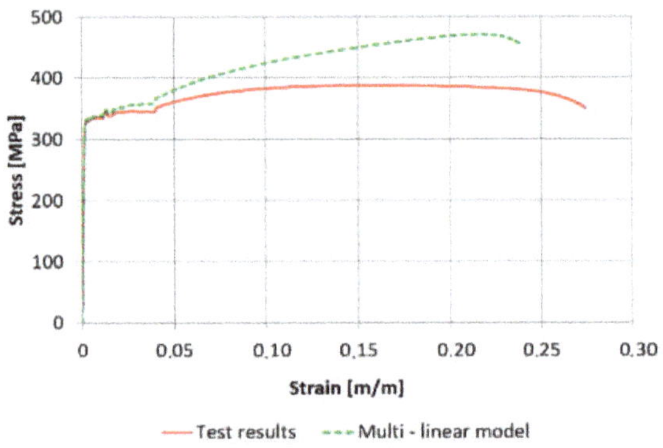

Figure 10. Stress–strain curves.

6.2. FEM Model of Full-Scale Construction

A model being the representation of the arch in the central part of the hall was used for calculations. The model consists of a single ABM 240 profile with the geometry adopted on the basis of the previous analyses. Material constants (modulus of elasticity, Poisson's modulus, shear modulus and coefficient of linear thermal expansion) have been adopted for structural analysis in accordance with EN 1993-1-1 standard [28].

The computation adopted an elastic-plastic multi-linear material model, determined according to the tests. The strength characteristics of steel were identified through a series of laboratory tests on 10 samples of steel sheet with a nominal thickness of 1.40 mm. The obtained mean yield strength amounted to $f_y = 340$ MPa, and ultimate strength was $f_u = 390$ MPa. A typical course of the stress–strain relationship in a single test, and elastic-plastic multi–linear material model is presented in Figure 10.

The test data were implemented via transformation functions in the σ_{true}–ε_{ln} system (elastic-plastic multi-linear material model) [3] according to the Equations (1), (2) presented below:

$$\varepsilon_{ln} = ln\left(1 + \varepsilon_{eng}\right) \tag{1}$$

$$\sigma_{true} = \sigma_{eng}\left(1 + \varepsilon_{eng}\right) \tag{2}$$

where:

ε_{ln}—relative logarithmic strain,

σ_{true}—true stress,

σ_{eng}—engineering stress (test result),

ε_{eng}—engineering strain (test result).

The free ends of the model are propped up in joints (a possibility to rotate towards X axis). The remaining degrees of freedom are blocked. On the side edges, boundary conditions are assumed that map the cooperation of the adjacent profiles (remote point). In the remote point system, displacement towards Z and Y are released, the remaining degrees of freedom are blocked. Due to lack of snow during the measurement period in winter, and the negligible influence of the pressure of wind on construction, only the thermal load was considered.

The purpose of the test was to determine the displacements and stress of the characteristic points located on the surface of the test object exposed to the action of thermal loads. For numerical calculations, the external and internal temperature of arch structure from three areas (corresponding to measured AOIs) was adopted. The temperature was measured precisely with the utilization of thermoelements.

6.3. Utilization of 3D DIC Measurements for Validation of FEM Model of Full-Scale Construction of Arch-Shaped Steel Sheets

Displacement measurements were taken in AOI1, AOI2 and AOI3 areas and 3 reference points were selected for each area (Figure 11) to validate the numerical model, 9 points in total. The tests lasted for several months, however validation of the numerical model was limited to a much shorter period. Below (Figures 12 and 13) we present exemplary results of measurements, which were performed in 12 h and 25 min. during a cold night and sunny day in April (largest external as well as internal temperature gradient occurred). The obtained results allowed to validate the numerical model, that can be used to calculate load capacity and stability of the coverings of the thin-walled arch-shaped sheet metals.

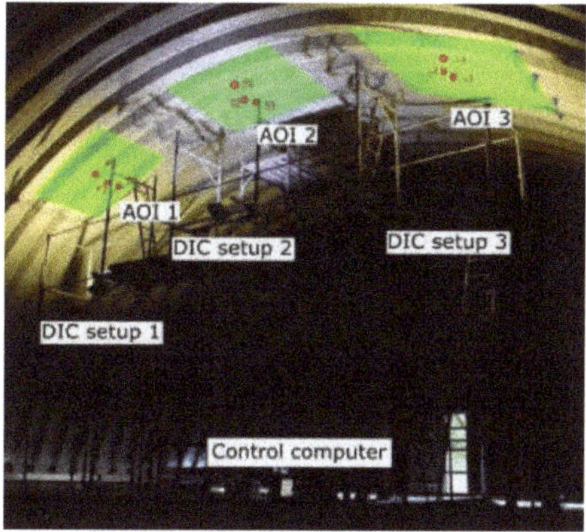

Figure 11. Measurement set-up.

Figure 13 presents the results of the displacements measurement and those of numerical calculations for the chosen reference points from the analysed measurement areas. The results were juxtaposed in relation to the change of temperature depending on the time of measurement. Displacements measured and determined on the basis of calculations are compatible in terms of increment directions and have similar values. Additionally, Figure 14 shows the analysis of the discrepancies of the calculation results and the test results depending on the value of temperature gradient. A certain regularity may be noticed in the distribution of discrepancies of the results. At high temperature gradients (over 18 °C), the discrepancies of neither test nor calculations results exceed 10%, at lower temperature gradients—the discrepancies of the tests results are much higher. This is probably due to the fact that numerical model did not take into account the boundary conditions related to the cooperation of the adjacent profiles, i.e., the friction between the cooperating profiles was not included in the calculations. Assuming that the friction at the jointing of the adjacent profiles has a constant value, its influence is much more visible at small displacements than at large displacements related to the effect of the higher temperature gradient.

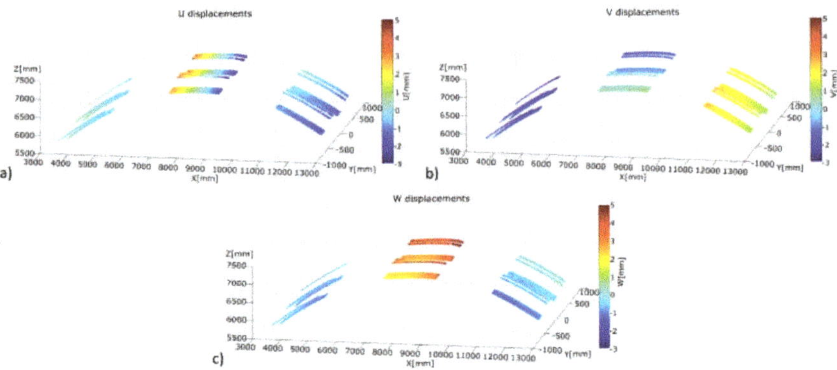

Figure 12. Exemplary results of the measurements stitched in the global coordinate system: (**a**) U displacement map, (**b**) V displacement map, (**c**) W displacement map [16].

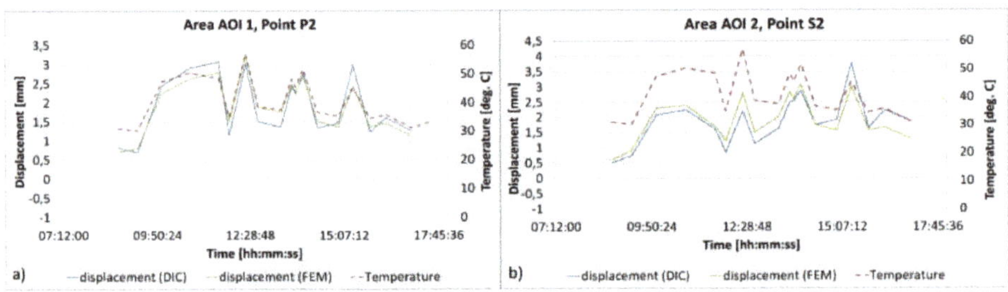

Figure 13. The comparison of displacement functions obtained for numerical models and experimental data, (**a**) point P2 (AOI 1), (**b**) point P2 (AOI 2).

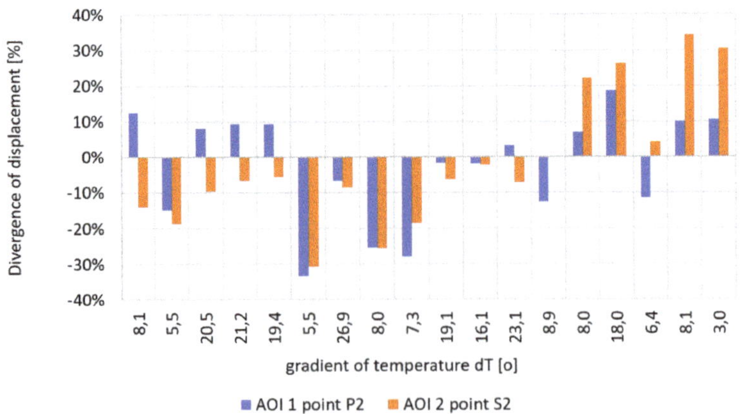

Figure 14. The analysis of the divergence of the measurement and calculations results.

7. Conclusions

So far, the DIC method applied in the construction industry has been considered a prototypical solution, more likely intended for testing/monitoring the elements of buildings, not for proper measurements used in a certification process in accordance with accepted standards. The methodology presented in this paper has shown the possibility and advantages of replacing conventional measuring methods based at point extensometers and applied at different stages of the analysis and testing of complicated building structures by applying the 3D DIC method. Its potential lies not only in the capability to measure in full field of view, but mostly in the possibility to digitally link the measurement with the numerical calculations, thus creating efficient hybrid experimental-numerical system with huge information resources useful, for instance, in the process of design optimization, FEM validation or diagnostics of complicated building structures. It should be pointed out that the presented methodology requires a complex measurement setup and is labor-intensive, therefore it should be utilized in the case of investigation of truly complex structures, in which stress concentrations can occurr in unexpected locations.

The procedure presented in Figure 2 and explained in detail in Sections 4–6 concerns the way to implement the particular stages when analysing self-supporting arch-shaped structures from profiled steel sheets with the use of numerical calculation methods supported by physical experiments. Each stage is used to determine and validate the optimal procedure to be applied in a FEM which, in the opinion of the authors, adequately indicates the solution to the most important problems related to the design of arch-shaped structures. The multistage research verification process enables the development of a reliable numerical model that is very useful and allows for the analysis of arch-shaped profiled steel sheets at diversified geometry and any load conditions.

The methodology presented in the paper may be used to determine the strength and functional properties of various varieties of the K-span system, which are required during the process of implementing the product for use in the construction industry (Polish and European technical assessments). The material presented in the paper may also be used in the future to develop an annex to the national standards in question regarding the design of thin-walled elements.

Author Contributions: Conceptualization, M.K.; methodology, K.M., A.P. and M.M.; investigation K.M., M.M. and P.W.; formal analysis, K.M., M.M. and A.P., validation, K.M. and A.P.; supervision, M.K.; funding acquisition, M.K. and A.P.; visualization, K.M. and A.P.; writing—original draft, K.M. and A.P.; writing—review and editing K.M., M.K. and A.P.

Funding: The authors gratefully acknowledge financial support from the project OPT4-BLACH (Grant No. PBS1/A2/9/2012) financed by the National Center for Research and Development and the statutory funds of the Faculty of Mechatronics, Warsaw University of Technology.

Conflicts of Interest: The authors declare no conflict of interest.

References

1. Piekarczuk, A.; Malesa, M.; Kujawinska, M.; Malowany, K. Application of hybrid FEM-DIC method for assessment of low cost building structures. *Exp. Mech.* **2012**, *52*, 1297–1311. [CrossRef]
2. Piekarczuk, A.; Malowany, K.; Więch, P.; Kujawińska, M.; Sulik, P. Stability and bearing capacity of arch-shaped corrugated shell elements: Experimental and numerical study. *Bull. Pol. Acad. Sci.* **2015**, *64*, 113–123. [CrossRef]
3. Piekarczuk, A.; Więch, P.; Malowany, K. Numerical investigation into plastic hinge formation in arched corrugated thin-walled profiles. *Thin Walled Struct.* **2017**, *119*, 13–21. [CrossRef]
4. Malowany, K.; Malesa, M.; Piekarczuk, A.; Kujawińska, M.; Skrzypczak, P.; Więch, P. Application of 3D digital image correlation for development and validation of FEM model of self-supporting metal plates structures. *Proc. Spie Int. Soc. Opt. Eng.* **2016**, *9803*, 98033W.
5. Cybulski, R.; Walentyński, R.; Cybulska, M. Local buckling of cold-formed elements used in arched building with geometrical imperfections. *J. Constr. Steel Res.* **2014**, *96*, 1–13. [CrossRef]
6. Laermann, K.H. Hybrid techniques in experimental solid mechanics. In *Optical Methods in Experimental Solid Mechanics*; Springer: Vienna, Austria, 2000; pp. 1–72.
7. Sebastian, C.; Patterson, E.; Ostberg, D. Comparison of numerical and experimental strain measurements of a composite panel using image decomposition. *Appl. Mech. Mater.* **2011**, *70*, 63–68. [CrossRef]

8. Lampeas, G.N.; Pasialis, V.P. A hybrid framework for nonlinear dynamic simulations including full-field optical measurements and image decomposition algorithms. *J. Strain Anal. Eng. Des.* **2013**, *48*, 5–15. [CrossRef]

9. Malesa, M.; Malowany, K.; Tomczak, U.; Siwek, B.; Kujawińska, M.; Siemińska-Lewandowska, A. Application of 3D digital image correlation in maintenance and process control in industry. *Comput. Ind.* **2013**, *64*, 1301–1315. [CrossRef]

10. Mechelke, K.; Kersten, T.P.; Lindstaedt, M. Comparative investigations into the accuracy behaviour of the new generation of terrestrial laser scanning systems. *Opt. 3D Meas. Tech. VIII* **2007**, *1*, 319–327.

11. Yang, H.; Xu, X.; Neumann, I. The benefit of 3D laser scanning technology in the generation and calibration of FEM models for health assessment of concrete structures. *Sensors* **2014**, *14*, 21889–21904. [CrossRef] [PubMed]

12. Xu, X.; Bureick, J.; Yang, H.; Neumann, I. TLS-based composite structure deformation analysis validated with laser tracker. *Compos. Struct.* **2018**, *202*, 60–65. [CrossRef]

13. Malesa, M.; Malowany, K.; Pawlicki, J.; Kujawinska, M.; Skrzypczak, P.; Piekarczuk, A.; Lusa, T.; Zagorski, A. Non-destructive testing of industrial structures with the use of multi-camera digital image correlation method. *Eng. Fail. Anal.* **2016**, *69*, 122–134. [CrossRef]

14. Shao, X.; Dai, X.; Chen, Z.; Dai, Y.; Dong, S.; He, X. Calibration of stereo-digital image correlation for deformation measurement of large engineering components. *Meas. Sci. Technol.* **2016**, *27*, 125010.

15. Dong, S.; Yu, S.; Huang, Z.; Song, S.; Shao, X.; Kang, X.; He, X. Target-based calibration method for multifields of view measurement using multiple stereo digital image correlation systems. *Opt. Eng.* **2017**, *56*, 124102.

16. Malowany, K.; Malesa, M.; Kowaluk, T.; Kujawinska, M. Multi-camera digital image correlation method with distributed fields of view. *Opt. Lasers Eng.* **2017**, *98*, 198–204. [CrossRef]

17. Pan, B. Digital image correlation for surface deformation measurement: Historical developments, recent advances and future goals. *Meas. Sci. Technol.* **2018**, *29*, 082001. [CrossRef]

18. Sutton, M.; Orteu, J.-J.; Schreier, H. *Image Correlation for Shape, Motion and Deformation Measurements: Basic Concepts, Theory and Applications*; Springer: New York, NY, USA, 2009.

19. Roye, K.L. *Metal Building Construction Using the MIC-240 ABM K-Span Machine*; Naval Postgraduate School: Monterey, CA, USA, 1996.

20. Wang, Y.-Q.; Sutton, M.A.; Ke, X.-D.; Schreier, H.W. Error Propagation in Stereo Vision: Part I: Theoretical Developments. *Exp. Mech.* **2011**, *51*, 405–422.

21. Ke, X.-D.; Schreier, H.W.; Sutton, M.A.; Wang, Y.-Q. Error propagation in stereo vision: Part II: Experimental validation. *Exp. Mech.* **2011**, *51*, 423–441.

22. *EN 1991-1-3 Eurocode 1: Actions on Structures—Part 1-3: General Actions—Snow Loads*; European Committee for Standardization, The European Union: Brussels, Belgium, 2003.

23. *EN 1991-1-4 Eurocode 1: Actions on Structures—Part 1-4: General Actions—Wind Action*; European Committee for Standardization, The European Union: Brussels, Belgium, 2005.

24. Orteu, J.-J.; Bugarin, F.; Harvent, J.; Robert, L.; Velay, V. Multiple camera instrumentation of a single point incremental forming process pilot for shape and 3D displacement measurements: Methodology and results. *Exp. Mech.* **2011**, *51*, 625–639. [CrossRef]

25. Chen, X.; Yang, L.; Xua, N.; Xie, X.; Sia, B.; Xu, R. Cluster approach based multi-camera digital image correlation: Methodology and its application in large area high temperature measurement. *Opt. Lasers Technol.* **2014**, *57*, 318–326. [CrossRef]

26. Muralikrishnan, B.; Phillips, S.; Sawyer, D. Laser trackers for large-scale dimensional metrology: A review. *Precis. Eng.* **2016**, *44*, 13–28. [CrossRef]

27. Besl, P.J.; McKay, N.D. A method for registration of 3-d shapes. *IEEE Trans. Pattern Anal. Mach. Intell.* **1992**, *14*, 239–256. [CrossRef]

28. *EN 1993-1-1: Eurocode 3: Design of Steel Structures—Part 1-1: General Rules and Rules for Buildings*; European Committee for Standardization, The European Union: Brussels, Belgium, 2005.

Article

Experimental Study on the Fracture Process Zone Characteristics in Concrete Utilizing DIC and AE Methods

Shuhong Dai [1,*], Xiaoli Liu [2,*] and Kumar Nawnit [2]

[1] School of Mechanics and Engineering, Liaoning Technical University, Fuxin 123000, China
[2] State Key Laboratory of Hydroscience and Engineering, Tsinghua University, Beijing 100084, China; nnet55@gmail.com
* Correspondence: dsh3000@126.com (S.D.); xiaoli.liu@tsinghua.edu.cn (X.L.)

Received: 1 February 2019; Accepted: 26 March 2019; Published: 30 March 2019

Abstract: The present work focuses on investigating the characteristics of the fracture process zone (FPZ) in concrete. The Single-edge notched (SEN) concrete beams under three-points bending are employed for conducting mode I fracture propagation. The displacement fields on the specimen surface and the internal AE signal of specimen are obtained simultaneously in real time by digital image correlation (DIC) and acoustic emission (AE) techniques. The experimental and analytical results indicated that the crack tip position, the crack extension length and the stress intensity factors (SIF) are obtained dynamically and quantitatively by DIC technique, and the length of FPZ is identified, respectively, by DIC and AE techniques in the crack extension process. The distribution of internal AE events is consistent with that of FPZ identified from surface deformation of specimens.

Keywords: fracture process zone; digital image correlation technique; acoustic emission technique; stress intensity factor

1. Introduction

Concrete is a quasi-brittle material. The micro-cracking region ahead of the real crack tip in concrete is defined as the fracture process zone (FPZ), and the characteristics of concrete FPZ have been a subject of massive experimental debates. Various techniques have been developed to track the evolution of micro-cracks in FPZ, such as the fiber optics technique [1], X-rays technique [2], fiber optic sensor [3], and optical microscopy [4]. Compared to the methods mentioned above, digital image correlation and acoustic emission techniques are increasingly popular recently.

The Digital image correlation (DIC) technique was firstly proposed by Peters and Sutton et al. [5,6]. It is an optical technique to determine surface deformations by matching the digital speckle images before and after deformation. DIC is a non-destructive and non-contact full field displacement measuring technique. The displacement and strain data obtained by DIC can be directly employed to theoretical analysis. Benefit from such advantages, the DIC has been widely applied to the study of concrete fracture. Choi and Shah [7] measured the lateral and axial deformations on a concrete specimen surface subjected to compression. Corr et al. [8] studied the interfacial transition zone between aggregates and cement paste in plain concrete. They also investigated its softening and fracture behavior. Wu et al. [9] studied the properties of the fracture process zone in concrete using DIC technique. Rouchier et al. [10] studied the whole process of concrete fracture using the DIC technique. The acoustic emission (AE) technique has been extensively applied in concrete engineering for approximately five decades since the 1960's [11]. The AE technique can continuously monitor the internal micro-cracks event and the failure process of concrete in real time, which makes it more popular than other methods. Wells [12] studied AE waveforms and determined the relationship between strain measurements and AE events. A frequency analysis and a source location analysis were reported to demonstrate the relationships [13]. These studies have produced practical applications for monitoring micro-cracks in concrete structures and are very useful in diagnostic applications. AE technique was utilized to assess the concrete fracture process, as the pioneering research works of Colombo [14]. The research works of Colombo et al. [15] have indicated that concrete micro cracks emit waves possessed smaller amplitudes, whereas macro cracks emit waves possessed larger amplitudes. Muralidhara [16] shed lights on the relationship between the AE event and the evolution of the fracture process in concrete. Recently, the relationship between the formation of FPZ in concrete and AE events has been extensively studied [17–20]. In order to cover their weaknesses and improve their advantages, DIC and AE techniques were combined to monitor the concrete crack evolution process [21–23].

In spite of the extensive work and many successes in characterizing FPZ, the comprehension of the characteristics of concrete FPZ requires further research. In this presentation, the concrete fracture tests are conducted on Single-edge notched beams through three-points bending conditions. The DIC technique is applied to measure the displacement and strain fields around the crack tip in real-time. Then the SIF, crack tip position and crack length are quantitatively and dynamically derived from the displacement fields measured by DIC technique. The AE technique is applied to monitor the internal fracture events during the crack propagation process, in particular the subcritical extension process. In short, the objective of this research is to investigate the characteristics of FPZ during the concrete crack propagation process.

2. Experimental Procedure

2.1. Digital Image Correlation Technique

Digital image correlation method relies on observations of random speckle patterns on the specimen surface. Image patterns are recorded before and after deformation of the specimens, then they are digitized and stored in a computer. The undeformed and deformed images are divided into small regions called "subsets", with each subset containing a group of pixels. Digital image correlation is used to match the subsets on the undeformed image with their corresponding subsets on the deformed image, as shown in Figure 1.

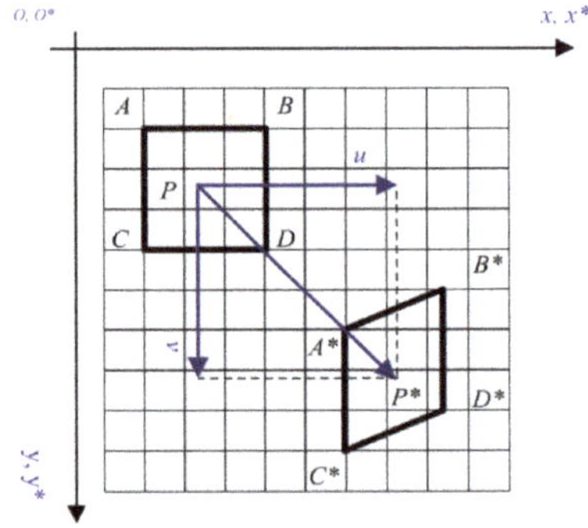

Figure 1. Initial subset (ABCD centered on *P*) and deformed subset (A*B*C*D* centered on *P**).

The cross-correlation function is chosen for this study, and it is defined as Equation (1):

$$F\left(x_m, y_n, u_x, u_y, \frac{\partial u_x}{\partial x}, \frac{\partial u_x}{\partial y}, \frac{\partial u_y}{\partial x}, \frac{\partial u_y}{\partial y}\right) = 1 - \frac{\sum_{i=-\frac{R}{2}}^{\frac{R}{2}} \sum_{j=-\frac{R}{2}}^{\frac{R}{2}} \left[\left(f(x_{m+i}, y_{n+j}) - \overline{f}\right) \cdot \left(g(x_{m+i}, y_{n+j}) - \overline{g}\right)\right]}{\left[\sum_{i=-\frac{R}{2}}^{\frac{R}{2}} \sum_{j=-\frac{R}{2}}^{\frac{R}{2}} \left(f(x_{m+i}, y_{n+j}) - \overline{f}\right)^2 \cdot \sum_{i=-\frac{R}{2}}^{\frac{R}{2}} \sum_{j=-\frac{R}{2}}^{\frac{R}{2}} \left(g(x_{m+i}, y_{n+j}) - \overline{g}\right)^2\right]^{\frac{1}{2}}} \tag{1}$$

The $f(x_i, y_j)$ represents the gray level value at coordinate (x_i, y_j) of the undeformed image, while the $g(x_j^*, y_j^*)$ represents the gray level value at coordinate (x_j^*, y_j^*) of the deformed image, the subset size is $N \times N$. The coordinates (x_i, y_j) and (x_j^*, y_j^*) are directly related by the deformation occurring between the two images. If the deformation occurs in two dimensions parallel to the camera, then the coordinates are related by Equation (2):

$$x^* = x + u_x + \frac{\partial u_x}{\partial x}\Delta x + \frac{\partial u_x}{\partial y}\Delta y$$
$$y^* = y + u_y + \frac{\partial u_y}{\partial x}\Delta x + \frac{\partial u_y}{\partial y}\Delta y \tag{2}$$

The u_x and u_y are the displacements of the subset center in the x and y directions, respectively, and Δx and Δy are distances from the subset center to any point in the subset (x, y).

2.2. Estimation of Stress Intensity Factors

The method for estimating mode I and mixed-mode I-II stress intensity factors is proposed here, based on the displacement fields determined via the DIC method. The displacement fields around a crack tip of the concrete specimen are expressed as Equation (3):

$$\begin{Bmatrix} u_x \\ u_y \end{Bmatrix} = \sum_{n=1}^{\infty} \frac{A_{\mathrm{I}n}}{2G} r^{n/2} \begin{Bmatrix} \kappa \cos \frac{n}{2}\theta - \frac{n}{2}\cos(\frac{n}{2}-2)\theta + \{\frac{n}{2}+(-1)^n\}\cos \frac{n}{2}\theta \\ \kappa \sin \frac{n}{2}\theta + \frac{n}{2}\sin(\frac{n}{2}-2)\theta - \{\frac{n}{2}+(-1)^n\}\sin \frac{n}{2}\theta \end{Bmatrix}$$
$$- \sum_{n=1}^{\infty} \frac{A_{\mathrm{II}n}}{2G} r^{n/2} \begin{Bmatrix} \kappa \sin \frac{n}{2}\theta - \frac{n}{2}\sin(\frac{n}{2}-2)\theta + \{\frac{n}{2}-(-1)^n\}\sin \frac{n}{2}\theta \\ -\kappa \cos \frac{n}{2}\theta - \frac{n}{2}\cos(\frac{n}{2}-2)\theta + \{\frac{n}{2}-(-1)^n\}\cos \frac{n}{2}\theta \end{Bmatrix} \tag{3}$$

The u_x and u_y are the displacement components in the specimen considered, and G is shear modulus. The κ is $(3 - \mu)/(1 + \mu)$ for plane stress, and r and θ are the polar coordinates around a crack tip, as shown in Figure 2.

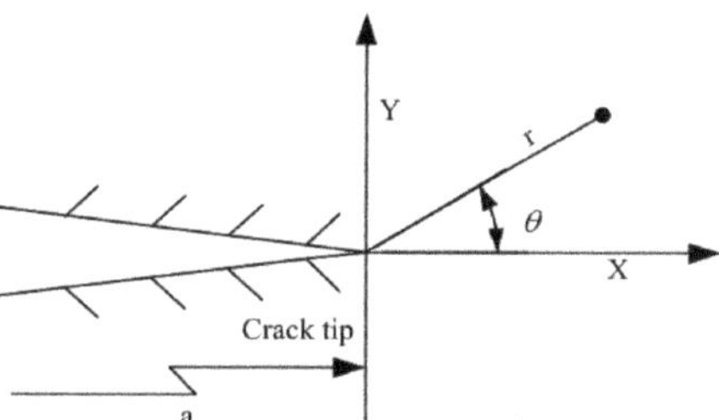

Figure 2. Coordinate system around a crack tip.

In the series solutions, the coefficient of the first terms $A_{\mathrm{I}1}$ and $A_{\mathrm{II}1}$ are related to stress intensity factors K_{I} and K_{II} of mode I and mode II through the relations of Equation (4).

$$A_{\mathrm{I}1} = \frac{K_{\mathrm{I}}}{\sqrt{2\pi}}, \; A_{\mathrm{II}1} = -\frac{K_{\mathrm{II}}}{\sqrt{2\pi}} \tag{4}$$

For the cases of mode I fracture, the displacement component of the specimen is perpendicular to the crack surface. This means that the u_y displacement field can be generally used to determine the stress intensity factors. In mixed-mode cases, however, the dominant displacement component for the crack can't be predicted before analysis. Therefore, radial u_r and circumferential displacement components u_θ on a polar coordinate system are used in this study to transform Cartesian displacements u_x and u_y. The displacements u_r and u_θ are obtained as Equation (5):

$$\left\{ \begin{array}{c} u_r \\ u_\theta \end{array} \right\} = \left[\begin{array}{cc} \cos\theta & \sin\theta \\ -\sin\theta & \cos\theta \end{array} \right] \left\{ \begin{array}{c} u_x \\ u_y \end{array} \right\} \tag{5}$$

The displacement fields in Equation (5) can, hence, be rewritten as Equation (6):

$$\left\{ \begin{array}{c} u_x \\ u_y \end{array} \right\} = \sum_{n=1}^{N} A_{\mathrm{I}n} \left\{ \begin{array}{c} f_{\mathrm{I}n}(r,\theta) \\ g_{\mathrm{I}n}(r,\theta) \end{array} \right\} - \sum_{n=1}^{N} A_{\mathrm{II}n} \left\{ \begin{array}{c} f_{\mathrm{II}n}(r,\theta) \\ g_{\mathrm{II}n}(r,\theta) \end{array} \right\} \tag{6}$$

The N is the number of terms of the series expansion of the displacement field. From Equations (5) and (6) with rigid-body displacements taken into account, the displacement fields can be expressed as Equations (7) and (8):

$$\begin{aligned} u_{rk} = & \left\{ \sum_{n=1}^{N} A_{\mathrm{I}n} f_{\mathrm{I}n}(r_k,\theta_k) - \sum_{n=1}^{N} A_{\mathrm{II}n} f_{\mathrm{II}n}(r_k,\theta_k) \right\} \cos\theta_k \\ & + \left\{ \sum_{n=1}^{N} A_{\mathrm{I}n} g_{\mathrm{I}n}(r_k,\theta_k) - \sum_{n=1}^{N} A_{\mathrm{II}n} g_{\mathrm{II}n}(r_k,\theta_k) \right\} \sin\theta_k + T_x \cos\theta_k + T_y \sin\theta_k \end{aligned} \tag{7}$$

$$\begin{aligned} u_{\theta k} = & -\left\{ \sum_{n=1}^{N} A_{\mathrm{I}n} f_{\mathrm{I}n}(r_k,\theta_k) - \sum_{n=1}^{N} A_{\mathrm{II}n} f_{\mathrm{II}n}(r_k,\theta_k) \right\} \sin\theta_k \\ & + \left\{ \sum_{n=1}^{N} A_{\mathrm{I}n} g_{\mathrm{I}n}(r_k,\theta_k) - \sum_{n=1}^{N} A_{\mathrm{II}n} g_{\mathrm{II}n}(r_k,\theta_k) \right\} \cos\theta_k - T_x \sin\theta_k + T_y \cos\theta_k + R r_k \end{aligned} \tag{8}$$

The T_x and T_y express the rigid-body translation in the x and y directions, respectively, and R is the rigid-body rotation. The subscript k ($k = 1, 2, \ldots , M$) denotes the index of the function evaluated at a point (r_k, θ_k) in the displacement field at which the displacement values are u_{rk} and $u_{\theta k}$. The M is the total number of displacement data points used in solving displacement equations. The unknown coordinates around a crack tip may thus be given in polar coordinates as Equation (9):

$$r_k = \sqrt{(x_k - x_0)^2 + (y_k - y_0)^2}, \quad \theta_k = \tan^{-1}\left(\frac{y_k - y_0}{x_k - x_0} \right) \tag{9}$$

The x_0 and y_0 are the location of the crack tip relative to an arbitrary Cartesian coordinate system. At any point in the displacement field, the coordinates r_k and θ_k, and displacements u_{rk} or $u_{\theta k}$ can be substituted into Equations (7) and (8). Therefore, stress intensity factors, higher-order terms and crack tip locations can be derived automatically from the displacement field determined by the DIC method.

2.3. Acoustic Emission Event Localization Technique

The AE technique has been extensively applied for the condition assessment and damage detection for concrete structures, as it was described by Grosse and Ohtsu in their book [13]. In particular, the AE technique can detect crack propagation that occurs not only on the surface but also deep inside the material. Therefore, a large number of AE analyses have been performed on concrete and concrete structures in Golaski's work [24]. One of the most important features of AE technique is the ability to localize the source of an AE event. Through the AE technology and signal localization method, the fracture process in concrete can be observed throughout the loading history. Signal localization is the basis of all analysis techniques used in AE technique. According to the 3-D localization problem, it is exactly determined when four or more travel times are available to calculate the three coordinates and the source time of an event. A least absolute deviation method is proposed as the AE location method in this presentation, and the objective function is defined as Equations (10) and (11)

$$F = \sum_{i=1}^{n} |C_i - C_m| \tag{10}$$

$$C_i = t_i - \left(\sqrt{(x_i - x_0)^2 + (y_i - y_0)^2 + (z_i - z_0)^2} / v \right) \tag{11}$$

The C_i is the source time to the sensor; and C_m is the median of $C_i (i = 1, 2, \ldots, n)$, which can be treated as the real source time. $C_i - C_m$ is the measurement error between calculated source time and real source time. The t_i is the onset time of the sensor, (x_i, y_i, z_i) is the coordinate of the sensor location, and (x_0, y_0, z_0) is the coordinate of the event source location.

2.4. Specimen, Loading Condition, and Measurement System

The concrete specimens are prepared with a standard P.O 32.5 Portland cement, which is crushed stone with a maximum diameter of 6 mm and river sand used as coarse and fine aggregate, respectively. The mix proportions are listed in Table 1. The modulus of elasticity and Poisson's ratio are measured through standard material test methods, the specimen size is $100 \times 50 \times 50$ mm^3. The measured values of the modulus of elasticity and the Poisson's ratio of the concrete material are $E = 35$ GP$_a$, $\mu = 0.26$, and the compressive strength is 27.5 MP$_a$.

Table 1. Concrete mix proportions.

Water/Cement Ratio	Cement (kg/m^3)	Sand (kg/m^3)	Aggregate (kg/m^3)	Water (kg/m^3)
0.48	446	593	1102	214

Six single-edge notched concrete beams under the three point bending are used for mode I fracture testing in the present study. Figure 3 shows a schematic representation of the concrete specimen and the locations of the three loading points of a three point bending test. The dimension of the SEN concrete beam is $210 \times 70 \times 23$ mm^3, and the span is 170 mm. A notch is made at the center of the concrete specimen edge using a diamond saw of 0.3 mm in thickness, with a length of $a_0 = 10$ mm. In implementing the DIC method, the interest regions of specimen surface are painted with white ink, and then covered with a black dot pattern with spray painting. An example of specimen prepared in this manner is shown in Figure 4.

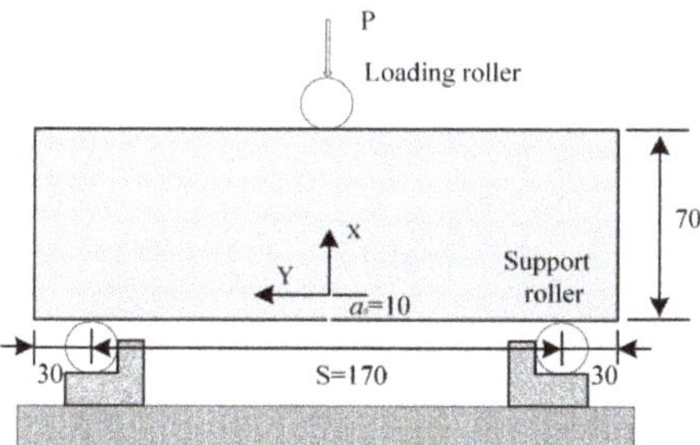

Figure 3. Concrete specimen and three-points bending conditions.

A servo hydraulic test machine with a capacity of 50 kN is applied for fracture tests, the displacement rate of the loading plane is set to 0.02 mm/min. In order to ensure contact between the loading system and specimens, a 50 N preload is applied before the testing. During the loading process, the painted area around the crack tip is recorded by a monochromatic charge-coupled device camera (Basler 404k, 2352×1720 pixels, BASLER, Ahrensburg, Germany) with a 105 mm focal lens. The specimen failure and crack propagation process is recorded by the camera in a rate of 15 frame pictures per second, then the pictures are stored in a computer automatically. The resolution of all captured images is 0.055 mm/pixel. As shown in Figure 4, sixteen AE sensors with a resonant frequency of approximately 150 kHz are attached to the specimen surface. Six sensors are in front surface and behind surface respectively, and four sensors are on the up and down edge of the specimen respectively.

AE signals are amplified 40 dB gain by a pre-amplifier. The sixteen sensor's signal data in 16-bit is recorded continuously and simultaneously in a frequency of 3 MHz.

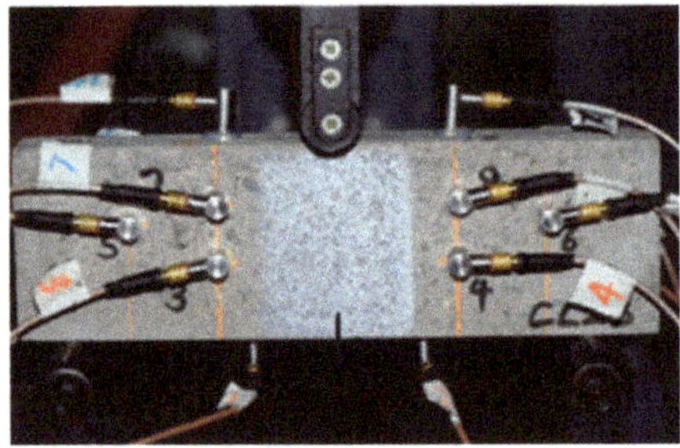

Figure 4. Single-edge notched concrete specimen sprayed with a dot pattern.

3. Results and Discussion

3.1. FPZ Evolution and SIFs

During a typical concrete fracture testing process, 13,089 images are captured by CCD in 872.6 s. A 55 × 30 mm^2 interest area covering the crack extension trace is shown in Figure 5, the displacement contour map of the interest area is obtained by the DIC technique. The crack tip position and displacement data around the crack tip are used as the initial value of the solution to the iterative Equation (8), then the SIFs and real crack tip position are obtained. As shown in Figure 5, 230 data points are selected around the crack tip, the data point position is marked by black dots. Although the accurate positions of the pre-crack tip can hardly be precisely ascertained at the outset, it is well known that the failure must start from the upper edge of the notch. Therefore, the center point of the notch's upper edge is set as the crack tip's initial position for the first iterative procedure. Then the obtained value of crack tip position is set as a new crack tip position for the consequent iterative procedures. As it is mentioned above, a series crack tip position can be derived from the displacement field around the crack tip during the crack propagation process. These crack tip positions are in the crack extension trace. These crack tip positions are used to calculate crack extension length δ_a during the FPZ evolution process. As shown in Figure 6, the crack extension length curve and loading curve are plotted together.

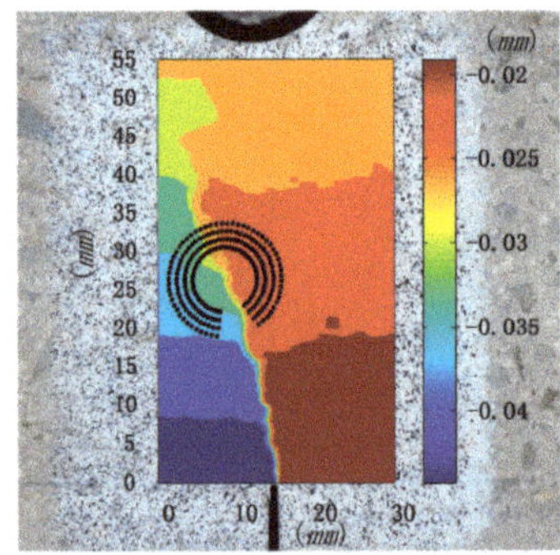

Figure 5. The speckle image captured by CCD and the displacement contour map of the interest area.

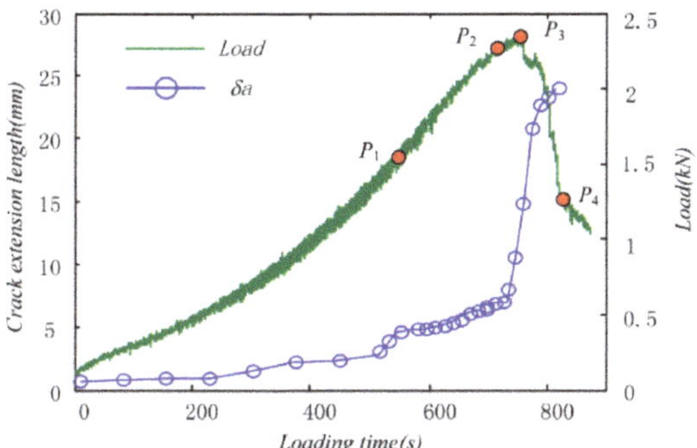

Figure 6. The curves of loading and crack extension length.

As shown in Figure 6, the crack extension length gradually increases from $t = 577.34$ s, the crack extension speed increases sharply after $t = 712.26$ s, the crack extension speed slows down after $t = 752.35$ s. Therefore, the first stage is usually called crack stable growth stage or subcritical growth stage. In this stage, micro-cracks are nucleating in front of the pre-crack tip along the notch direction, the crack extension length can be set as FPZ length. As shown in Figure 6, the concrete subcritical growth stage can be identified quantitatively and easily from the crack extension curves, and the FPZ length is about 8.35 mm. The second stage is unstable crack growth stage, micro-cracks coalesce into meso- and macro cracks, the crack extension length increases sharply, and the FPZ length is about 20.5 mm. In the third stage, the crack extension speed is mainly controlled by the test machine displacement rate. So, the concrete FPZ length is obtained from crack extension length curve derived from displacement field by DIC technique proposed in this presentation.

In Figure 7, the mode I SIF K_I curve and loading curve are plotted together. The SIF K_I is expected to be proportional to the applied load P and the square root of the length of preexisting crack a, as per the theory of fracture mechanics, while the value of K_I does not increase before $t = 577.34$ s. The SIF K_I begins to increase from $t = 577.34$ s, and reaches a critical value $K_I = 3.332$ MPa·m$^{1/2}$ at $t = 752.35$ s. Then the curve of K_I drops slightly and tends to a constant value. Hence, the critical value of $K_I = 3.332$ MPa·m$^{1/2}$ is defined as the mode I fracture toughness K_{IC}. As shown in Figure 8, the value of SIF K_{II} are much less than the value of K_I. It means that the beam is mainly cracked by tension force. Based on the analysis above, the typical contour maps of horizontal displacement during the loading process are shown in Figure 9.

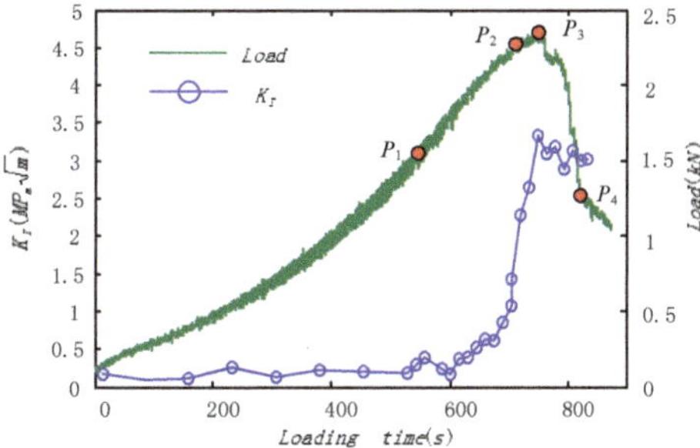

Figure 7. The curve of mode-I stress intensity factor K_I.

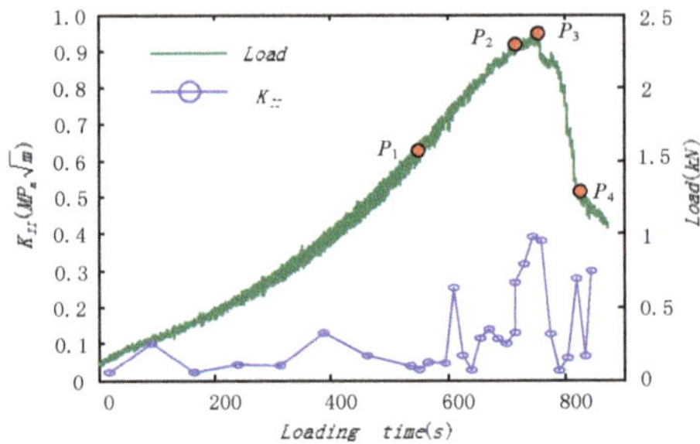

Figure 8. The curve of mode-II stress intensity factor K_{II}.

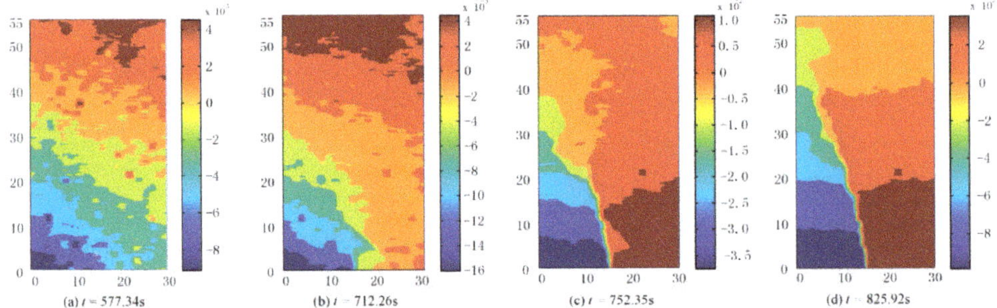

Figure 9. Horizontal displacement contour maps of the interest area (Unit: mm).

The displacement contour maps are corresponded with four spots marked on loading curve in Figure 6, respectively. Table 2 contained more details on specific values mentioned above, such as the loads, SIFs and crack extension lengths.

Table 2. Key experimental result.

Load Time (t)	Load (kN)	K_{I} (MPa·m$^{1/2}$)	K_{II} (MPa·m$^{1/2}$)	$\Delta\alpha$ (mm)	Load/Max Load (%)
577.34	1.58	0.206	0.018	4.65	67.2
712.26	2.29	2.545	0.281	8.35	97.4
752.35	2.35	3.332	0.390	15.02	100
825.92	1.24	3.0065	0.281	24.63	52.8

3.2. Internal AE Event of FPZ

Figure 10 shows the positions of 16 AE sensors and the last location distribution of the AE event. The 3rd to 8th AE sensors are attached onto the front specimen surface, and the 9th to 14th sensors are attached onto the back specimen surface. The first, second and the 15th and 16th AE sensor are attached onto the downside and upside of specimen edge respectively. A sufficient number of sensors can effectively suppress the influence on the events localization caused by the individual signals attenuation and sensors position. The definition of AE events is defined by a threshold of 100 millivolts. From Figure 10a, the pencil lead break position is near the 9th AE sensor. The total 9035 AE events are checked out from the signal waves. A total of 175 AE event positions are located effectively by the proposed method. It can be found that the location distribution of AE event in Figure 10b agreed with the deformation localization positions in Figure 9. Therefore, the location of AE events can be applied to study the micro-cracking events and the evolution of FPZ internal the concrete specimen.

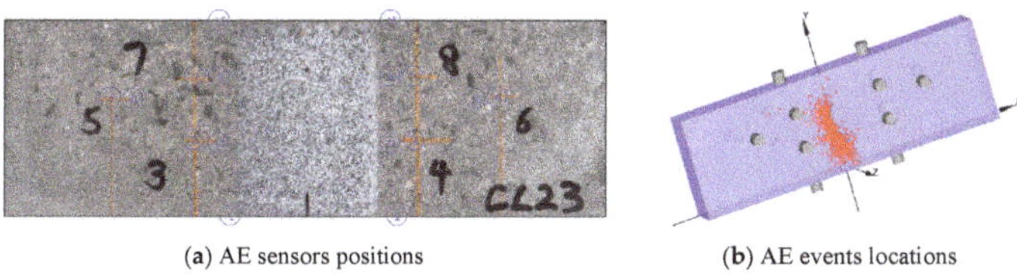

(**a**) AE sensors positions (**b**) AE events locations

Figure 10. The AE sensor position and the AE event location.

From Figure 11, the AE events are marked during the loading process. The marked positions on loading curve are the same as what have been mentioned in the Section 3.1. Therefore, it can be analyzed correspondingly for the internal micro-cracking characteristics and the surface deformation localizations of FPZ. The first AE event is caused by the pencil lead break. The following AE event is caused by the micro-cracking during the FPZ evolution process.

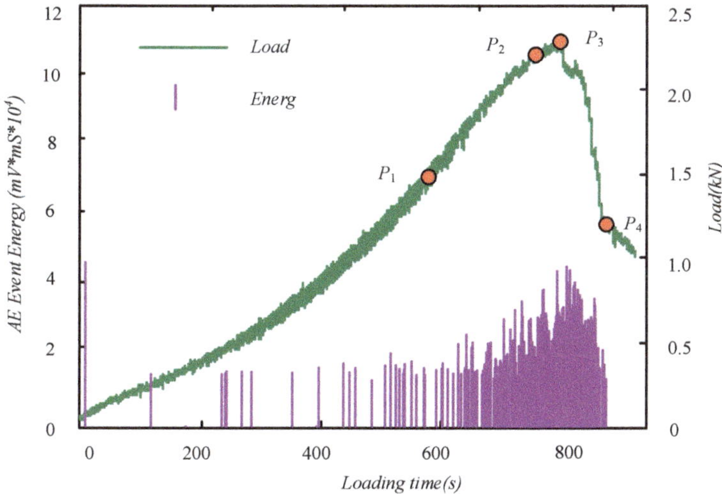

Figure 11. The curves of loading and AE event energy distribution.

It can be seen that the most AE events are concentrated between the main crack initial and unstable extension period. AE energy is sharply increased as macro crack forms in a FPZ. The emission energy of the event around the peak loading is higher than the other event. The AE energy in this analysis is defined as the product of amplitude and duration of the emission.

The AE event distribution in the FPZ at loading points P_1, P_2, P_3, and P_4 are shown in Figure 12. As it is shown in Figure 12a, the first AE event location is near the up loading roller, but the event is an isolated one. Conversely, the AE events which occurred from the pre-crack tip are a series of event in Figure 12b–d. Meanwhile, these events are distributed in a narrow band, the width of the band is about 10 mm. The band is in accordance with the displacement localization band in Figure 9. Therefore, it can be concluded that the internal AE event band is in accordance with the FPZ ahead of the pre-crack tip. The length of FPZ in Figure 12c,d are in agreement with the crack length shown in Figure 6 observed by the DIC method.

Corresponding to Figure 12, the surface horizontal strain fields are shown in Figure 13. The surface strain is obviously localized in a narrow band in Figure 13c,d. From Figures 12 and 13, it is observed that the FPZ corresponds to both the surface strain localization zone and the internal AE event zone.

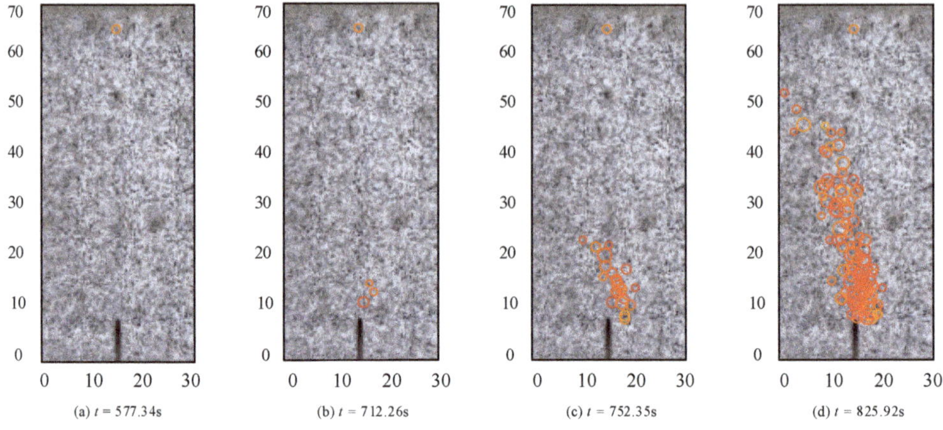

(a) $t = 577.34$s   (b) $t = 712.26$s   (c) $t = 752.35$s   (d) $t = 825.92$s

Figure 12. Location of AE events in FPZ (Unit: mm).

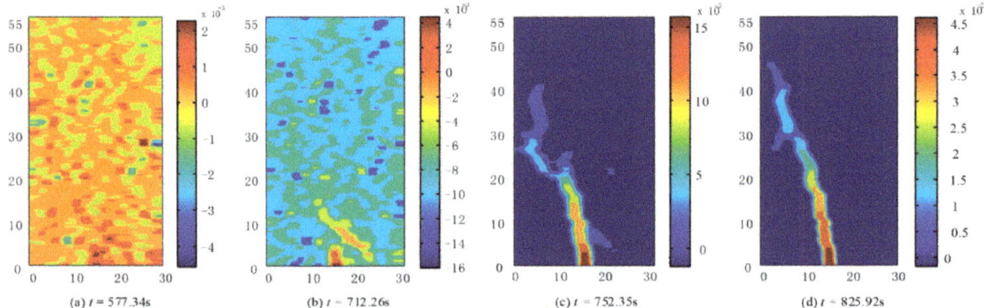

(a) $t = 577.34$s   (b) $t = 712.26$s   (c) $t = 752.35$s   (d) $t = 825.92$s

Figure 13. Strain localization ahead the crack tip (Unit: mm).

3.3. Concrete FPZ Characteristics

The internal and external characteristics of FPZ in concrete are identified by DIC and AE techniques quantitatively during the mode I fracture extension process. Although the measurement of FPZ length has attracted the attention of most researchers, the quantitative measurement of FPZ length has not yet been realized until now. The crack tip position and FPZ length is measured automatically and quantitatively by the DIC technique proposed in this presentation. The length of FPZ obtained by DIC and AE technique are in good agreement. The maximum of FPZ length is about 20 mm, the experimental results agreed with the hypothesis that the length of FPZ is about three times the maximum aggregate size [24].

The internal micro-cracking process is monitored by AE technique. The cracking source of AE events are located precisely by the proposed location methods. The results show that the internal AE event propagation process is completely consistent with the displacement and strain localization process obtained by DIC method on the surface of concrete specimen. During the concrete crack extension process, the FPZ corresponds to the surface deformation localization band and the internal AE event band. The width of fracture process zone is 3 mm on the surface of specimen and 10 mm in the interior of specimen. It means the internal influence area of crack in concrete is larger than that of FPZ observed on concrete surface.

The measurement of stress intensity factors at crack tip is usually a difficult problem during the FPZ extension process. The SIFs are derived from the displacement field around the FPZ tip by the DIC technique proposed in this presentation. The stress intensity factors indicate the distribution characteristics of the stress field around the FPZ. It should be mentioned that the emission energy monitored by AE sensor is a relative parameter, which can't be directly applied to quantify the real released energy of AE event. The comparative analysis of AE results based on deformation field and SIFs will help to explain and analyze the experimental results of AE in the FPZ of concrete.

4. Conclusions

The following conclusions can be drawn from experiments with Single-edge notched concrete beams under three-points bending using the digital image correlation and acoustic emission technique:

(1) The displacement fields, strain fields, crack tip position, crack extension length, and SIFs are obtained automatically and quantitatively by DIC technique proposed in this presentation, the FPZ is identified by the surface deformation of specimen during the crack extension process.

(2) The internal micro-cracking events of FPZ are localized automatically and dynamically by the proposed AE event location method. The distribution of internal acoustic emission events corresponds to the FPZ on the surface of the specimen.

(3) The length of the FPZ is obtained by DIC and AE technique, the results are consistent with each other.

(4) The internal and surface characteristics of FPZ evolution are identical during the concrete crack extension process. The comparative analysis based on DIC and AE technique is helpful to comprehend the characteristics of concrete FPZ.

Author Contributions: S.D. and X.L. conducted the experiments, wrote the paper. K.N. offered useful suggestions for the preparation and writing the paper.

Funding: This research was funded by the National Natural Science Foundation of China (41002075), the Open Research Fund Program of State Key Laboratory of Hydroscience and Engineering of Tsinghua University (sklhse-2015-C-06), and the Open Projects of Research Center of Coal Resources Safe Mining and Clean Utilization Liaoning (LNTU17KF15).

Conflicts of Interest: The authors declare that they have no conflicts of interest.

References

1. Denarié, E.; Saouma, V.E. Concrete fracture process zone characterization with fiber optics. *J. Eng. Mech.* **2001**, *127*, 494–502. [CrossRef]
2. Skarżyński, .; Tejchman, J. Experimental investigations of fracture process in concrete by means of X-ray Micro-computed tomography. *Strain* **2016**, *52*, 26–45. [CrossRef]
3. Zhao, J.L.; Bao, T.F.; Amjad, U. Optical fiber sensing of small cracks in isotropic homogeneous materials. *Sens. Actuators Phys.* **2015**, *225*, 133–138. [CrossRef]
4. Hadjab, H.; Chabaat, S.M. Use of scanning electron microscope and the non-local damage model to investigate fracture process zone in notched concrete beams. *Exp. Mech.* **2007**, *47*, 473–484. [CrossRef]
5. Peters, W.; Ranson, W. Digital imaging techniques in experimental stress analysis. *Opt. Eng.* **1982**, *21*, 427–431. [CrossRef]
6. Sutton, M.; Wolters, W. Determination of displacements using an improved digital correlation method. *Image Vis. Comput.* **1983**, *1*, 133–139. [CrossRef]
7. Choi, S.; Shah, S.P. Measurement of deformations on concrete subjected to compression using image correlation. *Exp. Mech.* **1997**, *37*, 307–313. [CrossRef]
8. Corr, D.; Accardi, M. Digital image correlation analysis of interfacial debonding properties and fracture behavior in concrete. *Eng. Fract. Mech.* **2007**, *74*, 109–121. [CrossRef]
9. Wu, Z.; Rong, H. An experimental investigation on the characteristics of FPZ in concrete using digital image correlation technique. *Eng. Fract. Mech.* **2011**, *78*, 2978–2990. [CrossRef]
10. Rouchier, S.; Foray, G. Damage monitoring in fibre reinforced mortar by combined digital image correlation and acoustic emission. *Constr. Build. Mater.* **2013**, *38*, 371–380. [CrossRef]
11. Rusch, H. Physical problems in the testing of concrete. *Zement Kalk Gips.* **1959**, *12*, 1–9.
12. Wells, D. An acoustic apparatus to record emissions from concrete under strain. *Nucl. Eng. Des.* **1970**, *12*, 80–88. [CrossRef]

13. Grosse, C.; Ohtsu, M. *Acoustic Emission Testing*; Springer Science & Business Media: Berlin/Heidelberg, Germany, 2008; pp. 211–212.
14. Colombo, S.; Forde, M.C.; Main, I.G.; Shigeishi, M. Predicting the ultimate bending capacity of concrete beams from the "relaxation ratio" analysis of AE signals. *Constr. Build. Mater.* **2005**, *19*, 746–754. [CrossRef]
15. Colombo, I.S.; Main, I.G.; Forde, M.C. Assessing damage of reinforced concrete beam using "b-value" analysis of acoustic emission signals. *J. Mater. Civ. Eng.* **2003**, *15*, 280–286. [CrossRef]
16. Muralidhara, S.; Prasad, B.K.R.; Eskandari, H. Fracture process zone size and true fracture energy of concrete using acoustic emission. *Constr. Build. Mater.* **2010**, *24*, 479–486. [CrossRef]
17. Sagar, R.V.; Prasad, B.R.; Singh, R.K. Laboratory investigations on concrete fracture using acoustic emission techniques. In *Acoustic Emission and Related Non-Destructive Evaluation Techniques in the Fracture Mechanics of Concrete*; Cambridge/Woodhead Publishing: Cambridge, UK, 2015; pp. 137–159.
18. Guo, M.; Alam, S.Y.; Bendimerad, A.Z. Fracture process zone characteristics and identification of the micro-fracture phases in recycled concrete. *Eng. Fract. Mech.* **2017**, *181*, 101–115. [CrossRef]
19. Xiangqian, F.; Shaowei, H.; Jun, L. Damage and fracture processes of concrete using acoustic emission parameters. *Comput. Concr.* **2016**, *18*, 267–278.
20. Ohno, K.; Uji, K.; Ueno, A. Fracture process zone in notched concrete beam under three-point bending by acoustic emission. *Constr. Build. Mater.* **2014**, *67*, 139–145. [CrossRef]
21. Verbruggen, S.; Aggelis, D.G.; Tysmans, T. Bending of beams externally reinforced with TRC and CFRP monitored by DIC and AE. *Compos. Struct.* **2014**, *112*, 113–121. [CrossRef]
22. Alam, S.Y.; Saliba, J.; Loukili, A. Study of evolution of fracture process zone in concrete by simultaneous application of digital image correlation and acoustic emission. In Proceedings of the 8th International Conference on Fracture Mechanics of Concrete and Concrete Structures, Toledo, Spain, 10–14 March 2013; pp. 1–9.
23. Alam, S.Y.; Loukili, A. Detecting crack profile in concrete using digital image correlation and acoustic emission. *Eur. Phys. J. Web Conf.* **2010**, *6*, 23003. [CrossRef]
24. Golaski, L.; Goszczynska, B.; Swit, G. System for the global monitoring and evalu ation of damage processes developing within concrete structures under service loads. *Balt. J. Road Bridge Eng.* **2012**, *7*, 237. [CrossRef]

Article

3D Strain and Elasticity Measurement of Layered Biomaterials by Optical Coherence Elastography based on Digital Volume Correlation and Virtual Fields Method

Fanchao Meng, Xinya Zhang, Jingbo Wang, Chuanwei Li, Jinlong Chen and Cuiru Sun *

Department of Mechanical Engineering, Tianjin Key Laboratory of Modern Engineering Mechanics, Tianjin University, No. 135 Yaguan Road, Tianjin 300354, China; 18602669349@163.com (F.M.); zhangxinya@tju.edu.cn (X.Z.); 15022145576@163.com (J.W.); licw16@tju.edu.cn (C.L.); jlchen66@tju.edu.cn (J.C.)
* Correspondence: carry_sun@tju.edu.cn

Received: 9 February 2019; Accepted: 21 March 2019; Published: 30 March 2019

Featured Application: 3D deformation measurement and elasticity quantification of biological tissue and biomaterials.

Abstract: The three-dimensional (3D) mechanical property characterization of biological tissues is essential for physiological and pathological studies. A digital volume correlation (DVC) and virtual fields method (VFM) based 3D optical coherence elastography (OCE) method is developed to quantitatively measure the 3D full-field displacements, strains and elastic parameters of layered biomaterials assuming the isotropy and homogeneity of each layer. The integrated noise-insensitive DVC method can obtain the 3D strain tensor with an accuracy of 10%. Automatic segmentation of the layered materials is realized based on the full field strain and strain gradient. With the strain tensor as input, and in combination with the segmented geometry, the Young's modulus and Poison's ratio of each layer of a double-layered material and a pork specimen are obtained by the VFM. This study provides a powerful experimental method for the differentiation of various components of heterogeneous biomaterials, and for the measurement of biomechanics.

Keywords: 3D deformation; digital volume correlation; optical coherence elastography; virtual fields method; layered material

---

1. Introduction

Optical coherence elastography (OCE) uses optical coherence tomography (OCT) as an imaging modality to measure the mechanical properties of biological tissue [1,2]. Thanks to the inherent high resolution and ~2 mm depth penetration of OCT, OCE has the potential for early disease diagnosis. To extract mechanical information such as strain and elastic modulus from OCT interference signals, phase-based [3,4] and speckle tracking based [5] methods can be used. Speckle tracking based OCE methods have an advantage over phase-based methods for simultaneous multi-dimensional static deformation measurement.

In 1998 Schmitt first proposed the concept of OCE or OCT elastography [5]. He used a normalized cross-correlation (NCC) speckle tracking method to measure the displacements and strains of gelatin phantoms and tissue specimens including pork meat, and intact skin in two dimesnions (2D). This work demonstrated the feasibility of applying speckle correlation to extract mechanical properties from OCT images, which are fundamentally laser speckle images [6]. In 2004, Rogowska et al. [7] improved the speckle tracking algorithm by using a zero-mean normalized cross-correlation (ZNCC) criterion. They found that the noise in axial and lateral displacement maps could be reduced by increasing the size of the subset (cross correlation kernel). Chan et al. [8] and Khalil et al. [9] then proposed a variation framework to reconstruct the Young's modulus of arterial tissue from the speckle correlation displacement measurement, using the inverse problem solving. Rogowska et al. later applied the speckle correlation method to measure the deformation and elastic modulus of atherosclerotic arterial samples [10]. However, only the average value of Young's modulus was obtained. Difficulties remain in measuring the Young's modulus of each layer of the aorta specimen. Sun et al. [11] conducted detailed experimental studies on the feasibility of combining a Newton-Raphson (NR) digital image correlation (DIC) algorithm [12] with OCT imaging for the strain measurement. They concluded that the speckle patterns were correlated well when the strain is less than 10%. Although measurement errors have not been well analysed and quantified from the above studies, speckle correlation methods have been proved to be applicable to integrate with OCT imaging for the deformation measurement.

Recently, with the development of digital volume correlation (DVC), 3D OCE techniques have been proposed. In 2013, Fu et al. [13] first combined OCT imaging with DVC for the 3D displacement and strain measurement by utilizing the commercial DVC software from LaVision (Göttingen, Germany). Later, they measured the full-field deformation of corneal considering the distortion induced by refraction in OCT reconstruction [14]. They also proposed to use the virtual fields method (VFM) to calculate the Young's modulus and Poisson's ratio of the homogeneous phantoms of two shapes [13]. The results were promising although there were fringe-like errors in the strain maps. In the same year, Nahas et al. [15] conducted proof-of-concept studies of combing full-field OCT with DVC for the strain tensor measurement of multi-layer phantoms and biological samples. The strains were calculated from the displacement gradient. This study demonstrated the feasibility of the method without a detailed evaluation of the accuracy of the results. Lately, Acosta Santamaría et al. measured the deformation of the porcine aorta immersed in tissue clearing agents under tensile loading by OCT and commercial DVC software [16]. These studies demonstrated the feasibility of using a DVC algorithm to process OCT images. However, the data format is often limited by the commercial software. Girard et al. conducted a series of studies. They developed a DVC algorithm based on a differential evolution method for the 3D displacement measurement [17], then proposed a VFM method, which is insensitive to rigid body motion, to recover the constitutive parameters of the optic nerve head [18]. This work provided a paradigm for combining OCT, DVC and VFM for the mechanical properties' measurement of heterogeneous tissue. However, using strains instead of displacements as the input for VFM calculation may be more accurate. Although there are not so many studies of combined OCT and DVC yet, it can be seen that it is completely feasible to combine them for the 3D displacement measurement. However, the noise insensitive method is required to calculate the strain tensors in dealing with the laser speckle noises in OCT images. In addition, when objects with complex structures are measured, a large amount of virtual displacements and large-scale linear equations maybe involved in VFM, which may result in unsolvable problems or wrong solutions. Thus, a method to simplify the VFM calculation is required when a nonhomogeneous sample is measured.

This study is to develop a DVC and VFM based OCE method for the quantitative characterization of mechanical properties of layered biological tissue. As the DVC algorithm has been advanced in all aspects since it was first proposed in 1999 [19], we picked out the steps that are insensitive to noise and integrated them into a 3D strain measurement method with OCT imaging. This included a zero-mean normalized sum of squared difference (ZNSSD) criterion [20], a 3D inverse compositional Gauss-Newton (IC-GN) matching method [21], and a local ternary quadratic polynomial fitting method [22] for strain calculation. We then chose a linear VFM method which is noise resistance [23]. The automatic segmentation of the layers is proposed based on the strain gradient acquired by DVC and OCT, so that the Young's modulus and Poisson's ratio of each layer can be solved individually by VFM. Strains are used as the input for the VFM calculation of material properties of double-layer materials under a compressive load. This quantitative 3D OCE method can obtain the elastic modulus of each layer of the specimen simultaneously. As many biological tissues, such as the skin and the vascular wall, are layered structures, this study will have a wide application potential in experimental biomechanics.

2. Equipment and Methods

2.1. 3D Strain Measurement by OCT Imaging and DVC

A swept source OCT (SSOCT) system and a compressive loading device were developed in our laboratory. A schematic diagram of the system is shown in Figure 1a. Laser emitted from a swept laser source (HSL-20-100-B, Santec, Aichi, Japan) goes to a Mach-Zahnder interferometer. The interference signal was detected by a balanced photodetector (EBR370006-02, Exalos, Schlieren, Switzerland) then acquired by a data acquisition card in a computer. The central wavelength of the laser is 1315 nm; the bandwidth is 88 nm, which results in an axial resolution of 9.93 μm. The refractive index of biological tissue is around 1.4, so the axial resolution in tissue is 7.09 μm. Sample beams are scanned by two Galvo mirrors. The line scan rate is 100 kHz, so it takes 0.46 s to scan a bulk of $214 \times 214 \times 1024$ voxels. A compressive loading device consisting of a base and an optical window was designed. A force sensor (FSH03221, FUTEK, Irvine, CA, USA) is set on the base to measure the applied load. An example of a 2D cross-sectional OCT image and a 3D volumetric image of a phantom are shown in Figure 1b,c.

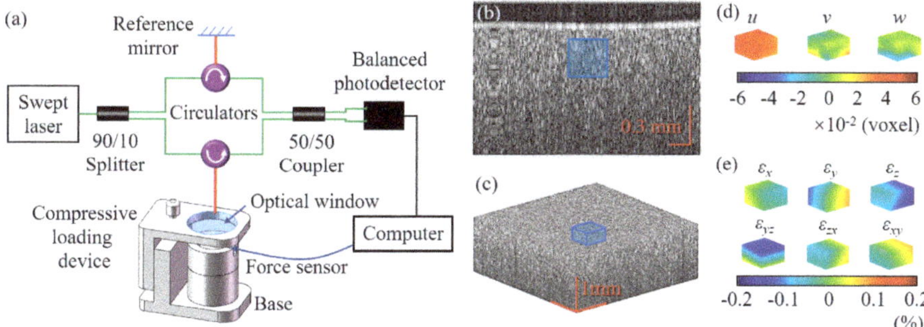

Figure 1. The SSOCT and compressive loading system. (**a**) A schematic diagram of the system; (**b**) a 2D OCT cross-sectional image (B-scan); (**c**) a 3D volumetric OCT image. Stationary experiment: (**d**) 3D displacements in the *x*, *y* and *z* directions of the ROI; (**e**) 3D strain tensors of the ROI. The ROI is marked out with a blue rectangle in (**b**) and with a blue cube in (**c**).

A DVC algorithm was developed by integrating a coarse search mechanism, an IC-GN based fine search algorithm and a local ternary quadratic polynomial fitting method for strain calculation. Firstly, because the ZNCC correlation function is not sensitive to noise, a ZNCC correlation function based coarse search was developed to find initial integer values of voxel shifts. Secondly, the IC-GN [21] based fine search algorithm was employed. The IC-GN algorithm changes the shape and location of target and reference sub-volumes simultaneously to find the maximum of the correlation function. Because the Hessian matrix does not need to update in each iteration, the computational speed can be ~1.7 times faster than the conventional DVC method [21]. The computing speed is ~4 s per point calculating OCT images by DVC. The convergence criterion was set as either the increment of each displacement component was less than 0.001 voxels, or the maximum iteration of 20 was reached. A cubic spline interpolation scheme with not-a-knot end condition was utilized for the calculation of sub-voxel intensities. More details about the IC-GN algorithm can be found in paper [21]. Through the coarse-fine search DVC calculation, the 3D displacements can be obtained with sub-voxel resolution. Thirdly, the local ternary quadratic polynomial fitting equations were used to fit $3 \times 3 \times 3$ voxels of the displacement data with the least square method. Then, the strains were calculated using Cauchy's formulae [24]. The displacement and strain resolution of this system are around 0.06 voxels (~1 μm) and 0.2%, respectively, tested by the stationary experiment as shown in Figure 1d,e, where the region of interest (ROI) is marked out with a blue rectangle in (b) and a blue cube in (c), which may become worse in deformation measurements due to the speckle blinking [25,26].

2.2. VFM

The VFM is based on the principle of virtual work, which has

$$-\int_V \sigma : \varepsilon^* dV + \int_{S_f} \mathbf{T} \cdot \mathbf{u}^* dS + \int_V \mathbf{b} \cdot \mathbf{u}^* dV = \int_V \rho \mathbf{a} \cdot \mathbf{u}^* dV \tag{1}$$

where σ is the real stress tensor; ε^* is the virtual strain tensor; $\mathbf{u}^*$ is the virtual displacement vector which should meets the boundary conditions; $\mathbf{T}$ is the external force vector on the surface of S_f; $\mathbf{b}$ is the body force vector; $\mathbf{a}$ is the acceleration vector; ρ is the density; V is the volume; S is the surface; ":" means the contraction operation of the second order tensor, i.e., $\sigma : \varepsilon^* = \sum_{i,j} \sigma_{i,j} \varepsilon^*_{i,j}$, $i, j = 1, 2, 3$. In this paper, the body force $\mathbf{b}$ can be neglected because it is relatively small compared with the external force $\mathbf{T}$; the samples are compressed statically, so the acceleration $\mathbf{a}$ can also be neglected. Hence, the Equation (1) can be simplified as

$$-\int_V \sigma : \varepsilon^* dV + \int_{S_f} \mathbf{T} \cdot \mathbf{u}^* dS = 0 \tag{2}$$

For convenience, the stress tensor and strain tensor are written as vectors:

$$\boldsymbol{\sigma} = \left(\sigma_x, \sigma_y, \sigma_z, \sigma_{yz}, \sigma_{zx}, \sigma_{xy}\right)^{\mathrm{T}} = \left(\sigma_1, \sigma_2, \sigma_3, \sigma_4, \sigma_5, \sigma_6\right)^{\mathrm{T}} \tag{3}$$

$$\boldsymbol{\varepsilon} = \left(\varepsilon_x, \varepsilon_y, \varepsilon_z, 2\varepsilon_{yz}, 2\varepsilon_{zx}, 2\varepsilon_{xy}\right)^{\mathrm{T}} = \left(\varepsilon_1, \varepsilon_2, \varepsilon_3, \varepsilon_4, \varepsilon_5, \varepsilon_6\right)^{\mathrm{T}} \tag{4}$$

According to elastic mechanics, the constitutive equations of 3D isotropic elastic materials are

$$\begin{aligned}
\sigma_1 &= Q_{11}\varepsilon_1 + Q_{12}(\varepsilon_2 + \varepsilon_3), & \sigma_4 &= \tfrac{1}{2}(Q_{11} - Q_{12})\varepsilon_4, \\
\sigma_2 &= Q_{11}\varepsilon_2 + Q_{12}(\varepsilon_1 + \varepsilon_3), & \sigma_5 &= \tfrac{1}{2}(Q_{11} - Q_{12})\varepsilon_5, \\
\sigma_3 &= Q_{11}\varepsilon_3 + Q_{12}(\varepsilon_1 + \varepsilon_2), & \sigma_6 &= \tfrac{1}{2}(Q_{11} - Q_{12})\varepsilon_6,
\end{aligned} \tag{5}$$

where the relationships between the two independent elastic parameters Q_{11}, Q_{12} and Young's modulus E and Poisson's ratio v are

$$Q_{11} = \frac{(1-v)E}{(1+v)(1-2v)}, \quad Q_{12} = \frac{vE}{(1+v)(1-2v)}. \tag{6}$$

Then, Equation (2) becomes

$$\int_V Q_{ij}\varepsilon_j \varepsilon^*_i dV = \int_{S_f} \mathbf{T} \cdot \mathbf{u}^* dS \quad i, j = 1, 2, \cdots, 6, \tag{7}$$

where Q_{ij} is elastic constant. If the material is homogenous, the Equation (7) can be expressed as

$$\begin{aligned}
&Q_{11} \int_V \left(\varepsilon_1 \varepsilon^*_1 + \varepsilon_2 \varepsilon^*_2 + \varepsilon_3 \varepsilon^*_3 + \tfrac{1}{2}\varepsilon_4 \varepsilon^*_4 + \tfrac{1}{2}\varepsilon_5 \varepsilon^*_5 + \tfrac{1}{2}\varepsilon_6 \varepsilon^*_6\right) dV \\
&+ Q_{12} \int_V \left(\begin{array}{c} \varepsilon_1 \varepsilon^*_3 + \varepsilon_3 \varepsilon^*_1 + \varepsilon_2 \varepsilon^*_3 + \varepsilon_3 \varepsilon^*_2 + \varepsilon_1 \varepsilon^*_2 + \varepsilon_2 \varepsilon^*_1 \\ -\tfrac{1}{2}\varepsilon_4 \varepsilon^*_4 - \tfrac{1}{2}\varepsilon_5 \varepsilon^*_5 - \tfrac{1}{2}\varepsilon_6 \varepsilon^*_6 \end{array}\right) dV = \int_{S_f} \mathbf{T} \cdot \mathbf{u}^* dS.
\end{aligned} \tag{8}$$

The integration can be approximated to summation by

$$\int_V \varepsilon_j \varepsilon^*_i dV = \sum_{k=1}^{M} \varepsilon_{j,k} \varepsilon^*_{i,k} v_k, \quad i, j = 1, 2, \cdots, 6, \tag{9}$$

where k represents the kth test point; M is the number of total points; v_k is the volume of the kth point or mesh; $\varepsilon^*_{i,k}$ is the virtual strain of the ith variable of the kth point; $\varepsilon_{j,k}$ is the real strain of the jth variable of the kth point. There are two unknown parameters Q_{11} and Q_{12} to solve, so two virtual fields are needed. Superscript (i) which means the ith virtual field and $i = 1, 2$ is used in order to build linearly independent equation:

$$\mathbf{A}\mathbf{Q} = \mathbf{B} \tag{10}$$

where the coefficient matrix $\mathbf{A}$ is a 2×2 square matrix; the constitutive parameter vector $\mathbf{Q}$ and the virtual work vector $\mathbf{B}$ are 2×1 vectors. The coefficient matrix $\mathbf{A}$ can be expressed as follows:

$$
\begin{aligned}
A_{i1} &= \int_V \left(\varepsilon_1 \varepsilon_1^{*(i)} + \varepsilon_2 \varepsilon_2^{*(i)} + \varepsilon_3 \varepsilon_3^{*(i)} + \tfrac{1}{2}\varepsilon_4 \varepsilon_4^{*(i)} + \tfrac{1}{2}\varepsilon_5 \varepsilon_5^{*(i)} + \tfrac{1}{2}\varepsilon_6 \varepsilon_6^{*(i)} \right) dV, \\
A_{i2} &= \int_V \left(\begin{array}{c} \varepsilon_1 \varepsilon_3^{*(i)} + \varepsilon_3 \varepsilon_1^{*(i)} + \varepsilon_2 \varepsilon_3^{*(i)} + \varepsilon_3 \varepsilon_2^{*(i)} + \varepsilon_1 \varepsilon_2^{*(i)} + \varepsilon_2 \varepsilon_1^{*(i)} \\ -\tfrac{1}{2}\varepsilon_4 \varepsilon_4^{*(i)} - \tfrac{1}{2}\varepsilon_5 \varepsilon_5^{*(i)} - \tfrac{1}{2}\varepsilon_6 \varepsilon_6^{*(i)} \end{array} \right) dV.
\end{aligned}
\tag{11}
$$

The constitutive parameter vector $\mathbf{Q}$ is

$$\mathbf{Q} = \left(\begin{array}{c} Q_{11} \\ Q_{12} \end{array} \right) \tag{12}$$

The virtual work vector $\mathbf{B}$ can be expressed as

$$B_i = \int_{S_f} \mathbf{T} \cdot \mathbf{u}^{*(i)} dS. \tag{13}$$

Hence, the two unknown elastic constants can be solved by Equations (14) and (15):

$$\mathbf{Q} = \mathbf{A}^{-1}\mathbf{B}, \tag{14}$$

$$E = \frac{(Q_{11} - Q_{12})(Q_{11} + 2Q_{12})}{Q_{11} + Q_{12}}, \quad v = \frac{Q_{12}}{Q_{11} + Q_{12}}. \tag{15}$$

For double-layered material, the principle of virtual work is still adaptable, while there will be two sets of linear equations, assuming each layer is homogeneous. Equation (2) can be written as:

$$^n\mathbf{A} \, ^n\mathbf{Q} = \, ^n\mathbf{B} \tag{16}$$

where n is the layer number; $^n\mathbf{A}$ is the coefficient matrix; $^n\mathbf{Q}$ is the constitutive parameters vector, and $^n\mathbf{B}$ is the virtual work vector done by external forces. The coefficient matrix $^n\mathbf{A}$ can be expressed as follows:

$$
\begin{aligned}
^n A_{i1} &= \int_{V_n} \left(\varepsilon_1 \varepsilon_1^{*(i)} + \varepsilon_2 \varepsilon_2^{*(i)} + \varepsilon_3 \varepsilon_3^{*(i)} + \tfrac{1}{2}\varepsilon_4 \varepsilon_4^{*(i)} + \tfrac{1}{2}\varepsilon_5 \varepsilon_5^{*(i)} + \tfrac{1}{2}\varepsilon_6 \varepsilon_6^{*(i)} \right) dV_n, \\
^n A_{i2} &= \int_{V_n} \left(\begin{array}{c} \varepsilon_1 \varepsilon_3^{*(i)} + \varepsilon_3 \varepsilon_1^{*(i)} + \varepsilon_2 \varepsilon_3^{*(i)} + \varepsilon_3 \varepsilon_2^{*(i)} + \varepsilon_1 \varepsilon_2^{*(i)} + \varepsilon_2 \varepsilon_1^{*(i)} \\ -\tfrac{1}{2}\varepsilon_4 \varepsilon_4^{*(i)} - \tfrac{1}{2}\varepsilon_5 \varepsilon_5^{*(i)} - \tfrac{1}{2}\varepsilon_6 \varepsilon_6^{*(i)} \end{array} \right) dV_n,
\end{aligned}
\tag{17}
$$

where n represents the nth layered material and $n = 1, 2$ for a double-layered material; V_n represents the volume of the nth layered material. The constitutive parameter vector $^n\mathbf{Q}$ is

$$^n\mathbf{Q} = \left(\begin{array}{c} ^nQ_{11} \\ ^nQ_{12} \end{array} \right) \tag{18}$$

where $^nQ_{11}$ and $^nQ_{12}$ represent the two independent elastic parameters of the nth layer. The virtual work vector $^n\mathbf{B}$ can be expressed as

$$^nB_i = \int_{S_{f_n}} {}^n\mathbf{T} \cdot \mathbf{u}^{*(i)} dS_n \tag{19}$$

where S_n means the entire surface, and $^n\mathbf{T}$ represents the external force vector of the nth layer. S_{fn} is the surface where the external force applies. $^n\mathbf{T}$ includes the forces on the interfaces of the layers, which becomes external forces after the layers are divided during calculation. $\mathbf{u}^{*(i)}$ is the ith virtual displacement vector. A flow chart of the layered VFM is shown in Figure 2.

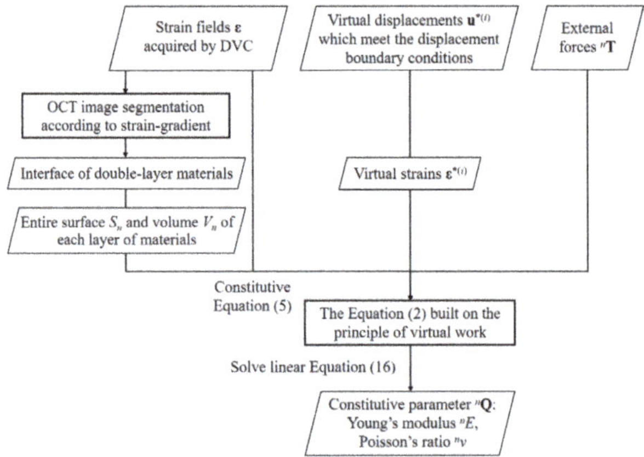

Figure 2. A flow chart of the layered VFM.

3. Experiments

3.1. Strain and Young's Modulus Measurement of a Homogeneous Phantom

3D displacements and strains of a phantom under compression were measured by the DVC-based 3D OCE technique. A homogenous phantom was made by mixing food-grade translucent 45-degree silica gel and titanium dioxide (TiO_2) scatterer with a diameter of 1 µm and density of 0.5 mg/mL. The phantom was cut into a $l_x = 4.24$ mm, $l_y = 4.24$ mm and $l_z = 11.60$ mm cube shown as Figure 3a and was set on the surface of the force sensor of the loading device shown in Figures 1a and 3b. The phantom was preloaded for about 5 min until the reading of the force sensor became stable at 1.072 N. Then, the first 3D OCT image was taken as shown in Figure 3d. The phantom was compressed by 0.046 N immediately after the first image was taken, and the second 3D OCT was taken. This procedure took less than 1 s, thus, tissue relaxation was neglected. The phantom was compressed for −41.0 µm reading from the translation of the optical window. The phantom was scanned 4 mm in both the width and length directions. The ROI was marked out with a blue cube shown in Figure 3d. It is 64.7 µm below the top surface of the phantom as shown in Figure 3c. The coordinate was built with the origin O at the bottom left corner of the ROI, and the three axes were along the three edges of the ROI as shown in Figure 3d. The number of DVC computed points was $M = 9 \times 9 \times 5 = 405$. The step lengths in the x, y and z directions were 17 voxels. By calibration, the dimensions of the ROI were $Lx = Ly = 2861.1$ µm, and $Lz = 844.1$ µm. The average of the correlation coefficient of the ROI was tested to be 0.659.

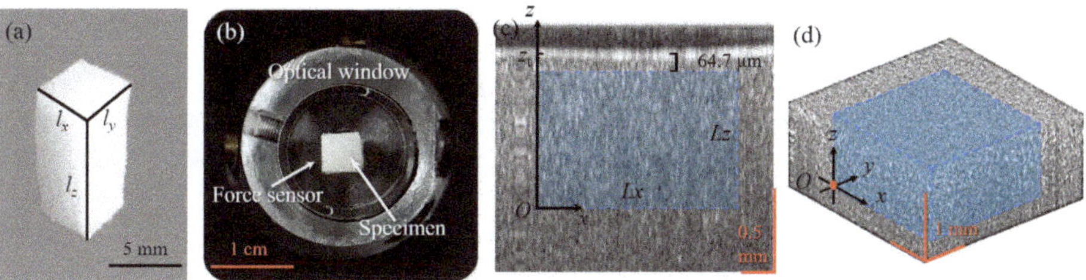

Figure 3. A sample of homogeneous phantom. (**a**) A photo of the phantom; (**b**) a photo of the phantom under compression; (**c**) a 2D OCT image of the phantom, where the ROI is marked with a blue rectangle; (**d**) a 3D OCT image of the phantom, where the ROI is marked with a blue cube.

All components of displacements and strains obtained by the DVC are shown in Figure 4. It can be seen from Figure 4c that the value of displacement in the z-direction decreases from the top to the bottom, which corresponds to theory. The mean value of the displacement on the top surface is -42.7 μm. So, the relative error for the z-direction displacement measurement is 4.1% compared with the imposed displacement of -41.0 μm. The displacements in the x and y directions shown in Figure 4a,b change diagonally, probably because the sample was slightly tilted and the OCT scanning direction was not strictly parallel with the edge of the sample. Each of the strains shown in Figure 4d–i is supposed to be a constant, and therefore the variations of the results demonstrate measurement errors. Figure 4f shows that the values are fairly constant except a thin top layer. The bigger values on the top portion, are possibly induced by the friction between the glass window and the top surface of the sample. The mean value of the ε_z of the whole ROI is -0.39% as shown in Figure 4f. While the applied strain is -41.0 μm/11.60 mm $= -0.35\%$, thus the relative measurement error is 11.4%. If the top portion with bigger values is excluded, the relative measurement error becomes 5.7%. The measurement error for shear strains shown in Figure 4g–i is within 0.2%.

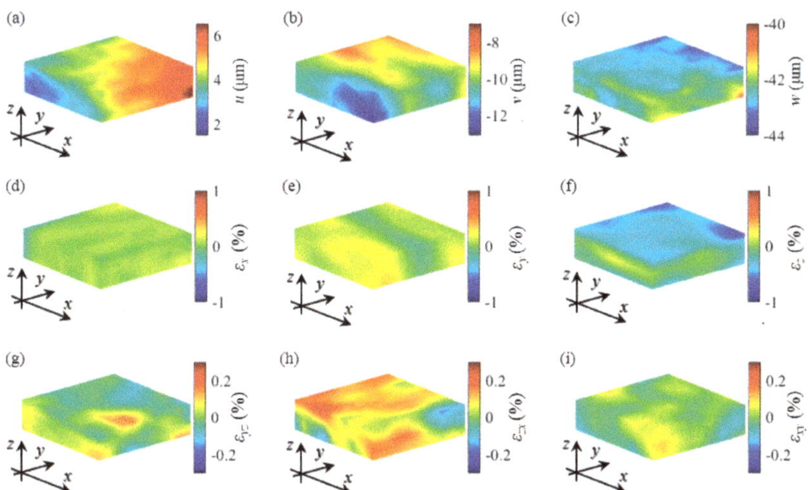

Figure 4. 3D deformation measurement of the homogeneous phantom. (**a–c**) 3D displacements in the x, y and z directions respectively; (**d–f**) 3D normal strains in the x, y and z directions respectively; (**g–i**) 3D shear strains.

The full-field strain ε obtained in Figure 4d–i and virtual strains $\varepsilon^{*(i)}$ were then utilized to form the coefficient matrix $\mathbf{A}$ of VFM in Equation (11). Considering the frictions on the top and bottom surface of the phantom, the boundary conditions were as follows:

$$\begin{cases} u|_{z=z_b,z=z_t} = 0 \\ v|_{z=z_b,z=z_t} = 0 \\ w|_{z=z_b} = 0 \\ T|_{z=z_t} = T_0 \end{cases} \tag{20}$$

where T_0 was the applied pressure, i.e., $T_0 = -0.046 \text{ N}/(4.24 \times 10^{-3} \text{ m} \times 4.24 \times 10^{-3} \text{ m}) = -2558.7$ Pa; z_t and z_b were the locations of the top and bottom surfaces of the phantom where the in-plane displacements were zeros. $z_t = 908.8$ μm was calculated according to the location of the white line in Figure 3c. $z_b = -10{,}691.2$ μm calculated by subtracting l_z from z_t.

According to the boundary conditions, the virtual displacement fields were chosen as follows:

$$\mathbf{u}^{*(1)} = \begin{pmatrix} 0 \\ 0 \\ z - z_b \end{pmatrix} \tag{21}$$

$$\mathbf{u}^{*(2)} = \begin{pmatrix} x(z - z_b)(z - z_t) \\ 0 \\ 0 \end{pmatrix} \tag{22}$$

Hence, the virtual strain fields were

$$\varepsilon^{*(1)} = \begin{pmatrix} 0 & 0 & 1 & 0 & 0 & 0 \end{pmatrix}^{\mathrm{T}} \tag{23}$$

$$\varepsilon^{*(2)} = \begin{pmatrix} (z - z_b)(z - z_t) \\ 0 \\ 0 \\ 0 \\ x(z - z_b + z - z_t) \\ 0 \end{pmatrix} \tag{24}$$

The coefficient matrix $\mathbf{A}$ in Equation (11) were written as

$$\begin{aligned} A_{11} &= \textstyle\sum_{k=1}^{M} \varepsilon_{3,k} v_k \,, \\ A_{12} &= \textstyle\sum_{k=1}^{M} (\varepsilon_{1,k} + \varepsilon_{2,k}) v_k, \\ A_{21} &= \textstyle\sum_{k=1}^{M} [\varepsilon_{1,k}(z_k - z_b)(z_k - z_t) + \varepsilon_{3,k} x_k(2z_k - z_b - z_t)] v_k \,, \\ A_{22} &= \textstyle\sum_{k=1}^{M} \left[(\varepsilon_{2,k} + \varepsilon_{3,k})(z_k - z_b)(z_k - z_t) - \tfrac{1}{2}\varepsilon_{5,k} x_k(2z_k - z_b - z_t) \right] v_k \,, \end{aligned} \tag{25}$$

where M is the number of the matched points in DVC and $M = 405$; v_k is the volume of the kth mesh (point) and $v_k = LxLyLz/M$. According to Equation (13) and static equilibrium, the virtual work vector $\mathbf{B}$ is

$$\begin{aligned} B_1 &= TLxLy(Lz - z_b) - TLxLy(0 - z_b), \\ B_2 &= 0. \end{aligned} \tag{26}$$

Then, coefficient matrix $\mathbf{A}$ and the virtual work vector $\mathbf{B}$ were input to Equation (14) to solve the elastic constants. Q_{11} and Q_{12} were calculated to be 2.2046×10^6 Pa and 1.1070×10^6 Pa. Young's modulus and Poisson's ratio were then calculated to be 1464.5 kPa and 0.334 respectively, by inputting

Q to Equation (15). The Young's modulus of the phantom was also measured by a uniaxial tensile test, which resulted in 1471.0 kPa. The relative discrepancy between the results obtained by the two methods is ~1%. The Poisson's ratio, calculated by dividing the average ε_z from the lateral strain obtained by DVC, was 0.306, where the the quadratic mean of ε_x and ε_y was taken as the lateral strain. The Poisson's ratio calculated by VFM was 0.334, which is 8% different to 0.306.

3.2. Strain and Young's Modulus Measurement of a Double-Layer Phantom

A double-layered phantom was made by mixing 1 μm and 0.5 mg/mL TiO_2 scatterer with 15-degree silica gel for the bottom layer and 0-degree silica gel for the top layer. The dimensions of the double-layer phantom were l_x = 4.20 mm, l_y = 4.00 mm and l_z = 1.90 mm. A compression experiment was conducted following the same procedure as the homogeneous phantom. The double-layer phantom was preloaded for 1.320 N, then compressed by 0.247 N as read from the force sensor. The phantom was scanned 4 mm along both the width and length directions, before and after deformation. A 3D and a 2D OCT image of the phantom are shown in Figure 5a, from which the interface of the two layers can be observed. The ROI is marked with a blue cube or rectangle, which is 99.3 μm below the top surface of the phantom as shown in Figure 5aii. The average of the correlation coefficient of the ROI was tested to be 0.581. The coordinate was built with the origin O at the bottom left corner of the ROI, and the three axes were along the three edges of the ROI as shown in Figure 5ai. The number of computed points was $M = 9 \times 9 \times 5 = 405$. The step lengths were 17 voxels in the x and y directions, and 8 voxels in the z-direction. The computed dimensions were $Lx = Ly = 2861.1$ μm and $Lz = 397.2$ μm. The 3D displacements of the ROI with the defined coordinate system are shown in Figure 5b–d. The 2D images are the cross-sections along the dashed lines on the 3D images. The distributions of 3D strains are shown in Figure 6. From the 3D display of ε_z in Figure 6c, the interface of the two layers can be observed. Strain-gradients were then calculated to segment the two layers automatically. Based on the abrupt change of the strain-gradient, the interface can be determined, which is drawn with dashed lines in Figure 6ci,cii. The strain-gradients of ε_z in the z-direction of the cross-section in Figure 6ci,cii are plotted on their cross-sectional OCT images as shown in Figure 7. For accuracy of computation in the VFM, broken lines simulated the physical interface between two layers, the blue curves shown in Figure 7. Although the physical boundary is recognizable in the phantom, the interfaces of most biological tissues are invisible. Hence, the automatic estimation of the boundary is necessary. The location of the minimum of the strain-gradient is where the interface of the two layers, i.e., the purple dashed lines as shown in Figure 7a,b, shows a good agreement with the physical boundary.

After the full-field strain ε of the double-layer phantom was obtained, the constitutive parameters were calculated by the VFM. Because the errors of ε_x are large on the left boundary as shown in Figure 6a, the values of the left two columns were dropped out during VFM calculation. Then Lx became: $Lx = 2225.3$ μm. The number of computed points became $M = (9 - 2) \times 9 \times 5 = 315$ in the VFM. The boundary conditions were the same as that of the homogenous phantom test, where the displacements in the x and y directions on the top and bottom surfaces were approximated as zeros, considering the friction on the contact surfaces. The value of z_t = 496.5 μm obtained from OCT imaging as shown in Figure 5aii; $z_b = z_t - l_z = -1404.0$ μm; $T_0 = -0.247$ N$/(4.20 \times 10^{-3}$ m $\times 4.00 \times 10^{-3}$ m$) = -14{,}702$ Pa.

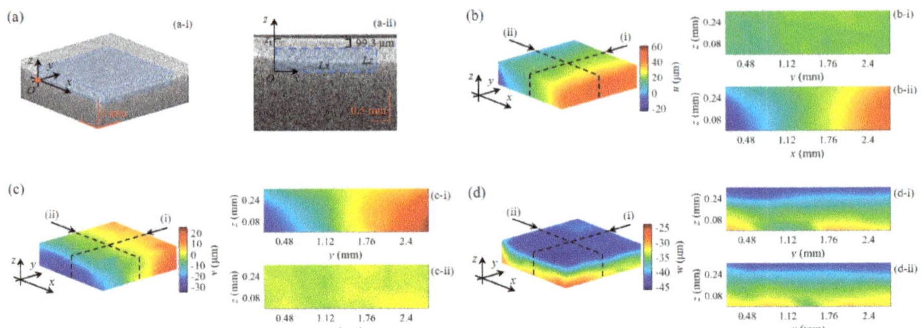

Figure 5. 3D displacement measurement of a double-layer phantom. (**a**) OCT images: (**a-i**) a 3D OCT image with the ROI marked by a blue cube; (**a-ii**) a 2D OCT image with the ROI marked by a blue rectangle; (**b–d**) 3D displacements in the x, y and z directions respectively. Two displacement maps on the right side of each 3D image are the cross-sections along the dashed lines on (**b–d**).

The full-field strain ε and virtual strains $\varepsilon^{*(i)}$ were input to Equation (17). The two coefficient matrixes $^{n}\mathbf{A}$ were

$$
\begin{aligned}
{}^{n}A_{11} &= \sum_{k=1}^{M_n} \varepsilon_{3,k} v_k \\
{}^{n}A_{12} &= \sum_{k=1}^{M_n} (\varepsilon_{1,k} + \varepsilon_{2,k}) v_k \\
{}^{n}A_{21} &= \sum_{k=1}^{M_n} \left[\varepsilon_{1,k}(z_k - z_b)(z_k - z_t) + \varepsilon_{3,k} x_k(2z_k - z_b - z_t) \right] v_k \\
{}^{n}A_{22} &= \sum_{k=1}^{M_n} \left[(\varepsilon_{2,k} + \varepsilon_{3,k})(z_k - z_b)(z_k - z_t) - \tfrac{1}{2}\varepsilon_{5,k} x_k(2z_k - z_b - z_t) \right] v_k
\end{aligned}
\tag{27}
$$

where the superscript n on the upper left corner represents the nth layer and $n = 1, 2$, where 1 means the bottom layer and 2 means the top layer; M_n is the number of meshes in the nth layer; v_k is the volume of the kth mesh and $v_k = LxLyLz/M$. One key step for the calculation is to segment the two layers according to the strain-gradient as described earlier, so as to determine the value of M_n. After the two layers were divided as shown in Figure 6c, the M_n were calculated to be $M_1 = 143$ and $M_2 = 172$ respectively. According to Equation (19) and the static equilibrium, the virtual work vectors $^{n}\mathbf{B}$ are

$$
\begin{aligned}
{}^{1}B_1 &= \frac{TLxLyLz}{M} \sum_{d=1}^{N_x N_y} N_d, \\
{}^{1}B_2 &= 0;
\end{aligned}
\tag{28}
$$

$$
\begin{aligned}
{}^{2}B_1 &= TLxLyLz \left(1 - \frac{1}{M} \sum_{d=1}^{N_x N_y} N_d \right), \\
{}^{2}B_2 &= 0,
\end{aligned}
\tag{29}
$$

where N_x and N_y are the number of points in the x and y directions respectively; N_d is the dth maximum number in the z-direction of the bottom layer, where $d = 1, 2, \ldots, N_x N_y$.

Then the coefficient matrixes $^{n}\mathbf{A}$ and the virtual work vectors $^{n}\mathbf{B}$ were substituted into the linear equations, Equation (16). The elastic constants $^{n}Q_{11}$ and $^{n}Q_{12}$ of the two layers were then solved as: $^{1}\mathbf{Q} = (1.1920 \times 10^6 \text{ Pa}, 8.8362 \times 10^5 \text{ Pa})^{\mathrm{T}}$; $^{2}\mathbf{Q} = (3.2683 \times 10^5 \text{ Pa}, 2.4166 \times 10^5 \text{ Pa})^{\mathrm{T}}$. The Young's moduli and Poisson's ratios were respectively $E_1 = 439.7$ kPa, $v_1 = 0.426$; $E_2 = 121.4$ kPa, $v_2 = 0.425$, where the subscript 1 means the bottom layer and the subscript 2 means the top layer. The Young's moduli were also measured by tensile tests. The comparison of the results between the VFM and the tensile tests is listed in Table 1, which shows that the relative discrepancy between the two measurements is less than

5%, which is 5% better than a simple plane boundary as we used before. The Poisson's ratio obtained by DVC was 0.422. The relative difference between the method of DVC and VFM is 1%.

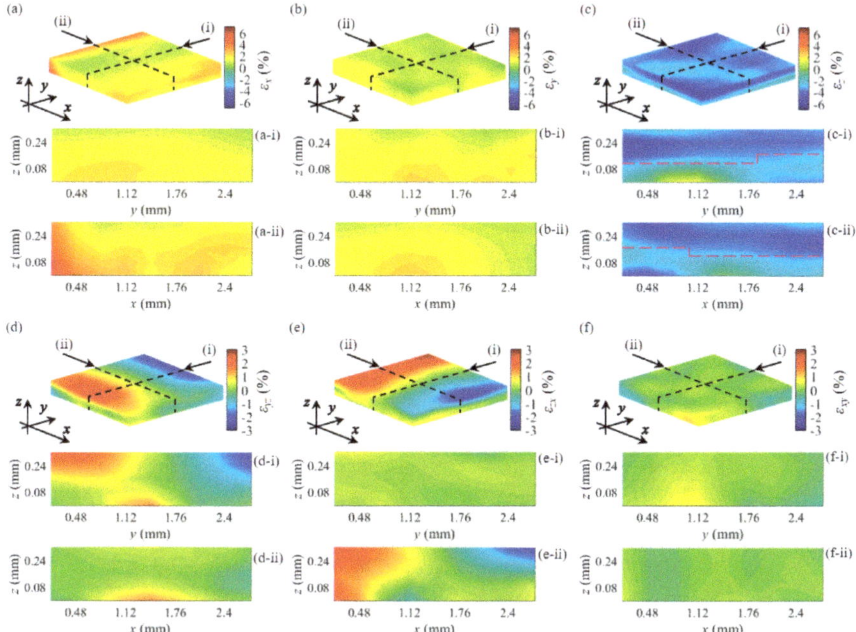

Figure 6. 3D strain fields of a double-layer phantom under compression. (**a–c**) Normal strains in the x, y and z directions respectively. The purple dashed lines in (**c-i**) and (**c-ii**) indicate the interface of the two layers. (**d–f**) Shear strains. Two strain maps shown underneath each of the 3D displays are the images of the cross-sections along the dashed lines.

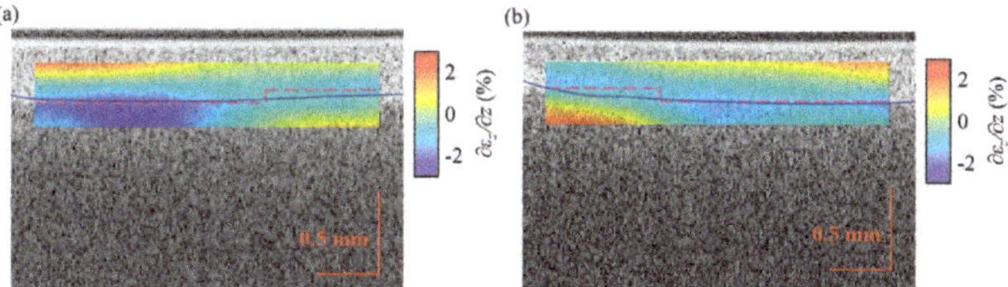

Figure 7. Strain-gradient of ε_z in the z-direction. (**a,b**) are 2D strain-gradient maps of the cross-section (**c-i**) and (**c-ii**) in Figure 6 respectively plotted on their 2D OCT cross-sectional images. The purple dashed line and blue solid line indicate the estimated interface and physical interface of the two layers respectively.

Table 1. The comparison between the tensile test and calculated constitutive parameters of a double-layer phantom by VFM.

Young's Moduli	Tensile Tests	VMF	Relative Errors
E_1 (kPa)	462.1	439.7	4.8%
E_2 (kPa)	120.8	121.4	0.5%

3.3. Strain and Young's Modulus Measurement of Pork

A piece of pork was bought from a grocery store. To test the DVC and VFM based method for the elasticity measurement of biological tissue, a fresh pork specimen was tested at room temperature. A small cube shown in Figure 8a was resected for measurement, which has a layer of pig's skin on the top. Dimensions of the pork sample were l_x = 10.36 mm, l_y = 10.30 mm and l_z = 7.50 mm. A reference volumetric OCT image was taken when 0.620 N preload was applied. Then another volumetric OCT image was taken when the specimen was compressed by 41.0 µm and the applied load was 0.082 N. 6 mm in both the width and length directions were imaged by the OCT. A cross-sectional OCT image of the specimen with the ROI marked by a blue rectangle is shown in Figure 8b. A reference 3D volumetric OCT image is shown in Figure 8c with the ROI marked by a blue cube, which is 461.8 µm below the top surface of the specimen as shown in Figure 8b. The average of the correlation coefficient of the ROI was tested to be 0.391. The coordinate was built with the origin O at the bottom left corner of the ROI, and the three axes were along the three edges of the ROI as shown in Figure 8c. The number of computed points was M = 31 × 31 × 8 = 7688. The step lengths were 7 voxels in the x, y and z directions. The computed dimensions were Lx = Ly = 4057.9 µm and Lz = 556.1 µm. The 3D displacements and strains of the ROI with the defined coordinate system are shown in Figure 9. The displacements in the x and y directions are less than 6 µm. The values of displacement and the normal strain in the z-direction shown in Figure 9c and f decrease with the increase of the depth. From Figure 9f the layered structure can be identified as the double-layered phantom. The other strain components are relatively small, mostly less than 1.0% but show heterogeneity of the tissue.

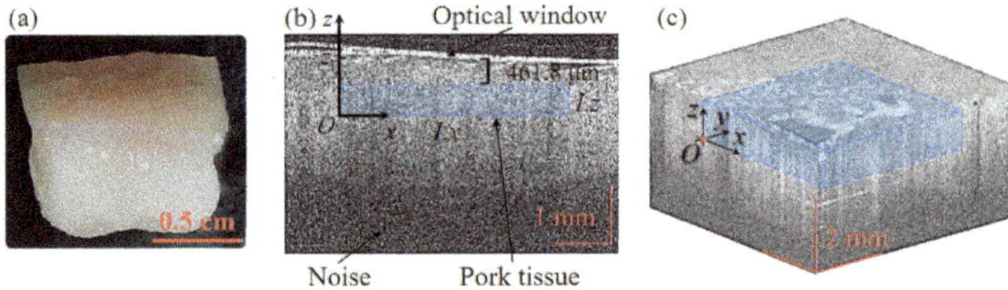

Figure 8. Pork sample. (**a**) A photo of the pork sample; (**b**) a central cross-sectional OCT image before deformation, where the ROI is marked with a blue rectangle; (**c**) a 3D OCT image before deformation, where the ROI is marked with a blue cube.

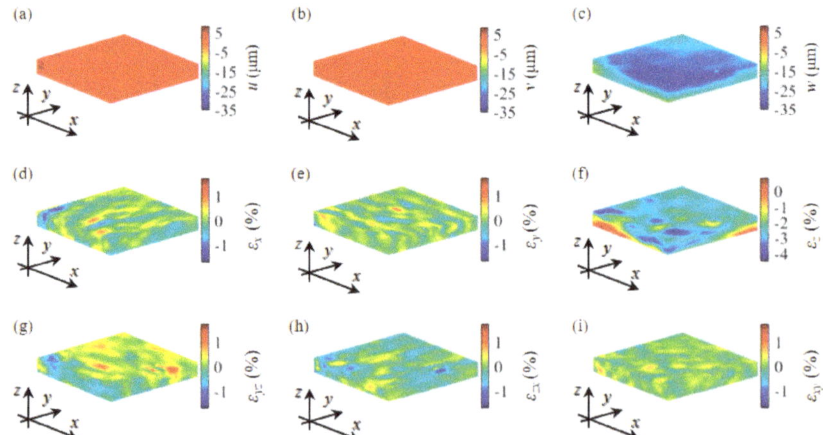

Figure 9. 3D deformation measurement of a pork specimen. (**a–c**) 3D displacements in the x, y and z directions respectively; (**d–f**) 3D normal strains in the x, y and z directions respectively; (**g–i**) 3D shear strains.

After the full-field strain ε of the pork specimen was obtained, the constitutive parameters were derived by the same procedure as the double-layer phantom using the VFM. Some of the geometric parameters in this experiment were $z_t = 1017.9$ µm as shown in Figure 8b; $z_b = z_t - l_z = -6482.1$ µm. $T_0 = -0.082$ N/(10.36 $\times$ 10^{-3} m $\times$ 10.30 $\times$ 10^{-3} m)$= -768$ Pa. By solving Equation (16), the Young's moduli and Poisson's ratios of the two layers were obtained. They were $E_1 = 123.9$ kPa, $v_1 = 0.083$ for the lower layer, and $E_2 = 31.62$ kPa, $v_2 = 0.123$ for the top layer. These values are in a similar range to the porcine skin's Young's modulus of 113 kPa measured by Yeung et al. [27].

4. Discussion

The 3D OCE method based on DVC and VFM has been developed in this study. Experimental results demonstrate that this method can obtain the 3D displacements, strains and constitutive parameters including Young's modulus and Poisson's ratio of layered materials.

The Young's modulus of the homogenous phantom obtained by the DVC and VFM based 3D OCE method is ~1% different from the value measured by a tensile test, which demonstrates the effectiveness of the new method. For the compressive loading experiments conducted in this paper, the theoretical strain distribution in the depth direction and the shear strains can be estimated. These distributions and analyses show the relative error is about 10%. Errors are also observable at the edges shown in Figure 4g–i. Reasons for the strain measurement errors include: (1) the influence of the glass window on the top of the specimen including the friction with the specimen and the slant compression; (2) the boundary effect of the polynomial fitting; (3) the influence of noise, which is unavoidable in laser interferometry. The measurement can be improved by improving the loading device, optimizing the virtual fields and applying data processing techniques to reduce noise. The Poisson's ratio can be easily calculated from the DVC results or by the VFM. The results obtained by the two methods are very close, which can be further verified by other experimental methods such as tensile testing.

A complete friction boundary condition was employed at both the top and bottom surfaces. This is reasonable, because the maximum ratio of shear stress to compressive normal stress on the top surface of the phantom is ~0.02 at the four corners, calculated by simulation. Generally, the static friction coefficient that is the ratio of the maximum static shear stress to normal stress between the glass and different rubbers is larger than 0.2 [28], which would be similar to the coefficient between the silica gel phantom and the glass. Because the shear stress is 10 times less than the maximum static shear stress, the complete friction boundary conditions can be applied in Equation (20).

The OCT noise will affect the gray value of speckle images and then affect the accuracy of DVC and VFM. Above all, the OCT noise will affect the correlation coefficient in DVC. The correlation coefficient decreased from 0.8 to 0.6 with the depth increasing in the experiment of the homogeneous phantom. Because the scattering light intensity decreases with the increase of depth, the information of the sample is reduced dramatically. The OCT noise will cause the results of DVC to no longer be reliable in deep regions. Hence, only the shallow regions were used. Furthermore, the OCT noise can mask the gray value of speckle images. It will result in matching errors even in the samples without any motion or deformation as the stationary experiment showed that the sensitivities of our system were 0.05 voxels and 0.2% for displacement and strain measurement, respectively. If the matching errors exist, the strains inputted into the VFM will lead to errors in the calculation of the constitutive parameters. By combining noise resistant methods, such as the ZNSSD criterion [20], the local ternary quadratic polynomial fitting method, and the linear VFM method [23], the experimental results will be insensitive to noise.

The 3D strain maps not only provide the necessary input, but also enable more automatic and accurate segmentation of the double-layer materials. This is important for the successful performance of the proposed VFM method. Although the interface was estimated as a series of broken lines, it is reliable because most of the purple dashed line is located at the blue curve in Figure 7b. The interface can also be described more accurately in the future work to improve the computing accuracy through finer meshing. The method of solving constitutive parameters of different components respectively has the merits of less virtual fields, large-scale linear equations and better solutions, because the solutions of the elastic constants acquired by the method of solving all components simultaneously had a negative value. From Figures 6c and 9f, it can be seen that the strain of the top layer is greater than the bottom layers, which indicates that the top layers are softer. Therefore, the image of strain distribution may be very useful for the diagnosis of diseases, as it shows the relative stiffness and boundaries of different components of soft tissue.

As a high-resolution laser interferometry imaging technique, the signal to noise ratio of OCT images decreases with the increases of depth as shown in Figure 8b, and the imaging is easy to be disturbed. In this experiment, only a thin portion of ~0.56 mm of the pork image was processed, as the correlation coefficient becomes too low anywhere below. As mentioned above, the OCT images are generated by laser interferometry. The scattering characteristic and anisotropy factor of the tissue influences the size and distribution of speckles of OCT images [29] and the DVC calculation. The size of OCT speckle was determined by OCT resolution, optical devices and microstructure of tissues [30,31]. To improve the accuracy of the DVC based 3D OCE deformation measurement, the influence of OCT speckle properties needs to be studied further. Kurokawa et al. [32,33] and Wijesinghe et al. [34] proposed a more sensitive displacement and strain detection method using complex cross-correlation. The sensitivity can be within tens of nanometers and tens of micro-strains, which improves about tenfold the DVC method. In [32,33] out of plane motion was measured from 2D images; in [34], particularly, 3D displacements were estimated from 3D OCT volumes. The using of complex data for correlation calculation may be also helpful for improving DVC calculation.

The pork tissue tested in this paper is assumed to be layered isotropic elastic and homogenous based on the strain maps shown in Figure 9. However, soft tissue is usually anisotropic. DVC has the advantage of obtaining the displacement and strain tensors of anisotropic materials, although the algorithm may need to be further improved. The VFM may become very complicated though. Image segmentation methods may be used to differentiate between various components of heterogeneous material, so that the VFM can be simplified. If the constitutive model of the tested tissue is too complicated or even unknown, finite element method updating or other inverse problem solving may be a good choice.

5. Conclusions

In conclusion, a DVC and VFM based OCE is developed to quantitatively measure the 3D displacements, strains and elasticity of double-layered biological tissue. The integrated noise-insensitive DVC method can obtain the 3D strain tensor with an accuracy of 10%. With the full-field strain as input, the VFM can recover the elastic modulus of homogeneous material with the accuracy of ~1%. The automatic segmentation method proposed based on the strain gradient can simplify the VFM calculation of layered samples. There are less virtual displacements and large-scale linear equations involved in the VFM, once a double-layered material is divided into two parts. The full-field strain tensor provides excessive information for elastic modulus quantification of the double-layered material by the VFM. The relative measurement error of the Young's modulus of the double-layered phantom is ~5% in this study, which can be improved by optimizing the experimental setup and the virtual fields. The simultaneous elasticity quantification of each of the two layers of the pork specimen, shows great promise for the proposed method for mechanical properties measurement of biological tissues. In all, the noise insensitive DVC and VFM based OCE method developed in this study can obtain the 3D displacements, strains, Young's modulus and Poison's ratio of layered materials simultaneously. It can be a powerful tool for the differentiation of various components of heterogeneous biomaterials and for biomechanics measurement.

Author Contributions: F.M. was mainly responsible for the design and implementation of experiments, the acquisition, processing, analyses and visualization of data. C.S. set the formulation or evolution of overarching research goals and guided the methodology of the experiments. She also acquired the funding and provided experimental equipment and materials. X.Z. and J.W. participated in the collection and processing of experimental data. C.L. provided the experimental device and gave some valuable opinions for the methodologies. The programming codes were written by J.W. and F.M. J.C. and C.S. administrated and supervised this project. The original draft of this paper was written by F.M., and the revision of the paper was completed by C.S., J.W., X.Z. and F.M.

Funding: This research was funded by [National Natural Science Foundation of China] grant number [11602166] and [Natural Science Foundation of Tianjin] grant number [16JCYBJC40500]. The APC was funded by [National Natural Science Foundation of China] grant number [11602166].

Acknowledgments: This work was supported by the National Natural Science Foundation of China (Grant #: 11602166); the Natural Science Foundation of Tianjin (Grant #: 16JCYBJC40500).

Conflicts of Interest: The authors declare no conflict of interest.

References

1.	Larin, K.V.; Sampson, D.D. Optical coherence elastography—OCT at work in tissue biomechanics [Invited]. *Biomed. Opt. Express* **2017**, *8*, 1172–1202. [CrossRef]
2.	Wang, S.; Larin, K.V. Optical coherence elastography for tissue characterization: A review. *J. Biophotonics* **2015**, *8*, 279–302. [CrossRef]
3.	Wang, R.K.; Ma, Z.S.; Kirkpatrick, J. Tissue Doppler optical coherence elastography for real time strain rate and strain mapping of soft tissue. *Appl. Phys. Lett.* **2006**, *89*, 144103. [CrossRef]
4.	Wang, R.K.; Kirkpatrick, S.; Hinds, M. Phase-sensitive optical coherence elastography for mapping tissue microstrains in real time. *Appl. Phys. Lett.* **2007**, *90*, 164105. [CrossRef]
5.	Schmitt, J.M. OCT elastography: Imaging microscopic deformation and strain of tissue. *Opt. Express* **1998**, *3*, 199–211. [CrossRef] [PubMed]
6.	Luo, Z.; Wang, Z.; Yuan, Z.; Du, C.; Pan, Y. Optical coherence Doppler tomography quantifies laser speckle contrast imaging for blood flow imaging in the rat cerebral cortex. *Opt. Lett.* **2008**, *33*, 1156–1158. [CrossRef]
7.	Rogowska, J.; Patel, N.A.; Fujimoto, J.G.; Brezinski, M.E. Optical coherence tomographic elastography technique for measuring deformation and strain of atherosclerotic tissues. *Heart* **2004**, *90*, 556–562. [CrossRef] [PubMed]
8.	Chan, R.C.; Chau, A.H.; Karl, W.C.; Nadkarni, S.; Khalil, A.S.; Iftimia, N.; Shishkov, M.; Tearney, G.J.; Mofrad, M.R.K.; Bouma, B.E. OCT-based arterial elastography: Robust estimation exploiting tissue biomechanics. *Opt. Express* **2004**, *12*, 4558–4572. [CrossRef]
9.	Khalil, A.S.; Chan, R.C.; Chau, A.H.; Bouma, B.E.; Mofrad, M.R.K. Tissue Elasticity Estimation with Optical Coherence Elastography: Toward Mechanical Characterization of In Vivo Soft Tissue. *Ann. Biomed. Eng.* **2005**, *33*, 1631–1639. [CrossRef] [PubMed]
10.	Rogowska, J.; Patel, N.; Plummer, S.; Brezinski, M.E. Quantitative optical coherence tomographic elastography: Method for assessing arterial mechanical properties. *Br. J. Radiol.* **2006**, *79*, 707–711. [CrossRef] [PubMed]
11.	Sun, C.; Standish, B.; Vuong, B.; Wen, X.Y.; Yang, V. Digital image correlation-based optical coherence elastography. *J. Biomed. Opt.* **2013**, *18*, 121515. [CrossRef]
12.	Chu, T.C.; Ranson, W.F.; Sutton, M.A. Applications of digital-image-correlation techniques to experimental mechanics. *Exp. Mech.* **1985**, *25*, 232–244. [CrossRef]
13.	Fu, J.; Pierron, F.; Ruiz, P.D. Elastic stiffness characterization using three-dimensional full-field deformation obtained with optical coherence tomography and digital volume correlation. *J. Biomed. Opt.* **2013**, *18*, 121512. [CrossRef]
14.	Fu, J.; Haghighi-Abayneh, M.; Pierron, F.; Ruiz, P.D. Depth-resolved full-field measurement of corneal deformation by optical coherence tomography and digital volume correlation. *Exp. Mech.* **2016**, *56*, 1203–1217. [CrossRef]
15.	Nahas, A.; Bauer, M.; Roux, S.; Boccara, A.C. 3D static elastography at the micrometer scale using Full Field OCT. *Biomed. Opt. Express* **2013**, *4*, 2138–2149. [CrossRef] [PubMed]
16.	Acosta Santamaría, V.A.; García, M.F.; Molimard, J.; Avril, S. Three-Dimensional Full-Field Strain Measurements across a Whole Porcine Aorta Subjected to Tensile Loading Using Optical Coherence Tomography–Digital Volume Correlation. *Front. Mech. Eng.* **2018**, *4*, 3. [CrossRef]

17. Girard, M.J.; Strouthidis, N.G.; Desjardins, A.; Mari, J.M.; Ethier, E.R. In vivo optic nerve head biomechanics: Performance testing of a three-dimensional tracking algorithm. *J. R. Soc. Interface* **2013**, *10*, 20130459. [CrossRef]
18. Zhang, L.; Thakku, S.G.; Beotra, M.R.; Baskaran, M.; Aung, T.; Goh, J.C.H.; Strouthidis, N.G.; Girard, M.J.A. Verification of a virtual fields method to extract the mechanical properties of human optic nerve head tissues in vivo. *Biomech. Model. Mechanobiol.* **2016**, *16*, 871–887. [CrossRef] [PubMed]
19. Bay, B.K.; Smith, T.S.; Fyhrie, D.P.; Saad, M. Digital volume correlation: Three-dimensional strain mapping using X-ray tomography. *Exp. Mech.* **1999**, *39*, 217–226. [CrossRef]
20. Pan, B.; Xie, H.; Wang, Z. Equivalence of digital image correlation criteria for pattern matching. *Appl. Opt.* **2010**, *49*, 5501–5509. [CrossRef]
21. Pan, B.; Wang, B.; Wu, D.; Lubineau, G. An efficient and accurate 3D displacements tracking strategy for digital volume correlation. *Opt. Lasers Eng.* **2014**, *58*, 126–135. [CrossRef]
22. Zauel, R.; Yeni, Y.N.; Bay, B.K.; Dong, X.N.D.; Fyhrie, P. Comparison of the linear finite element prediction of deformation and strain of human cancellous bone to 3D digital volume correlation measurements. *J. Biomech. Eng.* **2006**, *128*, 1–6. [CrossRef] [PubMed]
23. Pierron, F.; Grédiac, M. *The Virtual Fields Method*; Springer: New York, NY, USA, 2012; p. 57.
24. Wu, J. *Elasticity*; Higher Education Press: Beijing, China, 2001; p. 34.
25. Zaitsev, V.Y.; Matveev, L.A.; Matveyev, A.L.; Gelikonov, G.V.; Gelikonov, V.M. Elastographic mapping in optical coherence tomography using an unconventional approach based on correlation stability. *J. Biomed. Opt.* **2014**, *19*, 021107. [CrossRef]
26. Zaitsev, V.Y.; Matveev, L.A.; Matveyev, A.L.; Gelikonov, G.V.; Gelikonov, V.M. A model for simulating speckle-pattern evolution based on close to reality procedures used in spectral-domain OCT. *Laser Phys. Lett.* **2014**, *11*, 105601. [CrossRef]
27. Yeung, C.C.; Holmes, D.F.; Thomason, H.A.; Stephenson, C.; Derby, B.; Hardman, M.J. An ex vivo porcine skin model to evaluate pressure-reducing devices of different mechanical properties used for pressure ulcer prevention. *Wound Repair Regen.* **2016**, *24*, 1089–1096. [CrossRef]
28. Gu, Y.; Fei, G.; Zhang, H.; Liu, F. Test and analysis of friction coefficient on glass floor. *Glass* **2018**, *3*, 11–17.
29. Piederrière, Y.; Boulvert, F.; Cariou, J.; Le Jeune, B.; Guern, Y.; Le, G. BrunBackscattered speckle size as a function of polarization: Influence of particle-size and concentration. *Opt. Express* **2005**, *13*, 5030–5039. [CrossRef] [PubMed]
30. Lamouche, G.; Bisaillon, C.E.; Vergnole, S.; Monchalin, J.P. On the speckle Size in Optical Coherence Tomography. *Proc. SPIE* **2008**. [CrossRef]
31. Lamouche, G.; Bisaillon, C.E.; Maciejko, R.; Dufour, M.; Monchalin, J.P. Speckle size in optical coherence tomography. *Proc. SPIE* **2007**. [CrossRef]
32. Kurokawa, K.; Makita, S.; Hong, Y.J.; Yasuno, Y. Two-dimensional micro-displacement measurement for laser coagulation using optical coherence tomography. *Biomed. Opt. Express* **2015**, *6*, 170–190. [CrossRef]
33. Kurokawa, K.; Makita, S.; Hong, Y.J.; Yasuno, Y. In-plane and out-of-plane tissue micro-displacement measurement by correlation coefficients of optical coherence tomography. *Opt. Lett.* **2015**, *40*, 2153–2156. [CrossRef] [PubMed]
34. Wijesinghe, P.; Chin, L.; Kennedy, B.F. Strain tensor imaging in compression optical coherence elastography. *IEEE J. Sel. Top. Quantum Electron.* **2019**, *25*. [CrossRef]

Article

3D Strain Mapping of Opaque Materials Using an Improved Digital Volumetric Speckle Photography Technique with X-Ray Microtomography

Lingtao Mao [1,2,*], Haizhou Liu [2], Ying Zhu [2], Ziyan Zhu [2], Rui Guo [2] and Fu-pen Chiang [3]

[1]   State Key Laboratory of Coal Resources and Safe Mining, China University of Mining & Technology, Beijing 100083, China
[2]   School of Mechanics and Civil Engineering, China University of Mining & Technology, Beijing 100083, China; m18600616052@163.com (H.L.); zhuying_hy@163.com (Y.Z.); 13230110170@163.com (Z.Z.); Ray.Guo@student.cumtb.edu.cn (R.G.)
[3]   Laboratory for Experimental Mechanics Research and Dept. of Mechanical Engineering, Stony Brook University, Stony Brook, NY 11794-2300, USA; fu-pen.chiang@stonybrook.edu
*   Correspondence: mlt@cumtb.edu.cn

Received: 1 March 2019; Accepted: 26 March 2019; Published: 4 April 2019

Abstract: Digital volumetric speckle photography (DVSP) method has been used to strain investigation in opaque materials. In this paper, an improved DVSP algorithm is introduced, in which a multi-scale coarse–fine subset calculation process and a subvoxel shifting technique are applied to increase accuracy. We refer to the new algorithm as Multi-scale and Subvoxel shifting Digital Volumetric Speckle Photography (MS-DVSP). The displacement and strain fields of a red sandstone cylinder exposed to uniaxial compression and a woven composite beam under three-point bending are mapped in detail. The characteristics of the interior deformation of the specimens are clearly depicted, thus elucidating the failure mechanism of the materials.

Keywords: interior 3D deformation; digital volumetric speckle photography; X-ray microtomography; digital volume correlation; red sandstone; woven composite beam

---

1. Introduction

Speckle photography, a technique that uses a random speckle pattern to quantitatively measure displacement and strain, is a major milestone in the archive of experimental mechanics. Speckle photography has found application in many fields of science and engineering. The basic principle of the white light speckle technique was first proposed in a 1968 paper by J. Burch [1]. After the advent of laser, the laser speckle photography technique [2] was developed, followed by a major expansion of the white light speckle photography approaches [3]. The method's spatial resolution was significantly increased with the development of the electron speckle photography technique by Chiang et al., in 1997 [4], which employs submicron and nano-sized speckles. In addition, its ease of use and versatility were greatly enhanced by the development of the digital speckle photography (DSP) technique [5,6]. DSP is a 2-D method in that only the deformation of a plane (either a surface plane or interior plane [7]) can be mapped. Since the genesis of experimental mechanics field, one of the major goals has always been to develop a true 3D experimental stress/strain analysis technique whereby one can probe into solids for detecting the interior deformation with ease. The invention of techniques such as frozen stress photoelasticity [8,9], scattered light photoelasticity [10,11], integrated photoelasticity and photoelastic tomography [12–14], etc. were major developments in this field. However all of these techniques are rather tedious and time consuming, some of which are even error prone. These techniques require transparent birefringent materials to simulate the real object, which limits their application. Many important solid mechanics problems cannot be modeled by a transparent material, the most obvious of which are the 3D deformation of rock materials and fiber-reinforced composites. Aside from photoelasticity, moire, laser speckle, and white light speckle photography techniques have also been attempted to investigate the internal strain of 3D objects [15–18]. However, successes in such attempts have been restricted because only a few planes or sections can be probed. All of these techniques require the use of a transparent material.

With the rapid growth and application of high-resolution X-ray computed tomography (nano, microfocus and synchrotron), it becomes much easier to acquire volumetric images of opaque materials with high spatial resolution. Based on volumetric images, digital volume correlation (DVC) method, a 3D extension of 2D digital image correlation (DIC), has been proposed by Bay et al. for strain analysis of bone tissues exposed to compression loads [19]. This method has since been applied to many materials such as wood [20], compacted sugar [21], sand grains [22], cast iron [23], rock materials [24–26], concrete [27,28], and composites [29–31]. In the DVC method, surrounding an interrogated point, a cubic subset (subvolume) of voxels is selected in the reference volumetric image, its corresponding position is registered in the deformed volumetric image, from which its 3D displacement vector is retrieved. The registration process can be carried out either in the spatial domain or the frequency domain. In the frequency domain, fast Fourier transform (FFT) is used to increase the calculation efficiency.

As an extension of the 2D DSP (Digital Speckle Photography) method [5,6], we have developed a 3D interior full field deformation measurement technique called digital volumetric speckle photography (DVSP) with the help of X-ray microtomography [32]. This technique has been successfully applied to mapping the internal 3D strain fields of rocks [33–35], concrete [36], and composites [37–39]. In applying the DVSP method, the reference volumetric image and deformed volumetric image are divided into subsets of certain 3D voxel arrays. Each corresponding pair of the subsets are "compared" via a two-step 3D FFT analysis, which is computationally highly efficient. The result is a 3D map of displacement vectors, in terms of impulse functions, representing the collective displacement experienced by all the speckles within the subset of voxels. Mathematically the DVSP algorithm may be considered as an equivalent operation to the phase-only cross-correlation at the spectrum plane. The phase-only filter (POF) can result in a sharper impulse response function, and its signal is considerably stronger than that of the noise spectrum [6,40]. However, if the displacement between the subsets is large, a fair amount of noise is imposed in the correlation surface resulting from the nonoverlapping areas. This effect gives rise to a poor signal-to-noise ratio.

In this paper, a multi-scale and coarse-fine subset calculation process and a subvoxel shifting techniques are introduced into the DVSP algorithm to improve its accuracy. As a demonstration, we applied the new algorithm to analyzing the internal deformation of a red sandstone cylinder specimen exposed to uniaxial compression and a woven composite beam specimen under three-point bending.

2. Methodology of Multi-Scale Subset and Subvoxel Shifting in DVSP

2.1. Theory of Multi-Scale Subset and Subvoxel Shifting in DVSP

The theory of DVSP has been described previously [32] and is only briefly presented here for easy reference. Assume that a reference volumetric image and a deformed volumetric image of an object are acquired by a CT system (i.e., either in nano- or microscale), these two volumetric images are subsequently divided into volumetric subsets with arrays of $16 \times 16 \times 16$ voxels or $32 \times 32 \times 32$ voxels and then 'compared'. The cross correlation with the POF is operated in the frequency domain (ξ, η, ζ) as follows:

$$\overline{G}(\xi - u, \eta - v, \zeta - w) = \Im\left\{\frac{H_1\left(f_x, f_y, f_z\right) H_2^*\left(f_x, f_y, f_z\right)}{\sqrt{\left|H_1\left(f_x, f_y, f_z\right) H_2\left(f_x, f_y, f_z\right)\right|}}\right\}$$
$$= \Im\left\{\left|H_1\left(f_x, f_y, f_z\right)\right| \exp\left\{j\left[\phi_1\left(f_x, f_y, f_z\right) - \phi_2\left(f_x, f_y, f_z\right)\right]\right\}\right\} \tag{1}$$

where $\overline{G}(\xi - u, \eta - v, \zeta - w)$, an expanded impulse function located at (u, v, w), is the discrete 3D cross correlation between the two volumetric subsets. $\Im$ denotes the Fourier transform, with $H_1\left(f_x, f_y, f_z\right)$ being the Fourier transform of $h_1(x, y, z)$, representing the gray distribution function of a subset from the reference volumetric image, and $H_2\left(f_x, f_y, f_z\right)$ being the Fourier transform of $h_2(x, y, z)$, representing the gray distribution function of the corresponding subset from the deformed volumetric image. In addition, * denotes the complex conjugate. $\left|H\left(f_x, f_y, f_z\right)\right|$ and $\phi\left(f_x, f_y, f_z\right)$ are the spectral amplitude and phase fields, respectively.

In Equation (1), the term $\exp\left\{j\left[\phi_1\left(f_x, f_y, f_z\right) - \phi_2\left(f_x, f_y, f_z\right)\right]\right\}$ may be viewed as a POF, which can give a good balance between the peak sharpness and the noise tolerance in the correlation theory. As noted in reference [34], the random error of DVSP depends on subset size and micro structure pattern of objects. Furthermore, if the displacement between the corresponding subsets is large, the increase in the nonoverlapping area would cause an increase in decorrelation and thus result in an enhanced random error. On the other hand, in the case of images, $\overline{G}$ in Equation (1) is only displaying a delta-like function if u, v and w are integers. Non-integer translations between two subsets cause the peak in $\overline{G}$ to spread across neighboring voxels, subsequently degrading the quality of the displacement estimate. To identify sub-voxel investigation, the common approach is to apply cubic spline interpolation. The accuracy of these interpolation methods is highly dependent on the shape of the $\overline{G}$ function near the peak. In the 2D method, a multi-scale and coarse-fine subset calculation process and a subpixel shifting technique have been applied to decrease the error caused by decorrelation and non-integer translations [41,42]. These techniques are hereby introduced into DVSP, and we call this new algorithm Multi-scale and Subvoxel shifting Digital Volumetric Speckle Photography (MS-DVSP).

The essence of the multi-scale and coarse-fine calculation process is to minimize the nonoverlapping area under one voxel. In the coarse calculation process, the largest size of the subset ($2^n \times 2^n \times 2^n$) should not be larger than the size of the region-of-interest (ROI) in the reference volumetric image, and the integer voxel of displacement value (u_0, v_0, w_0) is obtained by the DVSP method. After that, the subset with the size ($2^{n-1} \times 2^{n-1} \times 2^{n-1}$) is applied, the corresponding subset in the deformed volumetric image is re-selected with reference to the first integer voxel predicted displacement (u_0, v_0, w_0), and the displacement value (u_1, v_1, w_1) of each subset is then calculated by the DVSP method. The subset size is gradually reduced according to above mentioned steps. When the subset size reaches a predetermined value and the obtained displacement values Δu, Δv, Δw are no more than 1 voxel in size, then surrounding an integral voxel of the crest a cubic subset with the size of $3 \times 3 \times 3$ voxels is selected and a cubic spline interpolation is used to assess the sub-voxel value δu, δv and δw. The reference subset can be moved with an amount δu, δv and δw by using the shifting property of Fourier transform. To weaken the edge effects and get good localization properties of Fourier transform, the shifted reference subset and the deformed subset are narrowed by a Kaiser window. Then, the subvoxel translation is calculated again. Some errors due to the interpolation require to reiterate by considering the "reference" subset with a new subvoxel shifting until a convergence criterion is matched. The final displacement value (u, v, w) of each subset can be obtained from the sum of the displacement values in the above steps. The flowchart of the MS-DVSP algorithm is shown in Figure 1. The procedures are programmed in MATLAB codes.

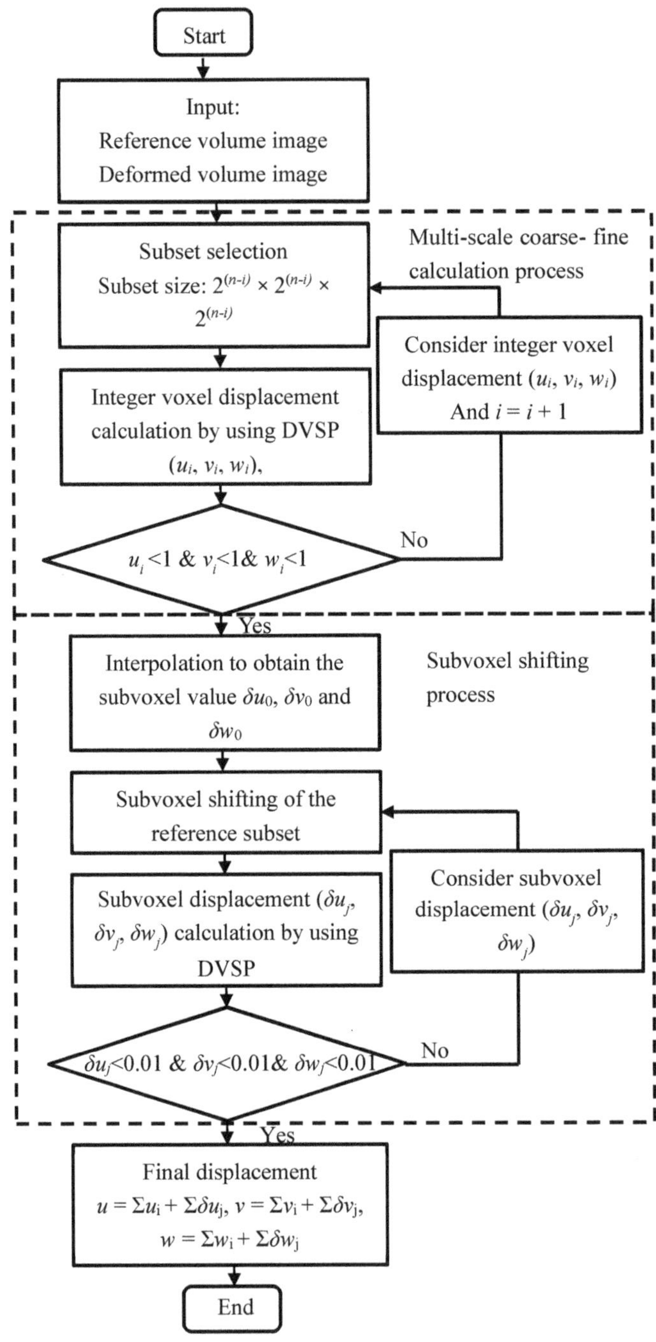

Figure 1. Flowchart of the (Multi-scale and Subvoxel shifting Digital Volumetric Speckle Photography) MS-DVSP algorithm.

2.2. X-ray Microtomography System and In Situ Loading Setup

The schematic of the X-ray microtomography system and a custom-made in situ loading setup is illustrated in Figure 2. The X-ray microtomography system mainly consists of a FXE 225 kV micro focus X-ray source from YXLON (Hamburg, Germany), an array detector (XRD 0822 AP14, Varex Imaging Corporation, Salt Lake City, UT, USA) of dimensions 1024 × 1024 pixels from PerkinElmer (Waltham, MA, USA), a turntable and its mechanical control unit, and a scanning control workstation [39]. The custom-made in situ loading setup is mounted on the turntable by using bolts, and rotates 360° with the turntable. In the loading experiment, the specimen is placed on the stage in the Poly(methyl methacrylate) (PMMA) chamber of the setup, and the mechanical loading is controlled by an electric motor. While the specimen is rotating step by step over 360°, a set of 720 radiograph projections are captured. When each projection is captured, the turntable does not move. It takes about 25 min for one 360° scan. After that, a Feldkamp cone-beam algorithm is used to reconstruct a sequence of 2D 16-bit slice gray images [43]. Based on these slice images, a 3D image is obtained.

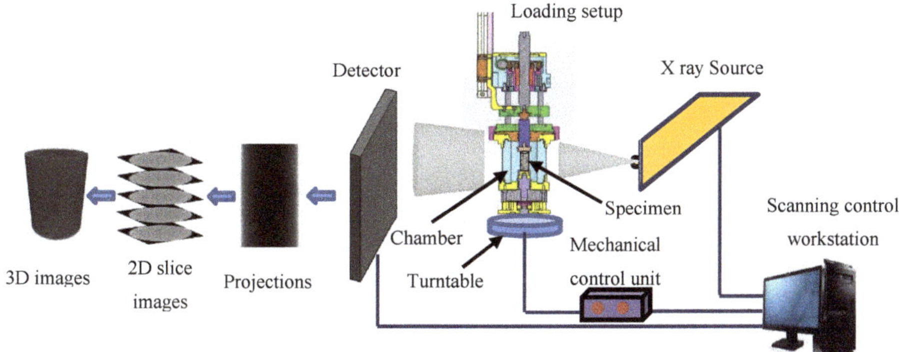

Figure 2. Schematic of the X-ray microtomography system and the in situ loading setup.

3. Experiment and Results

3.1. Internal Strain Investigation of Red Sandstone Exposed to Compression

A cylindrical specimen of red sandstone of 25 mm in diameter and 50 mm in length is scanned under different compression loadings in situ. The scan voltage is 120 kV, and the current is 200 μA. The experimental process is divided into 8 steps as shown in the stress–strain curve in Figure 3a. In each step, the specimen is scanned when the force is stable. In step 1, we take two consecutive scans of the specimen in identical settings and without moving (except the tomographic rotation) or deforming the specimen and designate them as Scans 1 and 2, respectively. Those two scans are used to evaluate artifacts and image noise influences in Section 4.2. In step 7, the loading value decreases due to the development of micro cracks, but the specimen still has loading capacity till the loading reaches the peak value of 12.08 MPa. In step 8, the broken specimen is scanned. In Figure 3, three sectional images at y = 6.20 mm, 12.50 mm and 18.80 mm of the specimen in step 6 and step 8 are shown, respectively. From the gray images shown Figure 3b–d, it is difficult to ascertain the damage. With reference to Figure 3e–h, the coalescence of tensile cracks mainly results in the failure of the specimen. At the lower left and right corners of the middle sectional image shown in Figure 3f, there are also some cracks due to shear stress. We select the volumetric image of step 1 as the reference image, and volumetric images of other steps as the deformed images, respectively. The size of the volumetric image has 560 × 560 × 801 voxels and the dimension of a unit voxel is 45 μm × 45 μm × 45 μm. By using the MS-DVSP algorithm, displacement fields are calculated. The final subset size is 32 × 32 × 32 voxels and the shift of the subset is 10 voxels. The calculated region of each section is the rectangular area marked with the dashed lines in Figure 3b–d. The u, v and w displacement fields of all sectional images of step 6 are plotted in Figure 4. As expected, the specimen mainly bore compression deformation along the z axis, and maximum displacements exist in the top region. Illustrated in Figure 4a,d,g, displacement fields of sectional images are divided into the negative and positive displacement regions, which make the specimen tensile along x axis. The main cracks shown in Figure 3e–h can be predicated by the zero displacement interface. This tensile deformation can also be investigated along y axis. In Figure 4b,e,h, the displacement of the sectional image at y = 6.20 mm is negative, while the displacement of most area of the sectional image at y = 18.80 mm is over zeros, especially in the lower region. These displacement distribution characteristics manifest that the expansion deformation occurs in circumference.

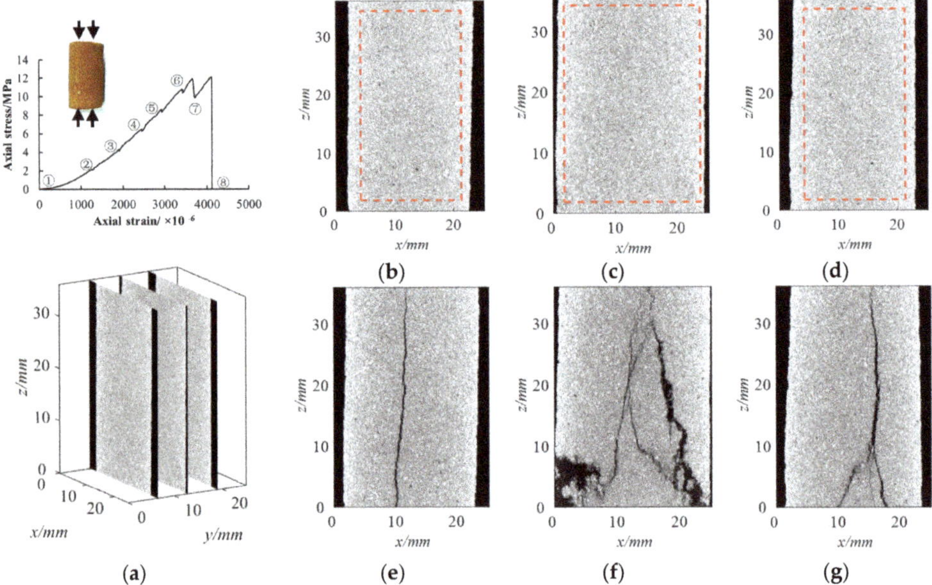

Figure 3. Stress–strain curve and reconstructed sectional images along y = 6.20 mm, 12.5 mm and 18.80 mm: (**a**) Stress–strain curve and three orthogonal sectional images of step 6; (**b**–**d**) sectional images of step 6, respectively; (**e**–**g**) sectional images of step 8, respectively.

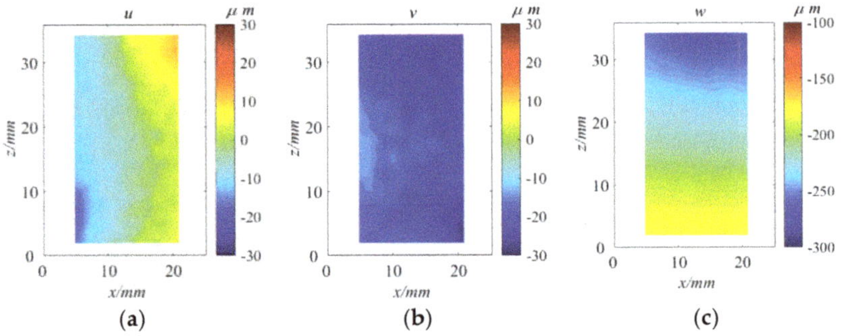

Figure 4. *Cont.*

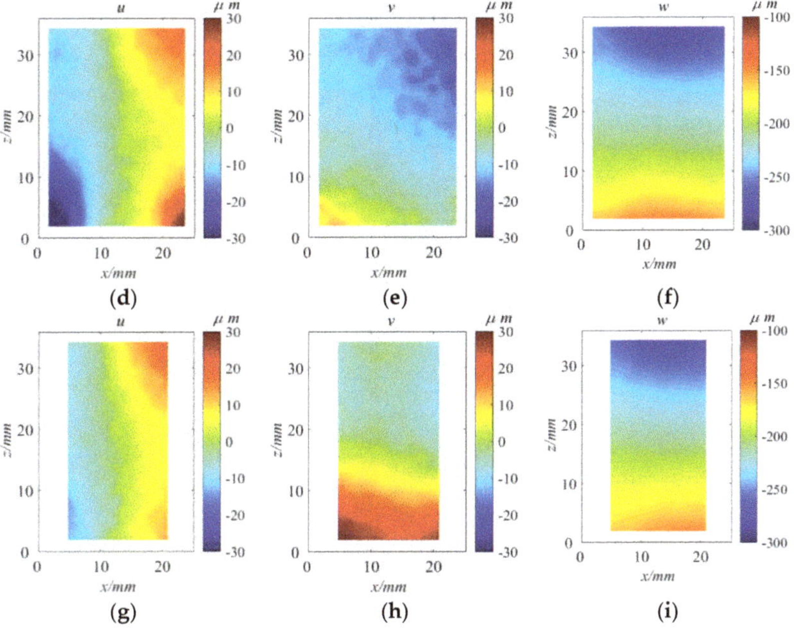

Figure 4. Contour plots of displacement values u, v and w of the sections of step 6: (**a**–**c**) along y = 6.20 mm, respectively; (**d**–**f**) along y = 12.50 mm, respectively; (**g**–**i**) along y = 18.80 mm, respectively.

Strain estimation can be derived from computed displacement fields by using numerical computation methods. During this procedure, the noise in computed displacement fields will result in decreasing the reliability of strain estimation. It is necessary to smoothing the calculated displacement dataset before strain estimation [44]. Here, we first apply multivariate kernel smoothing regression [45] to smooth the displacement fields. After that, a point-wise least-squares (PLS) [46] approach is used to compute the internal strain fields. The size of the strain calculation cubic element is $33 \times 33 \times 33$ voxels, and the distance between the center points of the neighbor element is 10 voxels. The normal strains and shear strains of these sections in step 4, 5 and 6 are plotted as shown in Figures 5–7, respectively. Here, in normal strain, we define positive corresponding to tension and negative corresponding to compression. From normal strain contours ε_{xx} shown in Figure 5, with the loading increase, it is noted that the deformation localization first occurs in both ends of the specimen, and develop to the middle. These localization regions correspond to the main cracks shown in Figure 3e–h. The distributions of normal strain ε_{zz} (shown in Figure 6) indicate that the compression deformation along the axis z is nearly uniform under small loading. With loading increase, deformations in both ends grow up more than ones in the middle. The maximum compression strain occurs at the bottom of the specimen. Due to friction among the specimen and two compression plates, the distributions of shear strain ε_{xz} take the shape of an 'x', as marked by the dashed lines in Figure 7, which depicts the characteristics of the shear deformations in the specimen. It is the shear deformation (shown in Figure 7f) cooperating with tensile deformation (shown in Figure 5f) that lead to the damage at the lower left and right corners of the specimen. These strain contours are useful in our efforts to understand the failure mechanism of the material.

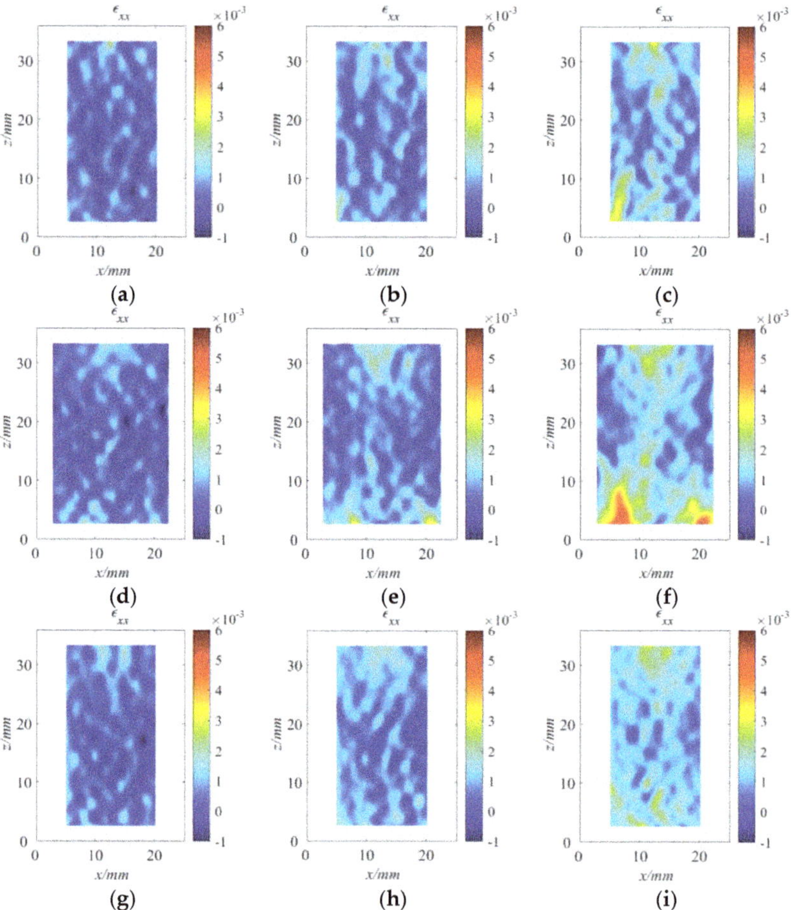

Figure 5. Contour plots of ε_{xx} strain fields of the sectional images in step 4, step 5 and step 6. (**a**–**c**) along $y = 6.20$ mm, respectively; (**d**–**f**) along $y = 12.50$ mm, respectively; (**g**–**i**) along $y = 18.80$ mm, respectively.

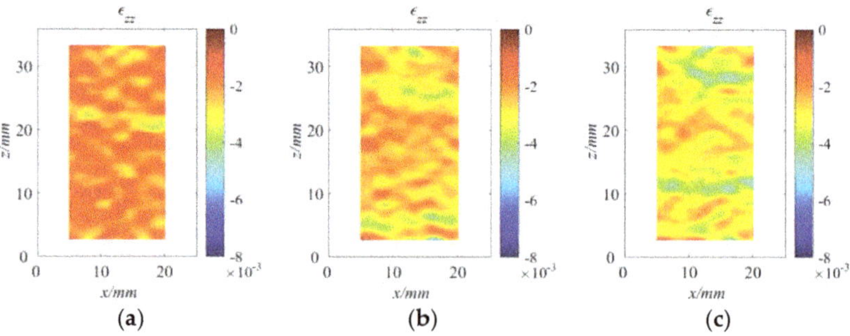

Figure 6. *Cont.*

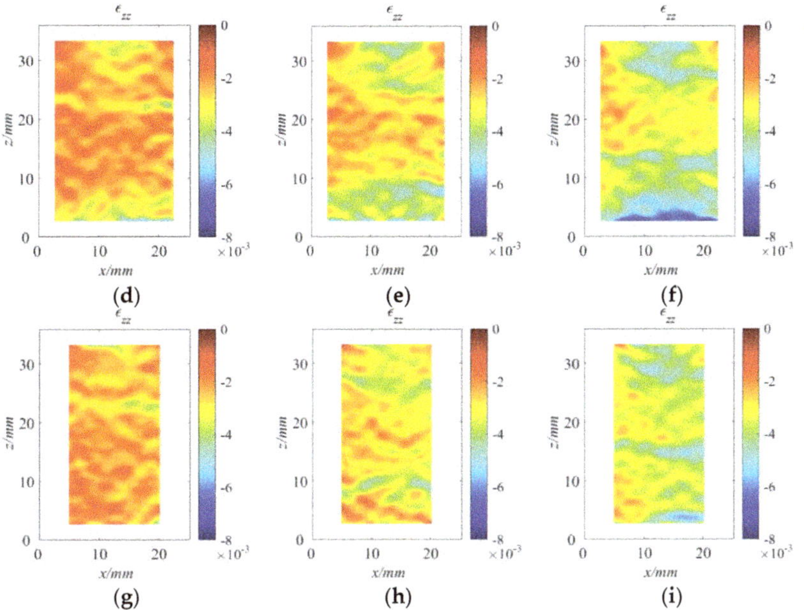

Figure 6. Contour plots of ε_{zz} strain fields of the sectional images in step 4, step 5 and step 6. (**a**–**c**) along $y = 6.20$ mm, respectively; (**d**–**f**) along $y = 12.50$ mm, respectively; (**g**–**i**) along $y = 18.80$ mm, respectively.

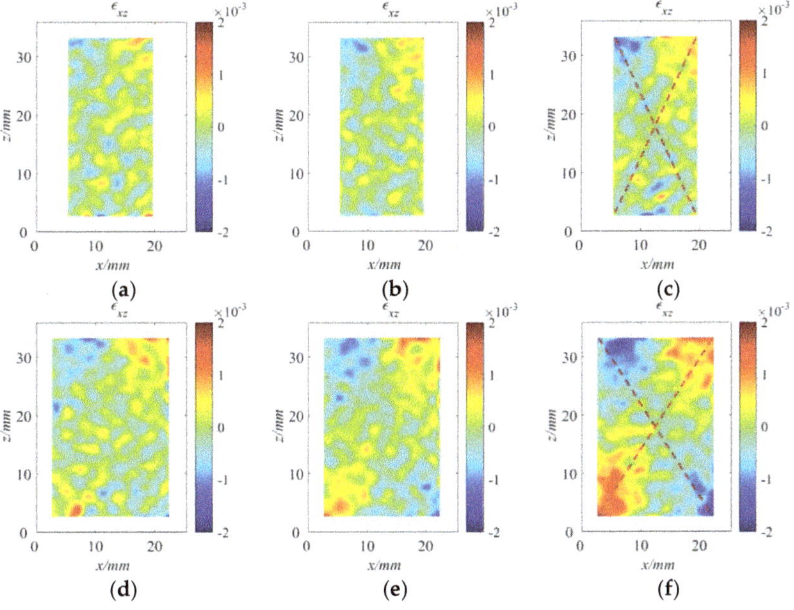

Figure 7. *Cont.*

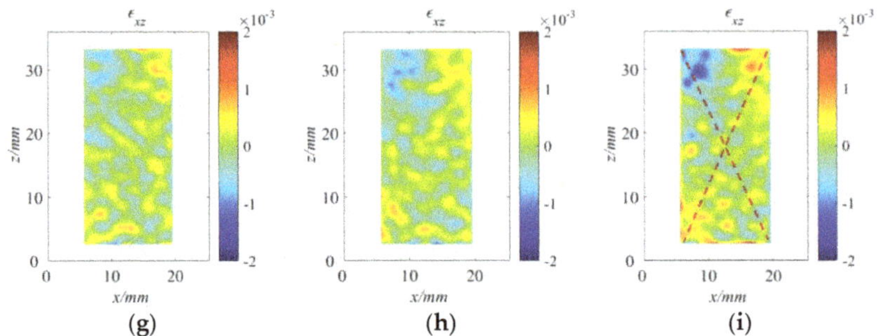

(g)   (h)   (i)

Figure 7. Contour plots of ε_{xz} strain fields of the sectional images in step 4, step 5 and step 6. (**a–c**) along $y = 6.20$ mm, respectively; (**d–f**) along $y = 12.50$ mm, respectively; (**g–i**) along $y = 18.80$ mm, respectively.

3.2. A Woven Composite Short Beam Under Three-Point Bending

A short woven composite short beam under three-point bending experiment has been described in reference [37,39]. The load-displacement curve is plotted in Figure 8, and Figure 11 scan steps are marked on the curve. The scan voltage is 130 kV, and the current is 200 μA. Five radiography projections are shown in Figure 8. The areas with high gray value correspond to pores in the specimen. From those projections before step 11, no much more damage information can be detected. In the projection of step 11, the specimen is broken, more areas with high gray value occur among layers on the left side of the specimen, which correspond to the cracks among layers. A 3D image of $900 \times 250 \times 361$ voxels is reconstructed in each scan step, and a unit voxel is 45 μm $\times$ 45 μm $\times$ 45 μm. Limited by the CT system resolution, fibers can not be distinguished from the matrix. The middle longitudinal sectional image and two transverse sectional images of the 10th scanning step are shown in Figure 9a–c, respectively. Pores have low gray value, which inverses to the projection. No discernable cracks are detected from these gray images. In the 11th scanning step, delamination and cracks are clearly visible on the left side of the specimen from the upper loading point to the bottom support shown in Figure 9e. We recalculate the internal displacement and strain fields with the MS-DVSP algorithm and the PLS approach, respectively. Due to heterogeneity in structures, we select the final subset size of $32 \times 32 \times 32$ voxels and the subset shift of 5 voxels to obtain much more displacement detail.

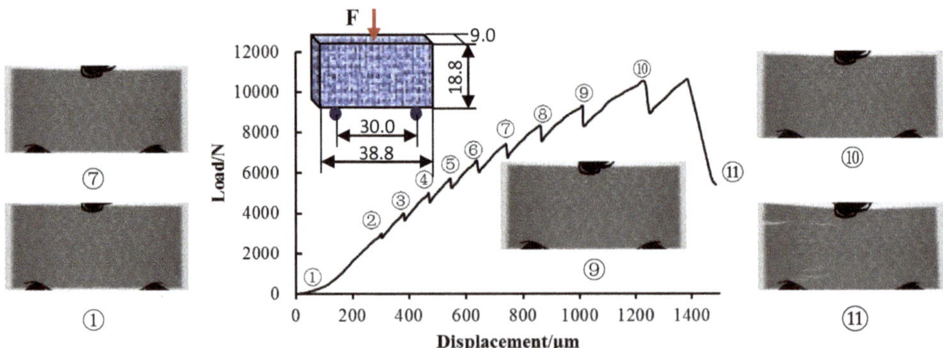

Figure 8. Loading–displacement curve and radiography projections of step 1, step 7, step 9, step 10 and step 11.

In Figure 10, v, w displacement fields and ε_{yy}, ε_{zz} and ε_{yz} strain fields of the middle longitudinal sectional image at step 7 (F = 6761 N), step 9 (F = 8333 N), and step 10 (F = 8982 N) are depicted, respectively. v, w displacement distributions are consistent to typical patterns of a short beam under three-point bending, but the periodic distribution of the internal structure in the woven composite specimen makes v displacement show periodic fluctuation in the longitudinal direction, which also be manifested by ε_{yy} strain distribution shown as Figure 10g–i. From ε_{zz} strain distribution illustrated in Figure 10j–l, the periodic fluctuation in the transverse direction is due to the layer structure of the specimen. From the contours of the shear strain ε_{yz}, the strain growth in regions from the loading point to two supports is evident with loading increase, and the high strain value is the prelude to the eventual delamination failure as shown in Figure 9e at Step 11. ε_{yz} strain fields of two transverse sectional images are depicted in Figure 11. Before the specimen is broken, it is also found that high strain occurs on the region which corresponds to the crack area shown in Figure 10f.

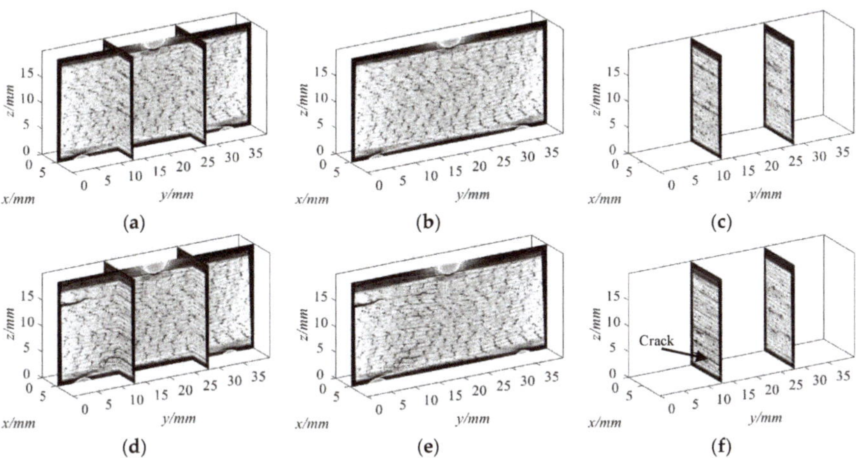

Figure 9. Reconstructed sectional images: (**a**) One middle longitudinal sectional image and two transverse sectional images of step 10; (**b**) The middle longitudinal sectional image at $x = 4.5$ mm of step 10; (**c**) two transverse sectional images of step at $y = 11.925$ and 27.000 mm of step 10; (**d**) One middle longitudinal sectional image and two transverse sectional images of step 11; (**e**) The middle longitudinalsectional image at $x = 5.625$ mm of step 11; (**f**) two transverse sectional images of step at $y = 11.925$ and 27.000 mm of step 11.

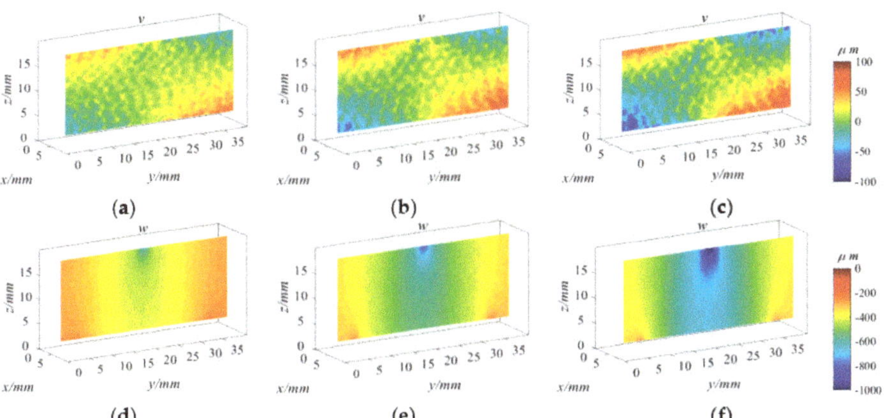

Figure 10. *Cont.*

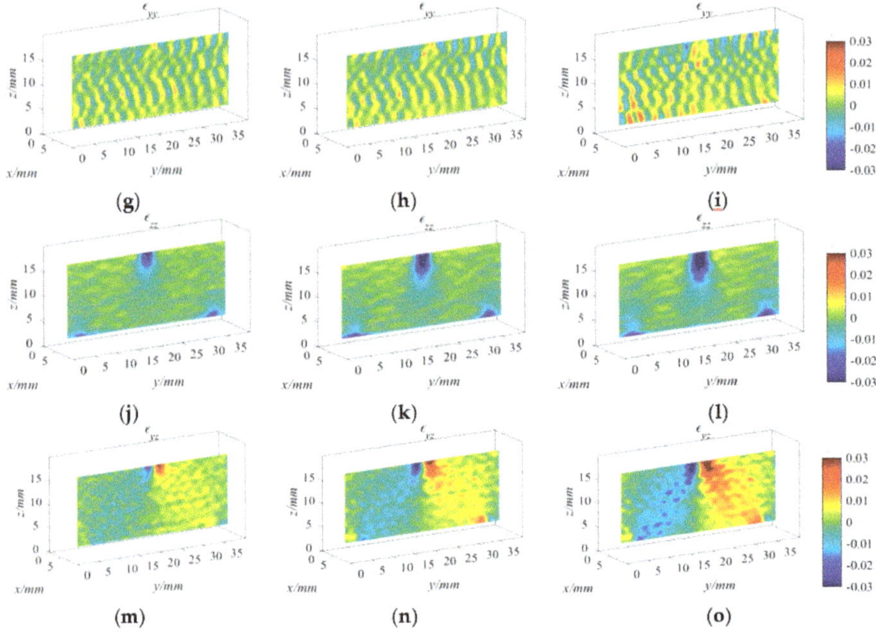

Figure 10. Displacement and strain fields of the middle longitudinal sectional image at step 7, step 9 and step 10, respectively; (**a–c**) v fields; (**d–f**) w fields; (**g–i**) ε_{yy} fields; (**j–l**) ε_{zz} fields; (**m–o**) ε_{yz} fields.

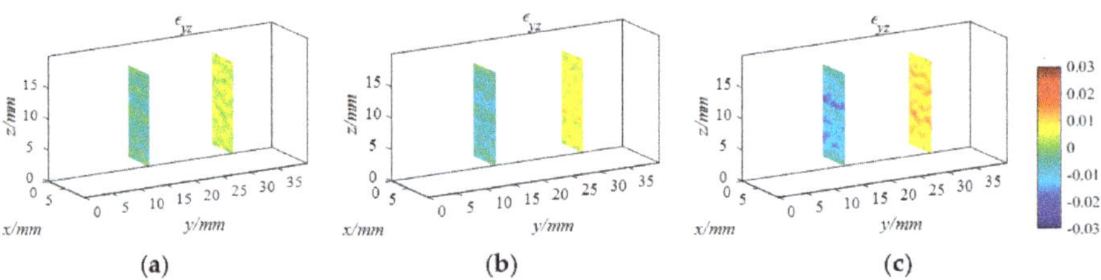

Figure 11. ε_{yz} strain contours of two transverse sectional images at step 7, step 9 and step 10, respectively; (**a**) step 7; (**b**) step 9 (**c**) step 10.

4. Discussion

4.1. Assessment of Accuracy of MS DVSP

For evaluating the robustness of an algorithm, numerical experiments are usually used. There are two methods to generate the reference and the deformed images. The first method is to apply computer simulation to digitally generating images. This method can isolate the possible errors caused by the image acquisition. In most DVC and DVSP applications, natural microstructures are seen as the speckle patterns. It is difficult to find a mathematic function to describe and simulate the natural microstructures. The second method is to employ the original image of a specimen as the reference image, and the deformed images can be generated by using synthetic shifts with Fourier transform. In this paper we use the second method. Two volumetric images of $200 \times 200 \times 200$ voxels are isolated from the reconstructed images of step 1 in the above two experiments, which are defined as the reference images, respectively. The "deformed" volumetric images are generated by imposing a rigid displacement ranging from 0.1 to 1.0 voxel with an increment of 0.1 voxel along the z direction. By using the MS-DVSP algorithm, pre-imposed sub-voxel translation displacements are calculated. The predetermined subset sizes are defined as $16 \times 16 \times 16$ voxels, $32 \times 32 \times 32$ voxels and $64 \times 64 \times 64$ voxels, respectively, and the shift of the subset is 10 voxels. Figure 12 shows both the mean bias error and the standard deviation error as a function of the imposed displacement for different subset sizes. With a predetermined subset size, the error reaches maximum at 0.5 voxels. In this situation the information between each subset in the reference and deformed images is the most biased. Furthermore the larger is the subset, the smaller the error. But, increasing the subset size would degrade the spatial resolution. We define the predetermined subset size as having $32 \times 32 \times 32$ voxels. With this subset size the maximum mean bias error of the red sandstone specimen is 0.0160 voxels and the corresponding standard deviation error is and 0.0060. For the woven composite specimen, the maximum mean bias error is 0.0062 voxels, and the corresponding standard deviation error is and 0.0089 voxels. The difference of accuracy between those two materials depends on their natural structures. The Shannon entropy values [47] of the red sandstone and the woven composite are 2.7606 and 3.0337, respectively, which means the woven composite has more feature information seen as speckle patterns resulting in a bit higher accuracy in the measurement.

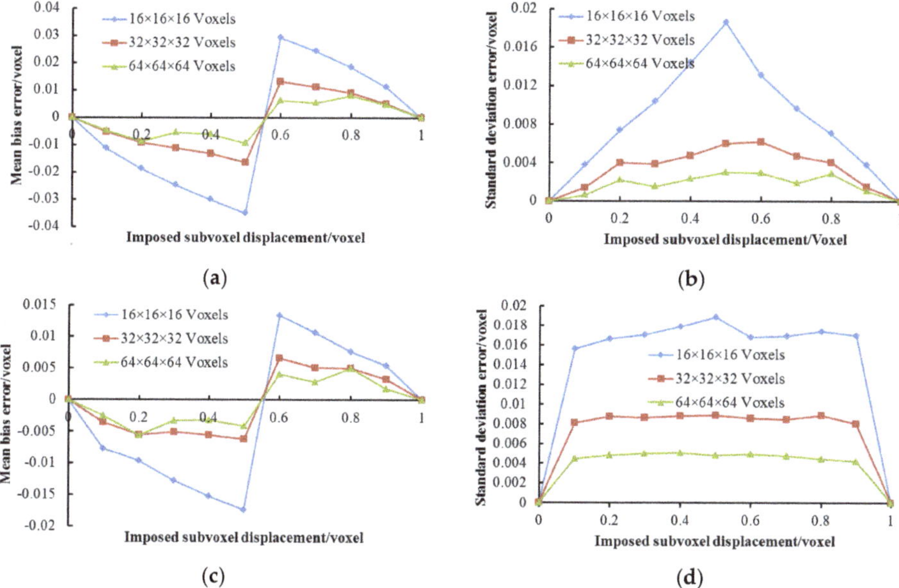

Figure 12. Comparison of displacement resolution among different sizes of subset. (**a**) Mean bias error of the red sandstone specimen; (**b**) Standard deviation error of the red sandstone specimen; (**c**) Mean bias error of the woven specimen; (**d**) Standard deviation error of the woven composite specimen.

In reference [39] a multi-scale coarse-fine subset calculation process was used in DVSP, which also increased the performance of DVSP. We call this algorithm M-DVSP. To compare the performance of MS-DVSP, M-DVSP, and DVSP, we calculate the displacements of these volumetric images with imposed sub-voxel translation using these three algorithms, respectively. The subset size is predetermined at $32 \times 32 \times 32$ voxels, and the shift of the subset is 10 voxels. The error curves are plotted as shown in Figure 13. The performance of M-DVSP algorithm is slightly better than that of DVSP, especially in standard deviation errors after 0.5 voxels. When the imposed displacement is increased continuously, the overlapping area between the reference subset and the deformed subset is decreased, and the noise caused by non-overlapping becomes stronger. As a result, it can be expected that the errors of DVSP will get progressively worse. Among these three algorithms, the performance of MS-DVSP is the best. Based on the mean bias error and its corresponding standard deviation error, we can find that the maximum error of the red sandstone specimen is 0.022 voxels of the MS-DVSP algorithm, 0.125 voxels of M-DVSP algorithm, and 0.124 of DVSP algorithm. The maximum error of MS-DVSP is only about one sixth of those errors of the other two algorithms, DVSP and M-DVSP. In the case of the woven composite specimen, the maximum errors are as follows: 0.015 voxels of the MS-DVSP algorithm, 0.150 voxels of the M-DVSP algorithm, and 0.151 of the DVSP algorithm. Thus, the maximum error of MS-DVSP is only about one tenth of those errors of the other two algorithms. These results indicate that MS-DVSP algorithm is an effective way to increase the accuracy of the technique, and the performance of MS-DVSP also depends on the internal micro-structures of objects.

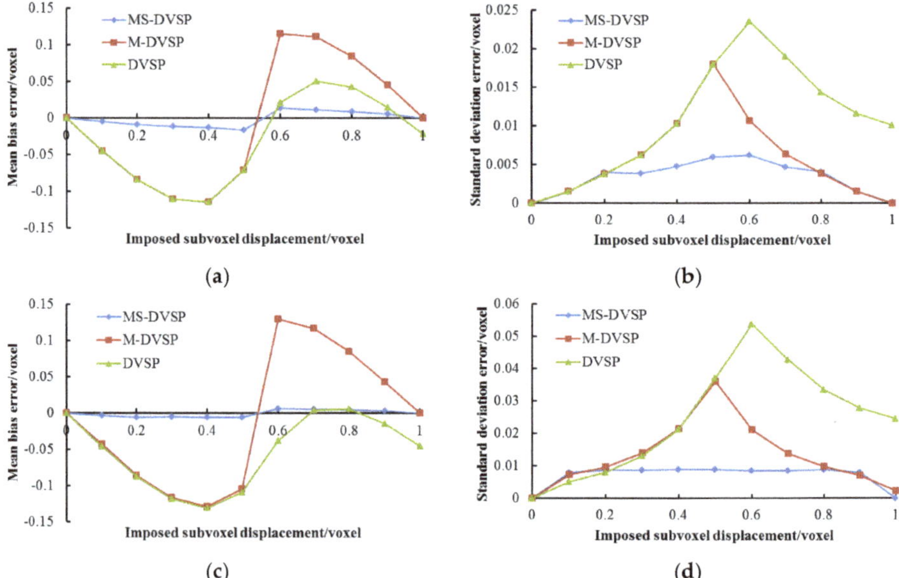

Figure 13. Accuracy comparison among MS-DVSP, M-DVSP and DVSP algorithm: (**a**) Mean bias error of displacement of the red sandstone specimen; (**b**) Standard deviation error of displacement of the red sandstone specimen; (**c**) Mean bias error of displacement of the woven composite specimen; (**d**) Standard deviation error of displacement of the woven composite specimen.

4.2. Influence of Artifacts in CT Image

During reconstruction of CT slice image, the non-uniformity of detector units and the polychromatic nature of the X-ray will give rise to ring and beam hardening artifacts. The self-heating effect of X-ray tube may result in distance variations that induce magnification changes and thus spurious dilatational strain. In addition, the image noise and contrast, the imperfect motion of the rotation stage, and the possible rigid body motion of the specimen will all influence the measurement result. Among all these effects, the spurious dilatational strain is definitely the most important to correct. More detailed information on the influence of these effects can be found in references [48–51]. Here, we just discuss the influence of the image noise and the imperfect motion.

To analyze the effect of artifacts of the micro X-ray CT system on the MS-DVSP algorithm, we take two consecutive scans in step 1 of the red sandstone specimen in Section 3.1 as the baseline experiment. Noise and artifacts of the system are included in the reconstructed images. For saving the computation time, we isolate two volumetric images having $200 \times 200 \times 200$ voxels from the center of reconstructed images of Scan 1 and Scan 2, define them as the reference image and the deformed image, respectively, and calculate displacements by using the MS-DVSP algorithm. The predetermined subset size is $32 \times 32 \times 32$ voxels, and the shift of the subset is 10 voxels. The average measured displacements are $u = -0.0059$ voxels, $v = -0.0147$ voxels, and $w = 0.0175$ voxels, and the standard deviation errors are 0.0107 voxels, 0.0115 voxels and 0.0105 voxels, respectively. The measured displacement along y and z axes is greater than the standard deviation error, indicating that some motion occurred along y and z axes between these two scans due to physical perturbations in the CT system. Because there are real noise and artifacts in the volumetric images of Scans 1 and 2, the above results indicate that the uncertainty of displacement measurement in all three directions is about 0.02 voxels with the MS-DVSP algorithm in the red sandstone specimen experiment.

For investigating the influence of noise and artifacts, two groups of deformed volumetric images are obtained by imposing subvoxel displacements on above two cropped volumetric images, and they are marked as Group 1, and Group 2, respectively. We define the cropped volumetric image of Scan 1 as the reference image, and the images of Group 1 and Group 2 as the deformed images, respectively. Then we calculate the displacements with the MS-DVSP algorithm. The predetermined subset size is $32 \times 32 \times 32$ voxels, and the shift of the subset is 10 voxels. Figure 14 illuminates the results of the displacement measurement. Due to the influence of noises and artifacts, the error increases almost two times. The uncertainty of displacement measurement is 0.015 voxel.

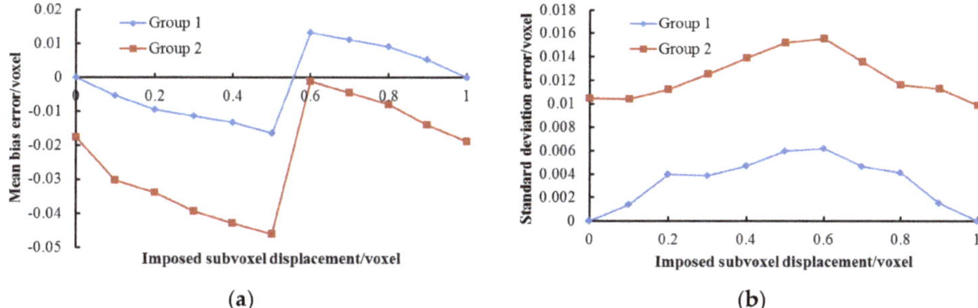

Figure 14. Influence of noise and artifacts on the imposed subvoxel displacement measurement (**a**) Mean bias error of displacement; (**b**) Standard deviation error of displacement.

As a more accurate alternative to fitting just rigid body movements, a least squares fit is used to calculate the rigid body translations and rotations, assuming small angles [52],

$$\begin{aligned}
u_{rigid} + \theta_z Y - \theta_y Z &\cong U \\
v_{rigid} - \theta_z X + \theta_x Z &\cong V \\
w_{rigid} + \theta_y X - \theta_x Y &\cong W
\end{aligned} \tag{2}$$

where θ_x, θ_y, and θ_z are the rotations about the x, y, and z axes, respectively, and X, Y, Z and U, V, W are the vectors of the x, y, z coordinates and u, v, w displacements, respectively, for all of the correlation points. The rigid body displacements are $u_{rigid} = -0.0082$ voxels, $v_{rigid} = -0.0204$ voxels and $w_{rigid} = 0.0167$ voxels, and the rotations are $\theta_x = 2.13 \times 10^{-5}$, $\theta_y = -3.15 \times 10^{-5}$ and $\theta_z = 3.43 \times 10^{-5}$. There are rigid body translations similar to the average measured displacements. Thus, when compared with the volumetric image of Scan 1, the volumetric image of Scan 2 incur very small rigid body translation and small rotation in all three directions. Compared with the deformation of the specimen in our study, the rigid body translation and rotation are very small, so we neglect the influence of the imperfect motion of the rotation stage in the data analysis.

To assess the uncertainty of strain measurement, we use the above calculated imposed subvoxel displacement results to compute strains with the PLS approach. Three sizes of strain calculation cubic box surrounding the point of interest are used, and they are $23 \times 23 \times 23$ voxels, $33 \times 33 \times 33$ voxels and $43 \times 43 \times 43$ voxels, respectively. Since only a rigid body translation is imposed, strains should be zero. However errors of displacement fields give rise to the error of strain calculation. The standard deviation errors of ε_{zz} is shown in Figure 15, which dictates that the uncertainty of strain measurement depends on the size of the calculation box, the larger is the box, the smaller the uncertainty. In PLS approach, too small calculation box is not enough to smooth the noise existing in local displacement fields, making strains having large variation. On the contrary, too large calculation box may result in local displacement fields oversmoothed. It is suggested that the size of the calculation box can be equal to or slightly larger than the size of the reference subset [46]. So in this study, we select the calculation box of $33 \times 33 \times 33$ voxels. With this size, the uncertainty of strain measurement increases more than two times due to the influence of noise and artifacts, and it is 2.34×10^{-4}.

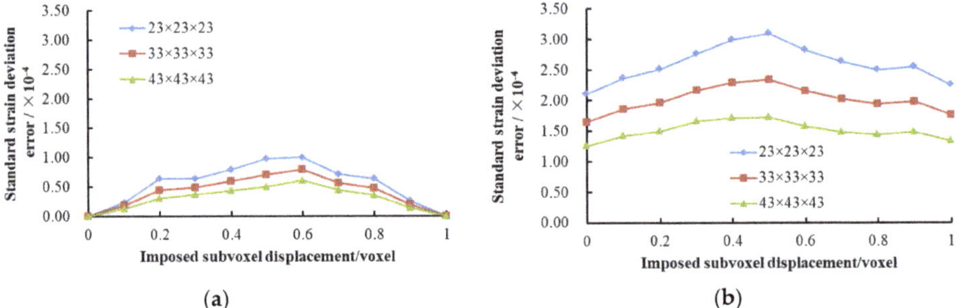

Figure 15. The standard deviation error of strain with different sizes of the strain calculation cubic box (**a**) Group 1; (**b**) Group 2.

5. Conclusions

This paper presents an improved algorithm called MS-DVSP to enhance the accuracy of the DVSP technique. The improvement in the algorithm lies in the fact that the two subsets that are correlated are shifted by nonintegral voxel values to obtain a maximum overlap between them. We tested the proposed scheme on red sandstone and woven composite specimens. Compared with DVSP, the accuracy and precision of MS-DVSP increase remarkably. In this study, with consideration of artifacts and noise in CT images, uncertainties of the displacement measurement and the strain measurement of the red sandstone specimen by using MS-DVSP algorithm, are 0.015 voxels and 2.34×10^{-4}, respectively.

The interior displacement and strain fields in a red sandstone specimen under compression and a woven composite specimen under 3-point bending are mapped using the MS-DVSP algorithm. From these plots, the characteristics of the interior deformation of the specimen are clearly depicted, thus elucidating the failure mechanism of the material. We believe the MS-DVSP algorithm will contribute to the further understanding of the physics and mechanics of opaque materials under stress.

Author Contributions: Conceptualization, L.M. and F.-p.C.; Methodology, L.M.; Software, L.M. and H.L.; Validation, H.L., Y.Z.; Z.Z. and R.G.; Formal Analysis, H.L., Y.Z.; Z.Z. and R.G.; Investigation, H.L., Y.Z.; Z.Z. and R.G.; Data Curation, L.M.; Writing—Original Draft Preparation, L.M.; Writing–Review & Editing, L.M. and F.-p.C.; Supervision, L.M.

Funding: This work was financially supported by the Major Program for Research and Development of Scientific Instrument of National Natural Science Foundation of China (51727807), the National Natural Science Foundation of China (51374211), the State Key Research Development Program of China (2016YFC0600705), the Fundamental Research Funds for the Central Universities (2009QM02), the US Office Of Naval Research's Solid Mechanics Program grant No: N0014-14-1-0419, and the Laboratory for Experimental Mechanic Research of the Department of Mechanical Engineering at Stony Brook University. F. P. Chiang wishes to thank Yapa Rajapakse, Director of the US Office of Naval Research's Solid Mechanics Program, for his support over the years for the advancement of the speckle technique.

Conflicts of Interest: The authors declare no conflicts of interest.

References

1. Burch, J.M.; Tokarski, J.M.J. Production of multiple beam fringes from photographic scatters. *Opt. Acta* **1968**, *15*, 101–111. [CrossRef]
2. Leendertz, J.A. Interferometric displacement measurements on scattering surface utilizing speckle effect. *J. Phys. E* **1970**, *3*, 214–218. [CrossRef]
3. Chiang, F.P.; Asundi, A. White light speckle method of experimental strain analysis. *Appl. Opt.* **1979**, *18*, 409–411. [CrossRef] [PubMed]
4. Chiang, F.P.; Jin, F. A new technique using digital speckle correlation for nondestructive inspection of corrosion. *Mater. Eval.* **1997**, *55*, 813–816.
5. Chen, D.J.; Chiang, F.P. Computer-aided speckle interferometry using spectral amplitude fringes. *Appl. Opt.* **1993**, *32*, 225–236. [CrossRef] [PubMed]
6. Chen, D.J.; Chiang, F.P.; Tan, Y.S.; Don, H.S. Digital speckle-displacement measurement using a complex spectrum method. *Appl. Opt.* **1993**, *32*, 1839–1849. [CrossRef] [PubMed]
7. Chiang, F.P.; Asundi, A. Interior displacement and strain measurement using white light speckles. *Appl. Opt.* **1980**, *19*, 2254–2256.
8. Burguete, R.L.; Patterson, E.A. A photoelastic study of contact between a cylinder and a half-space. *Exp. Mech.* **1997**, *37*, 314–323. [CrossRef]
9. Ju, Y.; Xie, H.P.; Zheng, Z.M.; Lu, J.B.; Mao, L.T.; Gao, F.; Peng, R.D. Visualization of the complex structure and stress field inside rock by means of 3D printing technology. *Chin. Sci. Bull.* **2014**, *59*, 5354–5365. [CrossRef]
10. Dupre, J.C.; Lagarde, A. Photoelastic analysis of a three-dimensional specimen by optical slicing and digital image processing. *Exp. Mech.* **1997**, *37*, 393–397. [CrossRef]
11. Kihara, T. Whole-field measurement of three-dimensional stress by scattered-light photoelasticity with unpoparized light. In Proceedings of the 14th International Conference on Experimental Mechanics, Poitiers, France, 4–9 July 2010; p. 32006.
12. Tomlison, R.A.; Patterson, E.A. The use of phase-stepping for the measurement of characteristic parameters in integrated photoelasticity. *Exp. Mech.* **2002**, *42*, 43–49. [CrossRef]
13. Wijerathne, M.L.L.; Oguni, K.; Hori, M. Stress field tomography based on 3D photoelasticity. *J. Mech. Phys. Solids* **2008**, *56*, 1065–1085. [CrossRef]
14. Aben, H.; Errapart, A. Photoelastic Tomography with linear and non-linear algorithms. *Exp. Mech.* **2012**, *52*, 1179–1193. [CrossRef]
15. Sciammarella, C.A.; Chiang, F.P. The moiré method applied to three-dimensional elastic problems. *Exp. Mech.* **1964**, *4*, 313–319. [CrossRef]
16. Chiang, F.P. A new three-dimensional strain analysis technique by scattered light speckle interferometry. In *The Engineering Uses of Coherent Optics*; Robertson, E.R., Ed.; Cambridge University Press: Cambridge, UK, 1976; pp. 249–262.
17. Asundi, A.; Chiang, F.P. Theory and application of white light speckle methods. *Opt. Eng.* **1982**, *21*, 570–580. [CrossRef]
18. Dudderar, T.D.; Simpkins, P.G. The development of scattered light speckle metrology. *Opt. Eng.* **1982**, *21*, 396–399. [CrossRef]
19. Bay, B.K.; Smith, T.S.; Fyhie, D.P.; Saad, M. Digital volume correlation: Three-dimensional. strain mapping using X-ray tomography. *Exp. Mech.* **1999**, *39*, 217–226. [CrossRef]
20. Forsberg, F.; Sjodahl, M.; Mooser, R.; Hack, E.; Wyss, P. Full three-dimensional strain measurements on wood exposed to three-point bending: Analysis by use of digital volume correlation applied to synchrotron radiation micro-computed tomography image data. *Strain* **2010**, *46*, 47–60. [CrossRef]
21. Forsberg, F.; Siviour, C.R. 3D deformation and strain analysis in compacted sugar using x-ray microtomography and digital volume correlation. *Meas. Sci. Technol.* **2009**, *20*, 1–8. [CrossRef]
22. Hall, S.A.; Bornert, M.; Desrues, J.; Pannier, Y.; Lenoir, N.; Viggiani, G.; Besuelle, P. Discrete and continuum analysis of localised deformation in sand using X-ray mCT and volumetric digital image correlation. *Geotechnique* **2010**, *60*, 315–322. [CrossRef]
23. Rethore, J.; Limodin, N.; Buffiere, J.-Y.; Hild, F.; Ludwig, W.; Roux, S. Digital volume correlation analysis of synchrotron tomographic images. *J. Strain Anal. Eng.* **2011**, *46*, 683–695. [CrossRef]

24. Lenoir, N.; Bornert, M.; Desrues, J.; Besuelle, P.; Viggiani, G. Volumetric digital image correlation applied to X-ray Microtomography images from triaxial compression tests on argillaceous rock. *Strain* **2007**, *43*, 193–205. [CrossRef]

25. Renard, F.; McBeck, J.; Cordonnier, B.; Zheng, X.J.; KandulaJesus, N.; Sanchez, J.R.; Kobchenko, M.; Noiriel, C.; Zhu, W.L.; Meakin, P.; et al. Dynamic in situ three-dimensional imaging and digital volume correlation analysis to quantify strain localization and fracture coalescence in sandstone. *Pure Appl. Geophys.* **2018**, 1–33. [CrossRef]

26. McBeck, J.; Kobchenko, M.; Stephen, A.H.; Tudisco, E.; Cordonnier, B.; Meakin, P.; Renard, F. Investigating the Onset of Strain Localization within Anisotropic Shale Using Digital Volume Correlation of Time-Resolved X-Ray Microtomography Images. *JGR Solid Earth* **2018**, *123*, 7509–7528. [CrossRef]

27. Chateau, C.; Nguyen, T.T.; Bornert, M.; Yvonnet, J. DVC-based image subtraction to detect microcracking in lightweight concrete. *Strain* **2017**, *54*. [CrossRef]

28. Bennai, F.; Hachem, C.E.; Abahri, k.; Belarbi, R. Microscopic hydric characterization of hemp concrete by X-ray microtomography and digital volume correlation. *Constr. Build. Mater.* **2018**, *188*, 983–994. [CrossRef]

29. Mendoza, A.; Schneiderc, J.; Parrab, E.; Obertb, E.; Rouxa, S. Differentiating 3D textile composites: A novel field of application for Digital Volume Correlation. *Compos. Struct.* **2019**, *208*, 735–743. [CrossRef]

30. Gonzalez, J.F.; Antartis, D.A.; Martinez, M.; Dillon, S.J.; Chasiotis, I.; Lambros, J. Three-Dimensional Study of Graphite-Composite Electrode Chemo-Mechanical Response using Digital Volume Correlation. *Exp. Mech.* **2018**, *58*, 573–583. [CrossRef]

31. Roux, S.; Hild, F.; Viot, P. Dominique Bernard. Three dimensional image correlation from X-Ray computed tomography of solid foam. *Compos. Part A* **2008**, *39*, 1253–1265. [CrossRef]

32. Chiang, F.P.; Mao, L.T. Development of interior strain measurement techniques using random speckle patterns. *Meccanica* **2015**, *50*, 401–410. [CrossRef]

33. Mao, L.T.; Zuo, J.P.; Yuan, Z.X.; Chiang, F.P. Full-field mapping of internal strain distribution in red sandstone specimen under compression using digital volumetric speckle photography and X-ray computed tomography. *J. Rock Mech. Geotech. Eng.* **2015**, *7*, 136–146. [CrossRef]

34. Mao, L.T.; Chiang, F.P. 3D strain mapping in rocks using digital volumetric Speckle photography technique. *Acta Mech.* **2016**, *227*, 3069–3085. [CrossRef]

35. Mao, L.T.; Hao, N.; An, L.Q.; Chiang, F.P.; Liu, H.B. 3D mapping of carbon dioxide-induced strain in coal using digital volumetric speckle photography technique and X-ray computer tomography. *Int. J. Coal Geol.* **2015**, *147–148*, 115–125. [CrossRef]

36. Mao, L.T.; Yuan, Z.X.; Yang, M.; Liu, H.B.; Chiang, F.P. 3D strain evolution in concrete using in situ X-ray computed tomography testing and digital volumetric speckle photography. *Measurement* **2019**, *133*, 456–467. [CrossRef]

37. Mao, L.T.; Chiang, F.P. Interior strain analysis of a woven composite beam using X-ray computed tomography and digital volumetric speckle photography. *Compos. Struct.* **2015**, *134*, 782–788. [CrossRef]

38. Mao, L.T.; Chiang, F.P. Mapping interior deformation of a composite sandwich beam using Digital Volumetric Speckle Photography with X-ray computed tomography. *Compos. Struct.* **2017**, *179*, 172–180. [CrossRef]

39. Mao, L.T.; Liu, H.Z.; Zhu, Z.Y.; Guo, R.; Zhu, Y.; Chiang, F.P. Digital volumetric speckle photography: A powerful experimental technique capable of quantifying interior deformation fields of composite materials. *Multiscale Multidiscip. Model. Exp. Des.* **2018**, *1*, 181–195. [CrossRef]

40. Sjodahl, M. Accuracy in electronic speckle photography. *Appl. Opt.* **1997**, *36*, 2875–2885. [CrossRef] [PubMed]

41. Perie, J.N.; Calloch, S.; Cluzel, C.; Hild, F. Analysis of a multiaxial test on a C/C composite by using digital image correlaiton and a damage model. *Exp. Mech.* **2002**, *42*, 318–418. [CrossRef]

42. Bergonnier, S.; Hild, F.; Roux, S. Digital image correlation used for mechanical tests on crimped glass wool samples. *J. Strain Anal.* **2005**, *40*, 185–197. [CrossRef]

43. Yang, M.; Liu, J.W.; Li, Z.C.; Liang, L.H.; Wang, X.L.; Gui, Z.G. Locating of 2π-projection view and projection denoising under fast continuous rotation scanning mode of micro-CT. *Neurocomputing* **2016**, *207*, 335–345. [CrossRef]

44. Pan, B.; Qian, K.M.; Xie, H.M.; Asundi, A. Two-dimensional digital image correlation for in-plane displacement and strain measurement: A review. *Meas. Sci. Technol.* **2009**, *20*, 062001. [CrossRef]

45. Hill, P.D. Kernel estimation of a distribution function. *Commun. Stat. Theory Methods* **1985**, *14*, 605–620. [CrossRef]

46. Pan, B.; Wu, D.; Wang, Z. Internal displacement and strain measurement using digital volume correlation: A least-squares framework. *Meas. Sci. Technol.* **2012**, *23*, 045002. [CrossRef]
47. Liu, X.Y.; Li, L.R.; Zhao, H.W.; Cheng, T.H.; Cui, G.J.; Tan, Q.C.; Meng, G.W. Quality assessment of speckle patterns for digital image correlation by Shannon entropy. *Optik* **2015**, *126*, 4206–4211. [CrossRef]
48. Limodin, N.; Rethore, J.; Adrien, J.; Buffiere, J.Y.; Hild, F.; Roux, S. Analysis and artifact correction for volume correlation measurements using tomographic images from a laboratory X-ray source. *Exp. Mech.* **2011**, *51*, 959–970. [CrossRef]
49. Wang, B.; Pan, B.; Lubineau, G. In-Situ Systematic Error Correction for Digital Volume Correlation Using a Reference Sample. *Exp. Mech.* **2018**, *58*, 427–436. [CrossRef]
50. Pan, B. Thermal error analysis and compensation for digital image/volume correlation. *Opt. Lasers Eng.* **2018**, *101*, 1–15. [CrossRef]
51. Buljac, A.; Jailin, C.; Mendoza, A.; Neggers, J.; Taillandier-Thomas, T.; Bouterf, A.; Smaniotto, B.; Hild, F.; Roux, S. Digital volume correlation: Review of progress and challenges. *Exp. Mech.* **2018**, *58*, 661–708. [CrossRef]
52. Gates, M.; Lambros, J.; Heath, M.T. Towards high performance digital volume correlation. *Exp. Mech.* **2011**, *51*, 491–507. [CrossRef]

Article

Enhanced Digital Image Correlation Analysis of Ruptures with Enforced Traction Continuity Conditions Across Interfaces

Yuval Tal [1,*], Vito Rubino [2], Ares J. Rosakis [2] and Nadia Lapusta [1,3]

[1] Seismological Laboratory, Division of Geological and Planetary Sciences, California Institute of Technology, Pasadena, CA 91125, USA; lapusta@caltech.edu

[2] Graduate Aerospace Laboratories, California Institute of Technology, Pasadena, CA 91125, USA; Vito.Rubino@caltech.edu (V.R.); arosakis@caltech.edu (A.J.R.)

[3] Division of Engineering and Applied Science, California Institute of Technology, Pasadena, CA 91125, USA

* Correspondence: yutal@caltech.edu; Tel.: +1-857-400-6173

Received: 1 March 2019; Accepted: 15 April 2019; Published: 18 April 2019

Abstract: Accurate measurements of displacements around opening or interfacial shear cracks (shear ruptures) are challenging when digital image correlation (DIC) is used to quantify strain and stress fields around such cracks. This study presents an algorithm to locally adjust the displacements computed by DIC near frictional interfaces of shear ruptures, in order for the local stress fields to satisfy the continuity of tractions across the interface. In the algorithm, the stresses near the interface are extrapolated by local polynomials that are constructed using a constrained inversion. This inversion is such that the traction continuity (TC) conditions are satisfied at the interface while simultaneously matching the displacements produced by the DIC solution at the pixels closest to the center of the subset, where the DIC fields are more accurate. We apply the algorithm to displacement fields of experimental shear ruptures obtained using a local DIC approach and show that the algorithm produces the desired continuous traction field across the interface. The experimental data are also used to examine the sensitivity of the algorithm against different geometrical parameters related to construction of the polynomials in order to avoid artifacts in the stress field.

Keywords: digital image correlation; dynamic interfacial rupture; traction continuity across interfaces

---

1. Introduction

Understanding the dynamics of shear rupture of interfaces is important for fields ranging from failure of composites and bimaterial structures, to earthquakes, car brakes, and even to pistons of internal combustion engines. Recent advances in high-speed camera technologies have enabled the development of an experimental setup that combines ultra-high-speed photography with digital image correlation (DIC) to quantify the behavior of propagating dynamic ruptures in the laboratory [1–3], including full-field measurements of displacements, velocities, strains, and stresses (Figure 1).

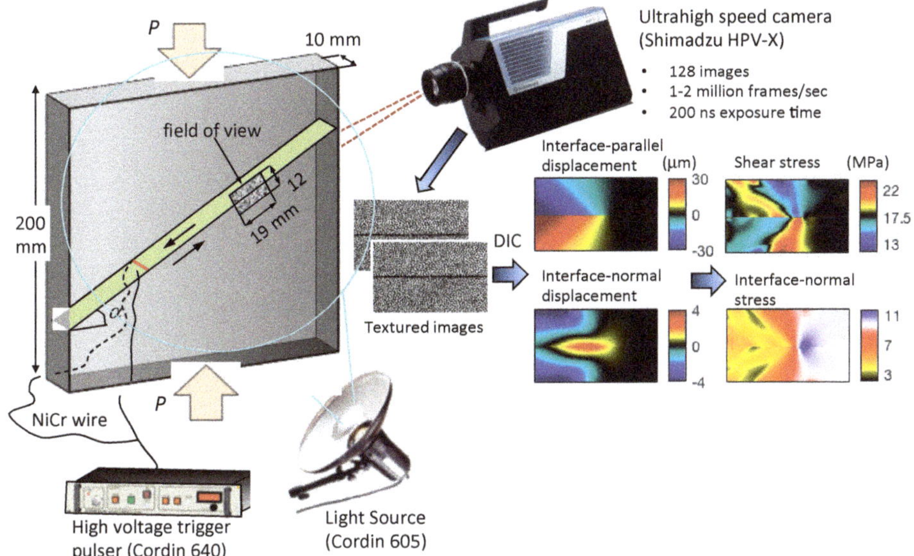

Figure 1. Schematics of the laboratory rupture experiment. Dynamic shear ruptures evolved spontaneously along the frictional interface, inclined at an angle α, of two Homalite plates under a compressional load P. Ruptures were initiated by the small burst of a NiCr wire placed across the interface and connected to a capacitor bank. The white light produced by a flash source was reflected by the specimen's surface and captured by a low-noise high-speed camera, typically at 1–2 million frames/s. The portion of the specimen to be imaged, the field of view, was coated by a flat white paint and decorated by a characteristic speckle pattern. Next, the textured images were processed by digital image correlation (DIC) algorithms to produce a temporal sequence of full-field displacement maps. The displacement fields were then post-processed to produce velocity, strain, strain rate, and stress maps.

DIC is an optical technique used to determine the full-field deformation between a reference and a deformed image [4–6]. The correlation algorithms use either global or local approaches to compute the desired field quantities [7,8]. In global approaches, the pattern matching is performed over the entire domain of analysis, typically using the finite element method [9,10], which allows enforcing compatibility of the displacement field. In local approaches, the image matching is performed with small windows, or "subsets", separated by a distance, referred to as "step", which can be less than half a subset size, (i.e., subsets can overlap) [4,11,12]. A number of recent investigations have focused on evaluating the accuracy and resolution of DIC [13,14]. The full-field measurements enabled by DIC have been used in a wide range of applications [4,6]. However, standard DIC techniques cannot capture displacement discontinuities associated with cracks or ruptures, as they assume a continuous displacement field [4]. Various approaches have been proposed to analyze discontinuous displacement fields [15–22], but they generally entail the application of constraints with a theoretical interpretation of the fields. This limits the implementation of such approaches to cases where the theoretical assumptions are valid.

In a previous study on dynamic ruptures conducted using an experimental setup similar to that employed here, it was possible to capture the static displacements associated with dynamic ruptures, after these ruptures traversed a part of a fault, because a low-frame-rate camera with high resolution (i.e., 4 megapixels) was employed [23]. In that study, the discontinuous displacement field was accurately mapped by simply having subsets over the interface as the high resolution and low noise of the images allowed for the subset size to be relatively small compared to the image size, so that smearing of the displacement fields caused by having subsets across the interface was minimal. Generally, local DIC approaches that provide the solution up to half a subset from the interface may also be sufficient when tracking the quasi-static propagation and opening of tensile crack faces and the associated strain field, using a large field of views and high-resolution cameras. However, these correlation approaches are not possible when the subset size is large compared to the field of view (i.e., when using low-resolution cameras), such as our ultra-high-speed camera (which has a resolution of 250×400 pixels^2) or high-resolution cameras, but with comparatively large subset sizes, situations typically arising in the study of dynamic problems [3]. To resolve the displacement discontinuities on the interface associated with the propagation of dynamic shear ruptures, the studies in [1–3] used the commercial DIC software Vic-2D (Correlated Solutions, Inc.) to separately correlate the domains above and below the interface. While standard local DIC methods calculate displacements up to half a subset away from the interface, the "fill boundary" algorithm of Vic-2D employed in [1–3] uses affine transformation functions to extrapolate the displacements from the center of the subset up to the interface. This enables the use of large subsets in each side of the interface, which are essential to overcoming the noise associated with ultra-high-speed photography and the analyzed rapid ruptures. An important limitation of this approach, however, is that the extrapolated displacements are less accurate than those obtained directly by the actual correlation, especially at pixels near the interface, which have the largest distance from the subset center. This leads to errors in the strains and stresses near the interface, which are obtained from the spatial derivatives of the displacement fields. In addition, the non-coupled correlation of the domains above and below the interface results in non-physical discontinuities in the tractions across the interface.

This study proposes a simple and fast method to supplement the DIC solution with a post-processing algorithm that enforces the continuity of normal and shear tractions across the interface. The algorithm uses a constrained inversion to construct local 2-D polynomials that satisfy traction continuity conditions at the interface while matching the displacements of pixels closer to the center of the subset, where the DIC solution is most accurate. The polynomials are then used to calculate all stress components near the interface. In Section 2, we describe the experimental setup, including the laboratory experiment, ultra-high-speed diagnostics, the digital image correlation approach, and the post-processing procedure to turn the displacement fields into strain and stress fields. In Section 3, we present the method to enforce traction continuity along the interface and study the effects of the parameters involved in this method on the full-field stresses, displacements and particle velocities. Implications for the analysis of friction using the stress fields produced by this approach and conclusions are given in Sections 4 and 5, respectively.

2. Monitoring Dynamic Shear Ruptures in the Laboratory

The algorithm developed in this paper is a post-processing algorithm that is designed to enforce the continuity of stresses measured near frictional interfaces during experimental dynamic ruptures. This section briefly summarizes the experimental setup, diagnostics, image analysis, and numerical methods that are used to obtain the displacement, strain, and stress fields before using the algorithm. This setup is the evolution of the Caltech " Laboratory Earthquake Setup " developed by Rosakis and his co-workers over a span of 15 years [1–3,24–29]. A more detailed description of the current form of the setup is given in [3].

The presentation focuses on the analysis of experimental dynamic ruptures, but the algorithm can be used for a wider range of frictional experiments that involve image analysis. The same methodology can also be used for the analysis of opening, shear, and mixed-mode cracks at both coherent and incoherent (frictional) interfaces separating both similar and dissimilar solids. Such interfaces are common in nature (e.g., faults), and are also important to a variety of engineering problems involving composites and bi-materials.

2.1. The Laboratory Setup

The laboratory setup is designed to study the dynamics of shear ruptures propagating along preexisting inclined frictional interfaces via full-field measurements of displacements, velocities, strains, and stresses associated with the rupture (Figure 1). Two Homalite-100 quadrilateral plates with a frictional interface inclined at an angle α are loaded under uniaxial compression P (Figure 1), resulting in initial shear and normal stresses on the interface of $\tau_0 = P\sin(\alpha)\cos(\alpha)$ and $\sigma_0 = P\cos^2(\alpha)$, respectively. The rupture is nucleated by a local pressure release provided by a rapid expansion of a NiCr wire filament due to an electrical discharge of Cordin 640 high-voltage capacitor (Cordin, Salt Lake City, UT, USA). A key aspect of the setup is that the low shear modulus of Homalite enables production of well-developed dynamic ruptures in samples of tens of centimeters.

Once the rupture initiates, a target area coated with a random black-speckle pattern is monitored using Shimadzu HPV-X ultra-high-speed camera system (Shimadzu, Kyoto, Japan), capable of recording up to 10 million frames per second, and Cordin 605 high-speed white-light source system with two light heads (Cordin, Salt Lake City, UT, USA) (Figure 1). In the experiment reported here, the camera recorded a sequence of 128 images of the patterns distorted by the propagating rupture, with a resolution of 400×250 pixels^2, at a temporal sampling of 2 million frames/second and an exposure time of 200 ns. Moreover, the HPV-X camera was equipped with a range of prime telephoto lenses, which allowed it to monitor fields of view of different sizes [3].

2.2. Digital Image Correlation to obtain Displacement Fields

We employed the local digital image correlation (DIC) software Vic-2D (Correlation Solutions Inc.) to analyze the sequence of images acquired with the HPV-X high-speed camera, and to produce evolving displacement maps, computed with respect to a selected reference configuration. In local approaches, the correlation is performed on local "subsets" separated by a distance, referred to as "step", which can be less than half the subset size (i.e., subsets can overlap) [4,11,12]. The 2D-DIC algorithms provide the two in-plane displacement components at each subset center. A parametric study performed in [3] showed that a subset size of 41 pixels and step size of 1 are needed to accurately resolve the spatial and temporal features of dynamic ruptures. In order to capture the discontinuous displacement field across the interface, the correlation is performed separately for the domains above and below the interface. While standard local DIC approaches are able to produce the displacement map up to half a subset away from the interface, the "fill boundary" algorithm of Vic-2D uses affine transformation functions to extrapolate the displacements from the center of the subset up to the interface.

2.3. Post-Processing of the Displacement Fields

Prior to the computation of strain fields, we filtered high-frequency noise from the displacement fields using a non-local-means (NL-means) filter [23,30,31]. This filter enables the displacement fields to be smoothed, while maintaining the original signal pattern. An example of full-field displacement maps of a laboratory rupture obtained using high-speed photography, DIC analysis, and a non-local filter is given in Figure 1. To facilitate comparisons, we report the same experimental rupture discussed in [1,3], produced with an applied vertical load $P = 23$ MPa and inclination angle $\alpha = 29°$, and monitored with a small field of view. This was a super-shear rupture propagating at a speed of 2200 m/s. Super-shear ruptures are interfacial ruptures whose tip propagates at speeds greater than the shear wave speed of the surrounding solid, as first observed in [24,26,32,33]. Note that the full-field maps are cropped and their size is slightly smaller than the reported field of view size. These displacement fields are used throughout the study to test the capability of new algorithms to enforce traction continuity (TC) at the interface.

Strains are computed from the filtered displacement fields using finite difference approximation. We employ a frame system x_1-x_2 parallel to the interface and use the central difference scheme for pixels away from the boundaries [23]:

$$\varepsilon_{11}(i,j,k) = \frac{u_1(i,j+1,k) - u_1(i,j-1,k)}{2\,s} \tag{1}$$

$$\varepsilon_{22}(i,j,k) = \frac{u_2(i+1,j,k) - u_2(i-1,j,k)}{2\,s} \tag{2}$$

$$\varepsilon_{12}(i,j,k) = \frac{1}{2}\left(\frac{u_1(i+1,j,k) - u_1(i-1,j,k)}{2\,s} + \frac{u_2(i,j+1,k) - u_2(i,j-1,k)}{2\,s}\right), \tag{3}$$

where $u_1(i,j,k)$ and $u_2(i,j,k)$ are the interface-parallel and interface-normal displacement components, respectively, for pixel (i,j) and frame k, and s is the step size. Note that the displacements in Equations (1)–(3) are expressed in pixels. Second-order backward and forward difference schemes are used to compute strains at the pixels immediately above and below the interface, respectively [1,3].

The stress changes with respect to the reference configuration (before rupture) are computed from the strain fields using the standard plane-stress linear elastic constitutive equations:

$$\begin{bmatrix} \Delta\sigma_{11}(x_1,x_2,t) \\ \Delta\sigma_{22}(x_1,x_2,t) \\ \Delta\sigma_{12}(x_1,x_2,t) \end{bmatrix} = \frac{E}{1-\nu^2} \begin{bmatrix} 1 & \nu & 0 \\ \nu & 1 & 0 \\ 0 & 0 & 1-\nu \end{bmatrix} \begin{bmatrix} \varepsilon_{11}(x_1,x_2,t) \\ \varepsilon_{22}(x_1,x_2,t) \\ \varepsilon_{12}(x_1,x_2,t) \end{bmatrix}, \tag{4}$$

where E is the Young's modulus and ν is Poisson's ratio. The actual stresses are given by:

$$\begin{aligned} \sigma_{11}(x_1,x_2,t) &= \sigma_{11(0)}(x_1,x_2) + \Delta\sigma_{11}(x_1,x_2,t) \\ \sigma_{22}(x_1,x_2,t) &= \sigma_{22(0)}(x_1,x_2) + \Delta\sigma_{22}(x_1,x_2,t) \\ \sigma_{12}(x_1,x_2,t) &= \sigma_{12(0)}(x_1,x_2) + \Delta\sigma_{12}(x_1,x_2,t), \end{aligned} \tag{5}$$

where $\sigma_{11(0)}(x_1,x_2)$, $\sigma_{22(0)}(x_1,x_2) = \sigma_0$, and $\sigma_{12(0)}(x_1,x_2) = \tau_0$ are the pre-stresses at the reference configuration. Since Homalite-100 is a strain-rate sensitive material at the strain rate levels developed during these dynamic ruptures, we use the dynamic values of the elastic constants to compute the stress changes [1,3]. An example of full-field maps of normal and shear stresses calculated from the displacement fields shown in Figure 1 are given in Figure 2a,b, respectively. The figures clearly show that the experimental technique enables capturing the spatial and temporal features of the ruptures. In particular, the more compressional and dilatational regions surrounding the crack are clearly imaged, as expected for shear ruptures (see [1,3] for more examples and physical insight). However, a significant limitation is that in the "fill boundary" algorithm of Vic-2D the domains above and below the interface are correlated and extrapolated towards the interface independently, without applying continuity constraints along the interface. Therefore, tractions are discontinuous at the interface, especially the normal component, where at some locations there is a difference of about 9 MPa between the normal tractions above and below the interface. Note that the increase of normal tractions towards the interface is the physical consequence of the shear rupture; however, close to the interface the tractions need to evolve towards being continuous. In the following section we present a post-processing algorithm to resolve this issue.

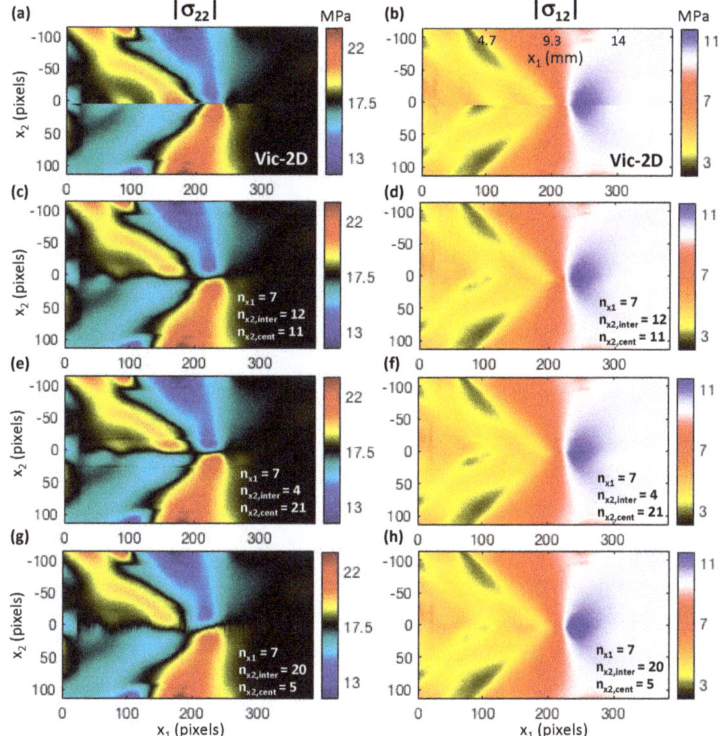

Figure 2. Full-field maps of normal (left) and shear (right) stresses measured for a super-shear crack-like rupture during a test with $\alpha = 29°$, $P = 23$ MPa ($\sigma_0 = 17.6$ MPa and $\tau_0 = 9.75$ MPa), and a field of view of 19 x 12 mm. (**a,b**) Full-field maps obtained using Vic-2D extrapolation near the interface. (**c,d**) Fulfilled maps obtained with the stress continuity constraints, using a good set of parameters n_{x1}, $n_{x2,inter}$, and $n_{x2,cent}$. (**e–h**) Enforcement of stress continuity with extreme values of $n_{x2,inter}$ and $n_{x2,cent}$ leads to discontinuities and spurious oscillations in the stress fields. Note that the full-field maps are cropped and their size is slightly smaller than the reported field of view size.

3. A Post-Processing Algorithm to Enforce Traction Continuity along the Interface

We present a method to supplement the DIC solution with a post-processing algorithm in which the displacements are modified in the proximity of the interface, such that the normal and shear tractions are continuous across the interface. The main idea is (i) to keep the displacement fields obtained with the local DIC up (or close) to the center of the subset touching the interface, and (ii) to use a portion of those fields to construct a polynomial extrapolation for the displacement so that stresses computed from the extrapolated displacements satisfy traction continuity conditions at the interface. The formulation presented here is for a coordinate frame x_1-x_2, such that x_1 is parallel to the interface. However, the method can easily be extended to any other frame. The procedure is described here for a given column of pixels at a given frame, and was implemented for all the columns and frames.

3.1. Traction Continuity Conditions

The condition for continuity of normal traction across the interface is given by

$$\sigma_{22(0)}^{0+} + \Delta\sigma_{22}^{0+} = \sigma_{22(0)}^{0-} + \Delta\sigma_{22}^{0-},$$

(6)

where the superscripts $(^{0+})$ and $(^{0-})$ denote the values immediately above and below the interface, respectively, and $\sigma_{22(0)}{}^{0+} = \sigma_{22(0)}{}^{0-}$ are the initial stresses. Assuming a linear elastic response for the materials above and below the interface, the condition in Equation (6) can be written in terms of strains as

$$\frac{E^+}{1-v^{+2}}\left(\varepsilon_{22}{}^{0+} + v^+{}_1\varepsilon_{11}{}^{0+}\right) = \frac{E^-}{1-v^{-2}}\left(\varepsilon_{22}{}^{0-} + v^-\varepsilon_{11}{}^{0-}\right), \tag{7}$$

where the superscripts $(^+)$ and $(^-)$ denote the material properties above and below the interface, respectively. For small strains the condition can be expressed in terms of displacement gradients as

$$\frac{E^+}{1-v^{+2}}\left(\frac{\partial u_2}{\partial x_2}{}^{0+} + v^+\frac{\partial u_1}{\partial x_1}{}^{0+}\right) = \frac{E^-}{1-v^{-2}}\left(\frac{\partial u_2}{\partial x_2}{}^{0-} + v^-\frac{\partial u_1}{\partial x_1}{}^{0-}\right). \tag{8}$$

Similarly, the continuity of shear traction is expressed via

$$\sigma_{12(0)}{}^{0+} + \Delta\sigma_{12}{}^{0+} = \sigma_{12(0)}{}^{0-} + \Delta\sigma_{12}{}^{0-}, \tag{9}$$

and

$$\mu^+\left(\frac{\partial u_1}{\partial x_2}{}^{0+} + \frac{\partial u_2}{\partial x_1}{}^{0+}\right) = \mu^-\left(\frac{\partial u_1}{\partial x_2}{}^{0-} + \frac{\partial u_2}{\partial x_1}{}^{0-}\right), \tag{10}$$

where μ is the shear modulus.

3.2. Approximating the Displacements with Local Polynomials

In order to modify the displacements near the interface for the jth column of pixels, we approximate the displacement fields u_2^+, u_1^+, u_2^-, and u_1^- with local polynomials in terms of both spatial coordinates. The polynomials are defined over rectangles of $n_{x1} \times n_{x2}$ pixels above and below the interface, which are centered at the jth column of pixels (Figure 3). The pixels within each rectangle are divided into a group of $n_{x1} \times n_{x2,inter}$ pixels that are closer to the interface, and a group of $n_{x1} \times n_{x2,cent}$ pixels that are closer to the center of the subset. To represent the spatial variations in stresses, we approximate the displacement fields with cubic polynomials with $n = 10$ coefficients as

$$\begin{aligned}
u_2^{p+}(x_1,x_2) &= a_2^{10+}x_1{}^3 + a_2^{9+}x_1{}^2x_2 + a_2^{8+}x_1{}^2 + a_2^{7+}x_1x_2{}^2 + a_2^{6+}x_1x_2 + a_2^{5+}x_1 + a_2^{4+}x_2{}^3 + a_2^{3+}x_2{}^2 \\
&\quad + a_2^{2+}x_2 + a_2^{1+} \\
u_1^{p+}(x_1,x_2) &= a_1^{10+}x_1{}^3 + \cdots + a_1^{1+} \\
u_2^{p-}(x_1,x_2) &= a_2^{10-}x_1{}^3 + \cdots + a_2^{1-} \\
u_1^{p-}(x_1,x_2) &= a_1^{10-}x_1{}^3 + \cdots + a_1^{1-}
\end{aligned} \tag{11}$$

Note that n_{x1} and n_{x2} should be chosen small enough to avoid smoothing local patterns in the displacement fields.

3.3. Inverting for the Polynomial Coefficients

To obtain the polynomial coefficients, we aim to minimize the misfit between the observed displacements u_2^{ob+}, u_2^{ob-}, u_1^{ob+}, and u_1^{ob-}, produced by the DIC analysis, and the displacements predicted by the polynomials u_2^{p+}, u_2^{p-}, u_1^{p+}, and u_1^{p-} at the pixels closer to the center of the subset. Recall that the DIC solution is computed up to half a subset away from the interface and extrapolated to the interface by the "fill Boundary" algorithm of Vic-2D. Since we expect such extrapolation to be reliable close to the center of the subset, but not close to the interface, we can use some of the extrapolated values by taking $n_{x2,inter} < 20$ (Figure 3c). We can also exclude the displacement extrapolated by Vic-2D and consider only the computed values (up to the subset center) by taking $n_{x2,inter} \geq 20$ (Figure 3d). The minimization condition leads to the following system of equations:

$$\left[\begin{bmatrix} X^+_{[m\times n]} & \cdots & \cdots & 0 \\ \vdots & X^-_{[m\times n]} & \ddots & \vdots \\ \vdots & \ddots & X^+_{[m\times n]} & \vdots \\ 0 & \cdots & \cdots & X^-_{[m\times n]} \end{bmatrix} \begin{bmatrix} A^+_{2[n]} \\ A^-_{2[n]} \\ A^+_{1[n]} \\ A^-_{1[n]} \end{bmatrix} \right] = \begin{bmatrix} u^{ob+}_{2[m]} \\ u^{ob-}_{2[m]} \\ u^{ob+}_{1[m]} \\ u^{ob-}_{1[m]} \end{bmatrix} \tag{12}$$

where

$$X^+_{[m\times n]} = \begin{bmatrix} x_{11}{}^3 & x_{11}{}^2 x_{12} & \cdots & 1 \\ \vdots & \vdots & \ddots & \vdots \\ x_{m1}{}^3 & x_{m1}{}^2 x_{m2} & \cdots & 1 \end{bmatrix}, \ x_2 > 0$$

and

$$X^-_{[m\times n]} = \begin{bmatrix} x_{11}{}^3 & x_{11}{}^2 x_{12} & \cdots & 1 \\ \vdots & \vdots & \ddots & \vdots \\ x_{m1}{}^3 & x_{m1}{}^2 x_{m2} & \cdots & 1 \end{bmatrix}, \ x_2 < 0$$

are the sensitivity sub-matrices above and below the interface, respectively,

$$A^+_{2[n]} = \begin{bmatrix} a_2^{10+} & \cdots & a_2^{1+} \end{bmatrix}^T$$
$$A^-_{2[n]} = \begin{bmatrix} a_2^{10-} & \cdots & a_2^{1-} \end{bmatrix}^T$$
$$A^+_{1[n]} = \begin{bmatrix} a_1^{10+} & \cdots & a_1^{1+} \end{bmatrix}^T$$
$$A^-_{1[n]} = \begin{bmatrix} a_1^{10-} & \cdots & a_1^{1-} \end{bmatrix}^T$$

are the assemblies of polynomial coefficients, and $m = n_{x1} \times n_{x2,cent}$ is the number of pixels within the group closer to the center of the subset.

The polynomials must also fulfill the constraints for continuity of tractions across the interface. The condition for continuity of normal tractions is enforced by

$$\frac{E^+}{1-v^{+2}}\left(\frac{\partial u_2^{p0+}}{\partial x_2} + v^+ \frac{\partial u_1^{p0+}}{\partial x_1}\right) = \frac{E^-}{1-v^{-2}}\left(\frac{\partial u_2^{p0-}}{\partial x_2} + v^- \frac{\partial u_1^{p0-}}{\partial x_1}\right) \tag{13}$$

and can be assembled into the following matrix form:

$$\left[\frac{E^+}{1-v^{+2}}\frac{\partial X^{0+}_{[n_{x1}\times n]}}{\partial x_2} - \frac{E^-}{1-v^{-2}}\frac{\partial X^{0-}_{[n_{x1}\times n]}}{\partial x_2} \frac{E^+}{1-v^{+2}}v\frac{\partial X^{0+}_{[n_{x1}\times n]}}{\partial x_1} - \frac{E^-}{1-v^{-2}}v\frac{\partial X^{0-}_{[n_{x1}\times n]}}{\partial x_1}\right]\begin{bmatrix} A^+_{2,[n]} \\ A^-_{2,[n]} \\ A^+_{1,[n]} \\ A^-_{1,[n]} \end{bmatrix} = 0, \tag{14}$$

where, for example,

$$\frac{\partial u_2^{p0+}}{\partial x_2} = \frac{\partial X^{0+}_{[n_{x1}\times n]}}{\partial x_2}A^+_{2,[n]} = \begin{bmatrix} 0 & x_{11}{}^2 & \cdots & 0 \\ \vdots & \vdots & \ddots & \vdots \\ 0 & x_{n_{x1}1}{}^2 & \cdots & 0 \end{bmatrix} A^+_{2,[n]}, \ x_2 = 0.$$

Similarly, the condition for continuity of shear stresses is enforced by

$$\left[\mu^+ \frac{\partial X^{0+}_{[n_{x1}\times n]}}{\partial x_1} \quad -\mu^- \frac{\partial X^{0-}_{[n_{x1}\times n]}}{\partial x_1} \quad \mu^+ \frac{\partial X^{0+}_{[n_{x1}\times n]}}{\partial x_2} \quad -\mu^- \frac{\partial X^{0-}_{[n_{x1}\times n]}}{\partial x_2} \right] \begin{bmatrix} A^+_{2,[n]} \\ A^-_{2,[n]} \\ A^+_{1,[n]} \\ A^-_{1,[n]} \end{bmatrix} = 0. \tag{15}$$

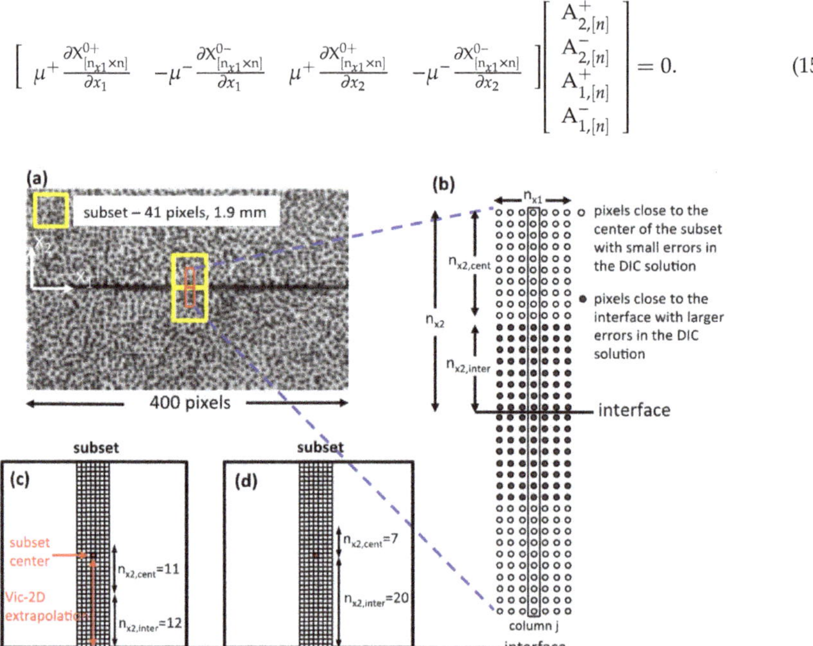

Figure 3. (**a**) A speckled image with two subsets of 41 × 41 pixels above and below the interface (yellow squares). The post-processing algorithm is performed on the regions inside the subsets marked by the red rectangles. (**b**) The geometry of the pixels over which the local polynomials approximating the displacement fields are defined. The polynomials approximating u_2^+, u_1^+ extend over a rectangle above the interface, while those approximating u_2^-, u_1^- extend over a rectangle below the interface; both rectangles include $n_{x1} \times n_{x2}$ pixels. The pixels within each rectangle are divided into a group of $n_{x1} \times n_{x2,inter}$ pixels that are closer to the interface, and a group of $n_{x1} \times n_{x2,cent}$ pixels that are closer to the center of the subset. The observed displacements in the latter group are used together with continuity of stresses constraints on the interface to construct the polynomials. (**c**,**d**) Two different combinations of $n_{x2,inter}$ and $n_{x2,cent}$ and their relationship with Vic-2D extrapolation. In the first combination, the group of pixels $n_{x1} \times n_{x2,cent}$, which are used as data in the inversion, include pixels where the displacement values were obtained by the "fill boundary" algorithm of Vic-2D.

Different methods can be used to solve the system of equations in Equation (12) together with the constraints in Equations (14) and (15) for the polynomial coefficients; here we use Matlab (The MathWorks, Natick, MA, USA) function *lsqlin*. Once the coefficients are obtained, the displacements at the pixels of the central column (n_{x2} pixels above and n_{x2} pixels below the interface) are replaced by those predicted by the corresponding polynomials, and the procedure is repeated for all columns and images. Note that the system of equations to be solved is small because the local polynomials should represent the local variations in displacements and stresses well. Thus, the computation time was small, and the modification of all pixel columns within an image was done within 2–3 s for the images analyzed here (400 pixels along the interface). Figure 4 shows the interface-parallel (a) and normal (b) displacements at the center of the field of view ($j = 190$) for the rupture of Figure 1. Both the displacements obtained by Vic-2D (with the "fill boundary" algorithm) and displacements modified by TC enforcement are shown. The maximum difference in displacements between the two solutions was smaller than 0.001 pixels, which is equivalent to 0.5% for u_1 and 1.5% for u_2, but it was enough to enforce the continuity of traction.

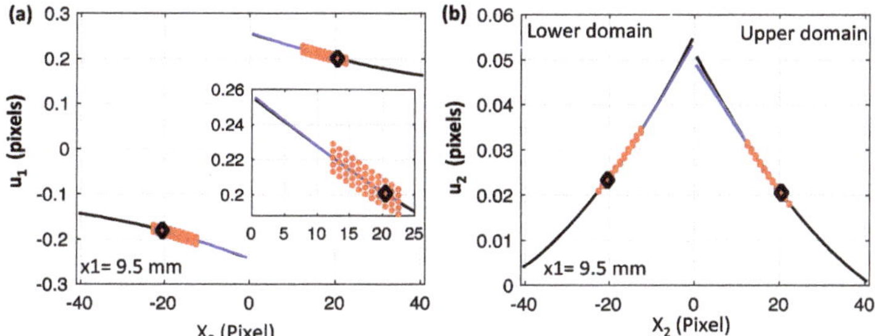

Figure 4. Interface parallel (**a**) and normal (**b**) displacements at the central columns of the subsets shown in Figure 3a. The black lines represent the displacements obtained with Vic-2D, while the blue lines are the displacements modified with the traction continuity algorithm, using $n_{x1} = 5$, $n_{x2,inter} = 12$, and $n_{x2,cent} = 11$. The red circles represent the observed displacements that are used to invert for the coefficients of the displacement polynomials u_2^{p+}, u_2^{p-}, u_1^{p+}, and u_1^{p-}. The black diamonds correspond to the centers of the subsets.

When the gradients in displacements in the direction parallel to the interface are large, as in the case of the rupture analyzed in this study, direct calculation of strains and stresses using the modified displacement fields may not result in completely continuous tractions at the interface. The calculation of strains and stresses for pixels in column j involves the displacements in columns $j-1$ and $j + 1$, as it requires displacement gradients with respect to x_1 (Equations (1)–(4)). When enforcing TC, these gradients are computed using the polynomial centered at column j. Once the displacement fields have been updated using the procedure described above, strains can be computed from the updated displacement fields, using finite difference schemes, as discussed in Section 2.3. However, the finite differences would be using the displacements of columns $j-1$ and $j + 1$, which are obtained from other polynomials than that used to enforce TC, those centered in columns $j-1$ and $j + 1$. As such, the spatial derivatives of the modified displacement field with respect to x_1 would be slightly different than those obtained during the enforcement of TC, using the polynomial centered at column j. Although the differences between polynomials centered at two neighboring columns are small, they may be enough to cause some discontinuities in tractions at the interface. To ensure the continuity of tractions, we computed strains and stresses within the $2 \times n_{x2}$ band, where displacements are updated directly as the algorithm goes over each column of pixels (i.e., we calculated the displacement derivatives for column j with respect to x_1 using the displacements in columns $j-1$ and $j + 1$ obtained from the polynomial centered at column j. Full-field maps of stresses modified by the algorithm using $n_{x1} = 7$, $n_{x2,inter} = 12$, and $n_{x2,cent} = 11$ (Figure 2c,d) demonstrated the capability of this approach to generate more realistic and continuous stress fields near the interface. Interestingly, because of the symmetry of the problem, the enforcement of TC led to normal stresses that are close to the background value ($\sim$17.6 MPa) at the interface.

3.4. The Effects of the Geometrical Parameters of the Polynomial

It is important to choose appropriate values for the geometrical parameters n_{x1}, $n_{x2,inter}$, and $n_{x2,cent}$ in order not to introduce unwanted field distortions. For instance, a choice of a small value of $n_{x2,inter}$ together with a large value of $n_{x2,cent}$ or vice versa can lead to stress fields that are continuous on the interface, but have sharp discontinuities (Figure 2e,f) or spurious oscillations (Figure 2g,h) in the bulk. In this section, we study the effects of different parameter combinations in more details, aiming to find the parameter sets that avoid these distortions in the stress field. We ran the algorithm 324 times, varying the geometrical parameters in the following ranges: $3 \leq n_{x1} \leq 9$, $4 \leq n_{x2,inter} \leq 20$, and $5 \leq n_{x2,cent} \leq 21$. Note that for $n_{x1} \geq 13$ the system was over constrained. Overall, we found that parameters in ranges of $8 \leq n_{x2,inter} \leq 14$ (1/5 to 1/3 of the subset size) and $7 \leq n_{x2,cent} \leq 15$ gave similar results, independent of the value of n_{x1}.

We found that the stress fields obtained with the algorithm were insensitive to the value of n_{x1}. For example, there was no observable difference between full-field maps of stresses obtained with $n_{x1} = 3$ and $n_{x1} = 9$ (Figure 5a–d). To further examine the effect of n_{x1}, we plot the normal and shear stresses vs position along two lines perpendicular to the interface at $x_1 = 170$ and 220 pixels (Figure 5e–h). For both lines there was a "jump" larger than 5.5 MPa in the normal stress at the interface when the calculation was done with the "fill boundary" algorithm of Vic-2D without TC enforcement. The TC algorithm enforced the normal stress to be continuous across the interface, with a normal stress at the interface that was about the average of the normal stresses at the pixels immediately above and below the interface when TC was not enforced. The results were not sensitive to n_{x1}. A small effect of n_{x1} on the shear stress was observed for the line located at x1 = 220 pixels (Figure 5h), with a difference of 0.12 MPa between the results for $n_{x1} = 3$ and 9. The shear stress at the interface when TC was enforced were smaller than the shear stresses at the pixels immediately above and below the interface when TC was not enforced.

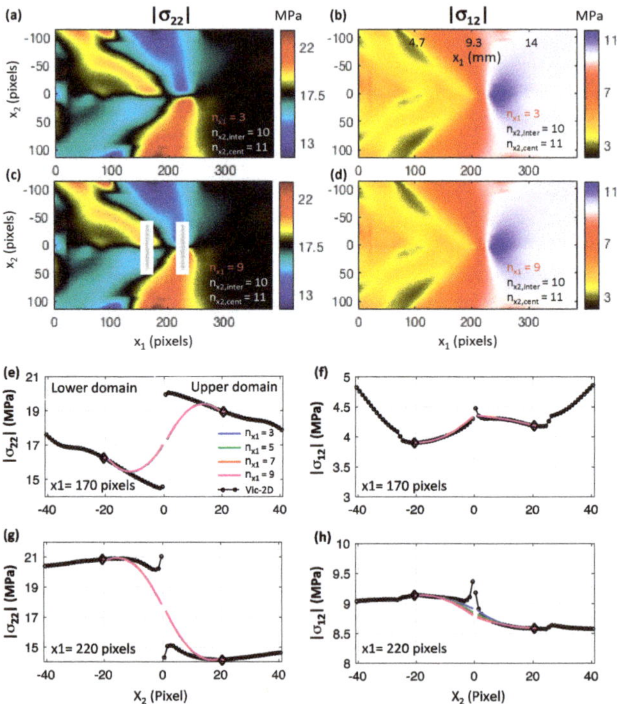

Figure 5. The effect of parameter n_{x1} on the stress field. (**a–d**) Full-field maps of normal (left) and shear (right) stresses for $n_{x1} = 3$ and 9. (**e–h**) Normal (left) and shear (right) stresses vs position along paths perpendicular to the interface at $x_1 = 170$ and 220 pixels (grey lines in (**c**)) for different values of n_{x1}. The stresses calculated with the "fill boundary" algorithm of Vic-2D without the traction continuity algorithm are also shown, where the black diamonds correspond to the centers of the subsets above and below the interface. In all the cases shown in this figure, $n_{x2,inter} = 10$ and $n_{x2,cent} = 11$.

As shown in Figure 2e–h, a large difference between the parameters $n_{x2,inter}$ and $n_{x2,cent}$ resulted in artifacts in the stress field. On the one hand, taking $n_{x2,inter} = 4$ and $n_{x2,cent} = 21$ (with $n_{x1} = 7$) resulted in discontinuities at the boundary between the Vic-2D solution and polynomial extrapolation (Figure 2e,f), $n_{x2,inter} + n_{x2,cent}$ pixels away from the interface due to a poor fit between the stresses obtained by the polynomial and the original stresses at this region. On the other hand, taking $n_{x2,inter} = 20$ and $n_{x2,cent} = 5$ (with $n_{x1} = 7$) resulted in spurious oscillations (Figure 2g,h). Some effects of $n_{x2,inter}$

were observed also when $n_{x2,cent}$ was fixed at a moderate value (e.g., $n_{x2,cent} = 11$), especially for the shear stress. A small value of $n_{x2,inter}$ (e.g., $n_{x2,inter} = 6$) led to large gradients in the normal stresses near the interface, as well as to small discontinuities at the pixels located n_{x2} pixels above and below the interface (Figure 6a,b). A large value of $n_{x2,inter}$ (e.g., $n_{x2,inter} = 16$) resulted in an artificial increase or decrease of stresses at localized regions around the interface compared with the domains above and below (Figure 6c,d). For example, the normal stress decreased locally at the interface at $x_1 = 370$ pixels (Figure 6c), while the shear stress increased locally at $x_1 = 170$ and 370 pixels (Figure 6d). Some issues are illustrated further in plots of the normal and shear stresses vs position along two lines perpendicular to the interface at $x_1 = 170$ and 220 pixels (Figure 6e–h). The value of normal stress on the interface was similar for $n_{x2,inter} \leq 14$, but diverged for higher values, with a difference of about 1 MPa for $n_{x2,inter} = 20$. For $n_{x2,inter} \leq 6$, the TC enforcement resulted in a poor agreement between the stresses obtained by the polynomial and the original stresses also at the pixels closer to the center of the subset. The shear stress at the interface also diverged, with an increase of about 1 MPa for $n_{x2,inter} = 16$, at $x_1 = 170$ pixels (Figure 6f).

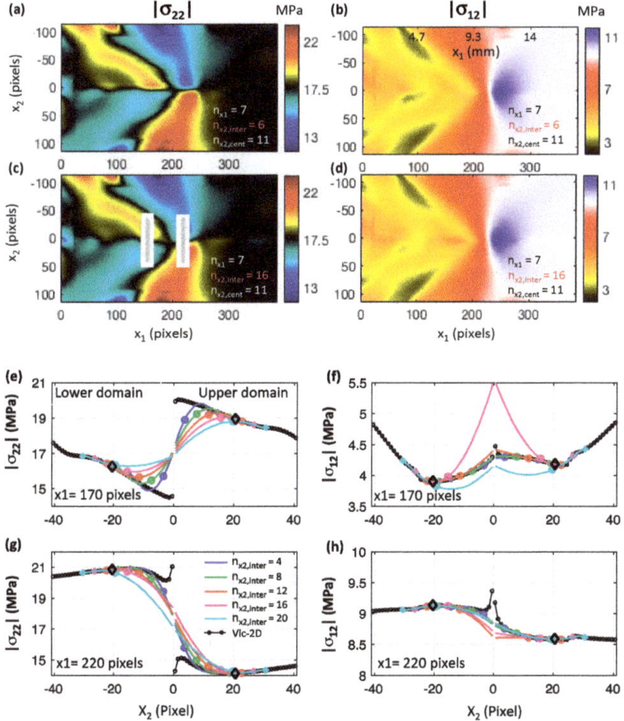

Figure 6. The effect of parameter $n_{x2,inter}$ on the stress field. (**a–d**) Full-field maps of normal (left) and shear (right) stresses for $n_{x2,inter} = 6$ and 16. (**e–h**) Normal (left) and shear (right) stresses vs position along paths perpendicular to the interface at $x_1 = 170$ and 220 pixels (grey lines in (**c**)) for different values of $n_{x2,inter}$. The circles on the curves correspond to $n_{x2,inter}$, while the filled diamonds correspond to n_{x2}. The stresses calculated with the "fill boundary" algorithm of Vic-2D without the traction continuity algorithm are also shown, where the black diamonds correspond to the centers of the subsets above and below the interface. In all the cases shown in this figure $n_{x1} = 7$ and $n_{x2,cent} = 11$.

The effect of $n_{x2,cent}$ was generally smaller than that of $n_{x2,inter}$. Full-field maps of stresses obtained with $n_{x2,cent} = 7$ and $n_{x2,inter} = 10$ (Figure 7a,b), were similar to those in Figure 2c,d. However, large value of $n_{x2,cent} = 19$ resulted in small discontinuities in normal stress at the pixels located n_{x2} pixels above and below the interface (Figure 7c,d). Both at $x_1 = 170$ and 220 pixels there was a very small effect of $n_{x2,cent}$ on the normal stress at the interface (Figure 7e,g). However, for $n_{x2,cent} \geq 17$, the fit between the stresses obtained by the polynomial and the original stresses was poor also for the pixels closer to the center of the subset. Similarly to $n_{x2,inter}$, the shear stress at the interface may significantly have differed for large values of $n_{x2,cent}$ (Figure 7f).

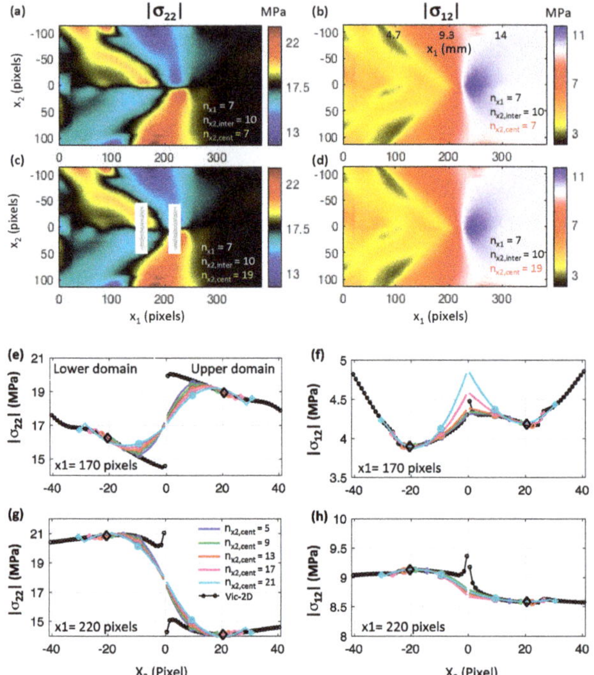

Figure 7. The effect of parameter $n_{x2,cent}$ on the stress field. (**a–d**) Full-field maps of normal (left) and shear (right) stresses for $n_{x2,cent} = 7$ and 19. (**e–h**) Normal (left) and shear (right) stresses vs position along paths perpendicular to the interface at $x_1 = 170$ and 220 pixels (grey lines in (**c**)) for different values of $n_{x2,cent}$. The circles on the curves correspond to $n_{x2,inter}$, while the filled diamonds correspond to n_{x2}. The stresses calculated with the "fill boundary" algorithm of Vic-2D without the traction continuity algorithm are also shown, where the black diamonds correspond to the centers of the subsets above and below the interface. In all the cases shown in this figure $n_{x1} = 7$ and $n_{x2,inter} = 10$.

4. Implications for Friction Analysis

The experimental setup developed in [1,3], which combines ultra-high-speed photography with digital image correlation, enables us to observe displacements, velocities, and stresses close to the interface. For the field of view shown in Figure 1, these quantities can be measured up to 25 μm (0.5 pixels) from the interface using the "fill boundary" algorithm of Vic-2D. This enables studying the evolution of friction, defined by the ratio $f = \sigma_{12}^0 / \sigma_{22}^0$, at any point along the interface, as well as its dependence on variables such as slip (relative displacement across the interface) and slip rate [1]. As a first approach to enforce traction continuity on the interface, the tractions σ_{12}^0 and σ_{22}^0 were calculated in [1] by averaging the stresses at the pixels immediately above and below the interface. Because of the anti-symmetric and symmetric patterns in the stress changes $\Delta\sigma_{22}$ and $\Delta\sigma_{12}$, respectively, the normal stress on the interface σ_{22}^0 was inferred to be nearly constant, as expected during the propagation of a dynamic shear rupture along a planar interface, and the shear stress σ_{12}^0 had a smoother evolution than the individual components σ_{12}^{0+} and σ_{12}^{0-} above and below the interface. The averaging enabled us to study friction, despite large differences between the normal stresses immediately above and below the interface [1], but it is important to examine whether those average values actually correspond to the tractions at the interface computed with the continuity condition.

In general, the averaged normal stresses (without TC enforcement) agreed with the normal stresses computed with TC enforcement for the pixels immediately above and below the interface (Figure 8a), with some small deviations. Note that, because we enforced the continuity at the interface, the stresses obtained with TC enforcement for the pixels located 0.5 pixels above and below the interface did not completely overlap, with a maximum difference of 0.3 MPa. Because of its symmetry with respect to the interface, the discontinuity in shear stress across the interface for the original Vic-2d results (Figure 8b) was significantly smaller than in the normal stress. Overall, there was a good agreement between the shear tractions obtained by the averaging procedure and those obtained with TC enforcement, again, with some minor differences. Interestingly, the differences were consistently smaller in the cohesive zone (the zone where the shear traction drops from a maximum to a residual value), with a maximum difference of about 1 MPa at $x_1 = 9$ mm. This is also shown locally in Figure 5h. Because the stresses were computed from displacement gradients, small changes in the displacements were enough to enforce the traction continuity on the interface, and there were only minor differences between the interface-parallel displacements and velocities obtained with TC enforcement and those obtained without TC enforcement (Figure 8c,d). The small differences in stresses resulted in some differences in the evolution of local friction with slip and slip rate (Figure 8e,f). However, the important characteristics of the local friction, such as the peak, residual levels, and weakening distance, were similar.

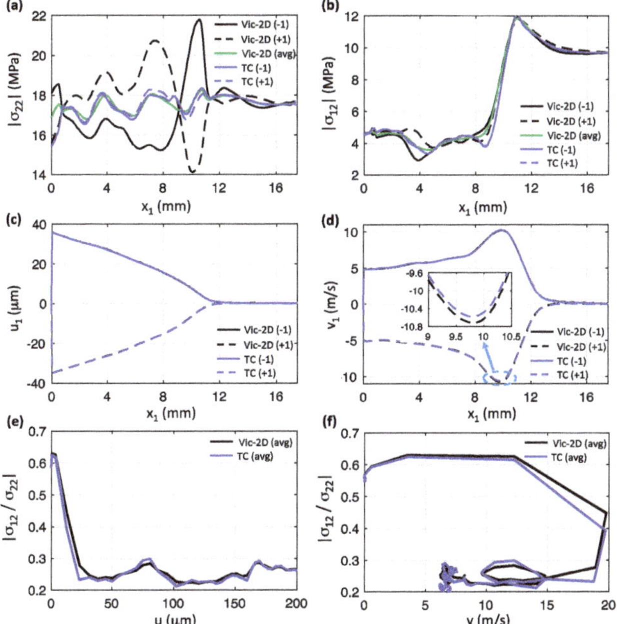

Figure 8. (**a**,**b**) normal and shear stresses along the interface. The black curves represent the stresses ±0.5 pixels from the interface obtained without traction continuity (TC) enforcement, while the green curve represents the average of the two. The blue curves represent the stresses obtained with TC enforcement ±0.5 pixels from the interface. (**c**,**d**) Interface-parallel displacements and velocities ±0.5 pixels from the interface, obtained with (blue) and without (black) TC enforcement. (**e**,**f**) Resolved friction vs slip and slip rate at $x_1 = 8.5$ mm, obtained with (blue) and without (black) TC enforcement. Both curves represent the average of the values immediately above and below interface. The polynomial geometrical parameters are $n_{x1} = 7$, $n_{x2,inter} = 10$, and $n_{x2,cent} = 11$.

5. Conclusions

In this work, the local DIC solution is supplemented with a fast post-processing algorithm to enforce the continuity of tractions across the interfaces of shear ruptures. This procedure allows us to obtain more physically meaningful stress fields near the interface, which is important when DIC is applied to study the dynamics of laboratory frictional ruptures. In the algorithm, the stresses near the interface are calculated from local polynomials that are constructed using a constrained inversion. This inversion is such that the resulting displacements match the displacements of pixels closer to the center of the subset, where the DIC solution is most accurate, while the resulting stresses satisfy traction continuity conditions at the interface.

We applied the algorithm to the displacement fields of a laboratory shear rupture obtained with a local DIC procedure, to show that the algorithm indeed produces the desired continuous traction fields across the interface. A sensitivity study provided a constraint on the parameters $n_{x2,inter}$, $n_{x2,cent}$, and n_{x1} involved in the construction of the polynomials, such that undesired artifacts in the stress fields were eliminated. We found that parameter ranges of $8 \leq n_{x2,inter} \leq 14$ (1/5 to 1/3 of the subset size) and $7 \leq n_{x2,cent} \leq 15$ gave similar results, independent of the value of n_{x1}. Relatively minor changes in displacement fields around the interface can produce non-negligible gradient changes, resulting in some non-uniqueness regarding the exact stress evolution towards the interface, even if the traction continuity is enforced. Future progress can be made by combining the results inferred by the analysis presented in this work with dynamic rupture modeling that matches the observed full fields. The dynamic solutions can then be interrogated for the spatial dependencies of stress fields next to the interface.

The averaging procedure for stresses above and below the interface, employed in the previous study [1] to study friction, works well to mimic the traction continuity across the interface due to the special nature of this rupture problem that exhibits certain symmetries across the interface. However, the averaging procedure may not work as well in cases where the symmetries are disrupted, as in the case of ruptures approaching a free surface [27]. We will examine such cases in future studies.

Author Contributions: Conceptualization, Y.T., V.R., A.J.R., and N.L.; Methodology, Y.T.; Investigation, Y.T. and V.R.; Data curation, V.R.; Writing—original draft preparation, Y.T.; Writing—review and editing, V.R., A.J.R., and N.L.; Supervision, A.J.R. and N.L.

Funding: This study was supported by the US National Science Foundation (NSF) (grant EAR-1651235), the US Geological Survey (USGS) (grant G16AP00106), and the Southern California Earthquake Center (SCEC), contribution No. 18131. SCEC is funded by NSF Cooperative Agreement EAR-1033462 and USGS Cooperative Agreement G12AC20038.

Acknowledgments: We thank Zefeng Li for fruitful discussions about the constrained inversion.

Conflicts of Interest: The authors declare no conflict of interest.

References

1. Rubino, V.; Rosakis, A.J.; Lapusta, N. Understanding dynamic friction through spontaneously evolving laboratory earthquakes. *Nat. Commun.* **2017**, *8*, 15991. [CrossRef] [PubMed]
2. Gori, M.; Rubino, V.; Rosakis, A.J.; Lapusta, N. Pressure shock fronts formed by ultra-fast shear cracks in viscoelastic materials. *Nat. Commun.* **2018**, *9*, 4754.
3. Rubino, V.; Rosakis, A.J.; Lapusta, N. Full-field ultrahigh-speed quantification of dynamic shear ruptures using digital image correlation. *Exp. Mech.* **2019**, in press. [CrossRef]
4. Sutton, M.A.; Orteu, J.J.; Schreier, H. *Image Correlation for Shape, Motion and Deformation Measurements: Basic Concepts, Theory and Applications*; Springer: New York, NY, USA, 2009.
5. Hild, F.; Roux, S. *Digital Image Correlation*; Wiley-VCH: Weinheim, UK, 2012.
6. Sutton, M.A.; Matta, F.; Rizos, D.; Ghorbani, R.; Rajan, S.; Mollenhauer, D.H.; Schreier, H.W.; Lasprilla, A.O. Recent Progress in Digital Image Correlation: Background and Developments since the 2013 W M Murray Lecture. *Exp. Mech.* **2017**, *57*, 1–30. [CrossRef]

7. Hild, F.; Roux, S. Comparison of Local and Global Approaches to Digital Image Correlation. *Exp. Mech.* **2012**, *52*, 1503–1519. [CrossRef]

8. Wang, B.; Pan, B. Subset-based local vs. finite element-based global digital image correlation: A comparison study. *Theor. Appl. Mech. Lett.* **2016**, *6*, 200–208. [CrossRef]

9. Besnard, G.; Hild, F.; Roux, S. "Finite-element" displacement fields analysis from digital images: Application to Portevin-Le Châtelier bands. *Exp. Mech.* **2006**, *46*, 789–803. [CrossRef]

10. Sun, Y.; Pang, J.H.; Wong, C.K.; Su, F. Finite element formulation for a digital image correlation method. *Appl. Opt.* **2005**, *44*, 7357–7363. [CrossRef]

11. Sutton, M.A.; Wolters, W.J.; Peters, W.H.; Ranson, W.F.; McNeill, S.R. Determination of displacements using an improved digital correlation method. *Image Vis. Comput.* **1983**, *1*, 133–139. [CrossRef]

12. Sutton, M.A.; Cheng, M.; Peters, W.H.; Chao, Y.J.; McNeill, S.R. Application of an optimized digital correlation method to planar deformation analysis. *Image Vis. Comput.* **1986**, *4*, 143–150. [CrossRef]

13. Reu, P.L.; Toussaint, E.; Jones, E.; Bruck, H.A.; Iadicola, M.; Balcaen, R.; Turner, D.Z.; Siebert, T.; Lava, P.; Simonsen, M. DIC Challenge: Developing Images and Guidelines for Evaluating Accuracy and Resolution of 2D Analyses. *Exp. Mech.* **2018**, *58*, 1067–1099. [CrossRef]

14. Rossi, M.; Lava, P.; Pierron, F.; Debruyne, D.; Sasso, M. Effect of DIC spatial resolution, noise and interpolation error on identification results with the VFM. *Strain* **2015**, *51*, 206–222. [CrossRef]

15. Jin, H.; Bruck, H.A. Theoretical development for pointwise digital image correlation. *Opt. Eng.* **2005**, *44*, 067003. [CrossRef]

16. Réthoré, J.; Hild, F.; Roux, S. Shear-band capturing using a multiscale extended digital image correlation technique. *Comput. Methods Appl. Mech. Eng.* **2007**, *196*, 5016–5030. [CrossRef]

17. Réthoré, J.; Hild, F.; Roux, S. Extended digital image correlation with crack shape optimization. *Int. J. Numer. Methods Eng.* **2008**, *73*, 248–272. [CrossRef]

18. Poissant, J.; Barthelat, F. A novel "subset splitting" procedure for digital image correlation on discontinuous displacement fields. *Exp. Mech.* **2010**, *50*, 353–364. [CrossRef]

19. Nguyen, T.L.; Hall, S.A.; Vacher, P.; Viggiani, G. Fracture mechanisms in soft rock: Identification and quantification of evolving displacement discontinuities by extended digital image correlation. *Tectonophysics* **2011**, *503*, 117–128. [CrossRef]

20. Tomicevic, Z.; Hild, F.; Roux, S.; Tomicevic, Z.; Hild, F.; Roux, S.; Image, M.D. Mechanics-Aided Digital Image Correlation. *J. Strain Anal. Eng. Des.* **2013**, *48*, 330–343. [CrossRef]

21. Hassan, G.M.; Dyskin, A.; Macnish, C.; Pasternak, E.; Shufrin, I. Discontinuous Digital Image Correlation to reconstruct displacement and strain fields with discontinuities: Dislocation approach. *Eng. Fract. Mech.* **2018**, *189*, 273–292. [CrossRef]

22. Hassan, G.M.; Macnish, C.; Dyskin, A. Extending Digital Image Correlation to Reconstruct Displacement and Strain Fields around Discontinuities in Geomechanical Structures under Deformation. In Proceedings of the IEEE Winter Conference on Applications of Computer Vision, Waikoloa, HI, USA, 5–9 January 2015; IEEE: Piscataway, NJ, USA, 2015; pp. 710–717.

23. Rubino, V.; Lapusta, N.; Rosakis, A.J.; Leprince, S.; Avouac, J.P. Static Laboratory Earthquake Measurements with the Digital Image Correlation Method. *Exp. Mech.* **2015**, *55*, 77–94. [CrossRef]

24. Xia, K.; Rosakis, A.J.; Kanamori, H. Laboratory Earthquakes: The Sub-Rayleigh-to-Supershear. *Science* **2004**, *303*, 1859–1861. [CrossRef]

25. Lu, X.; Lapusta, N.; Rosakis, A.J. Pulse-like and crack-like ruptures in experiments mimicking crustal earthquakes. *Proc. Natl. Acad. Sci. USA* **2007**, *104*, 18931–18936. [CrossRef] [PubMed]

26. Rosakis, A.J.; Xia, K.; Lykotrafitis, G.; Kanamori, H. Dynamic shear rupture in frictional interfaces: Speeds, directionality, and modes. *Treatise Geophys.* **2007**, *4*, 153–192.

27. Gabuchian, V.; Rosakis, A.J.; Bhat, H.S.; Madariaga, R.; Kanamori, H. Experimental evidence that thrust earthquake ruptures might open faults. *Nature* **2017**, *545*, 336–339. [CrossRef]

28. Mello, M.; Bhat, H.S.; Rosakis, A.J.; Kanamori, H. Identifying the unique ground motion signatures of supershear earthquakes: Theory and experiments. *Tectonophysics* **2010**, *493*, 297–326. [CrossRef]

29. Mello, M.; Bhat, H.S.; Rosakis, A.J.; Kanamori, H. Reproducing the supershear portion of the 2002 Denali earthquake rupture in laboratory. *Earth Planet. Sci. Lett.* **2014**, *387*, 89–96. [CrossRef]

30. Buades, A.; Coll, B.; Morel, J. The Staircasing Effect in Neighborhood Filters and its Solution. *IEEE Trans. Image Process.* **2006**, *15*, 1499–1505. [CrossRef] [PubMed]

31. Buades, A.; Coll, B.; Morel, J. Nonlocal Image and Movie Denoising. *Int. J. Comput. Vis.* **2008**, *76*, 123–139. [CrossRef]

32. Rosakis, A.; Samudrala, O.; Coker, D. Cracks faster than the shear wave speed. *Science* **1999**, *284*, 1337–1340. [CrossRef] [PubMed]

33. Rosakis, A.J. Intersonic shear cracks and fault ruptures. *Adv. Phys.* **2002**, *51*, 1189–1257. [CrossRef]

 *applied sciences*

Article

Application of the Non-Contact Video Gauge on the Mechanical Properties Test for Steel Cable at Elevated Temperature

Yong Du [1,2,*] and Zhang-ming Gou [1]

[1] College of Civil Engineering; Nanjing Tech University, Nanjing 211800, China; gouzhangming@njtech.edu.cn
[2] Department of Civil and Environmental Engineering, National University of Singapore, Singapore 117575, Singapore
* Correspondence: yongdu_mail@njtech.edu.cn

Received: 18 March 2019; Accepted: 18 April 2019; Published: 23 April 2019

Abstract: As the limit of traditional contact measurement, it is difficult to precisely measure the steel cables twisted by a branch of wires especially at elevated temperature. In this paper the strain-stress relationships of S355 and S690 structural steel, 1860 MPa steel cable twisted by seven wires have been measured by the strain gauge, extensometer and non-contact video gauge at ambient temperature and elevated temperature, respectively. Comparison of the stress-strain curves gotten by different measuring technology, it indicates that the non-contact video gauge can provide a more efficient and reliable database than the strain gauge as well as extensometer, especially at an elevated temperature. It is worth noting that the non-contact video gauge can capture not only the full range of stress-strain curves of steel cables, but is also efficient for the specimens with a complex shape.

Keywords: non-contact video gauge; measurement; stress-strain relationship; uniaxial tensile test; elevated temperature

---

1. Introduction

In order to improve the reality of a structural response for pre-tensioned steel structures exposed to fire by using numerical simulation, it is very important to capture the full range of the stress-strain relationship at elevated temperature for steel cables, which are always with a complex shape. It is well known that the strain gauge can accurately capture the stress-strain relationship for steel material at ambient temperature, but cannot at an elevated temperature. In previous test, most of the stress-strain relationships of steel material have been tested by extensometers at an elevated temperature. The brittle ceramic arms of an extensometer had to be removed before the ultimate strength in order to prevent being broken due to the limit measuring range. Thus, it can only capture the partial range of the stress-strain curve before the ultimate strength at an elevated temperature [1]. Although the external mechanical device with linear variable displacement transducer can trace the full range of the stress-strain curve at an elevated temperature, the strain resulted from this kind of measurement device is the average strain, which is different from the engineering strain. Furthermore, it is difficult to grip a group of wires in a tensile test due to the complex surface of a steel cable twisted by a branch of wire. The displacement between the front end of an arm and the surface of the specimen always occurred, especially for the test object with a complex geometrical surface such as a steel cable [2].

Recently, a charge-coupled device camera (CCDC) system, which has been improved by the digital image correlation method (DICM), has been used to measure the strain in the test coupon subjected to tension. A computer program based on this correlation method was developed for the calculation of the displacement components and the deformation gradients of an object surface due to deformation [3,4]. Several experiments were performed to demonstrate the viability and accuracy of the correlation method.

To form a gauge length is the primary step to collect test data by CCDC. Hu et al. [5] presented a novel laser-engraving technology to create speckles on the surface of a metallic specimen, which will form gauge length and be kept up till the melting temperature of metallic materials. This kind of speckles was also applied to investigate the elastic modulus of tungsten material up to 1000 °C. However, it is not economical and convenient to mark gauge length for steel cable specimens compared with a high temperature paint sparkler.

In 1996, Lyons et al. [4] proposed that the thermal expansion tests of standard inconel specimens up to 650 °C was conducted by a combined dual element optical fiber lamp and digital camera. The normalized cross interaction function was developed to determine the value of thermal expansion of specimens. In 2010, Pan et al. [6] reported that the blue LED light and band-pass filter imaging system can overcome the decorrelation effect due to high bright light at elevated temperature. This kind of technology was used by CCDC to investigate the thermal expansion of Cr-Ni austenitic stainless steel up to 1200 °C. In 2012, Gales et al. [7] carried out a series of tensile tests on pre-stressing steel at an elevated temperature by a steady method and transient method, respectively. It indicates that the novel digital image correlation method can measure the strain of tendon at an elevated temperature. In 2015, Yang et al. [8] investigated the mechanical properties of Gh738 material at an elevated temperature using the digital image correlation method.

The CCDC is also used to investigate the mechanical response of welded joints, concrete members and structural members at ambient temperature and elevated temperature respectively [9–14].

In order to calibrate the CCDC system, which can be used to measure the steel cables with a complex shape till rupture at an elevated temperature, a different measurement method was compared with each other in the present study as there is no previous study that can be referenced directly for this proposal. Finally, the present study conducted the test to capture the full range of stress-strain curves at an elevated temperature, which can improve the accuracy of the fire resistance analysis for pre-tensioned steel structures in the civil engineering area.

2. Basic Principle of Digital Image Correlation Method

From then, now, the previous studies have contributed to update the algorithm of the digital image correlation method (DICM), which is the core technology of the CCDC system, and improved the accuracy of the measurement by CCDC system. Hereby, the basic principle of the CCDC system was introduced as below.

Digital image correlation method is an optical measurement based on digital image processing and numerical calculation. According to the basic principle of DIMC, speckle pattern is taken as the information resource of the deformation for an object [15,16]. Figure 1a displays the initial speckle pattern and Figure 1b shows the speckle pattern after deformation. Comparison of Figure 1a,b, the geometric coordination of a moving point, P, can be traced by using the digital image correlation method.

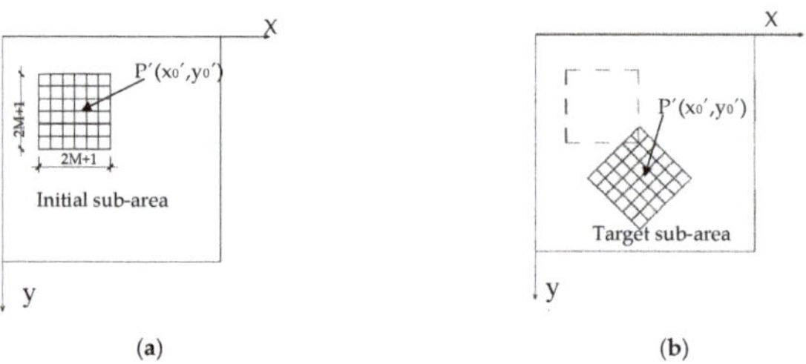

Figure 1. Displacement information for the digital image correlation method. (**a**) Initial information; (**b**) Target information.

Based on the digital image correlation method, shown in Figure 2, a charge-coupled device camera has been used by the CCDC system to capture a series of speckle patterns and transferred to a data collector. In the meantime, the computer software, which developed on the base of the digital image correlation method, analyzed the database efficiently, and output the transient strain of a specimen. The measurement range of the non-contact video gauge is up to a strain of 100%, which can capture the full range of strain till specimens fails. The high quality video also can be recorded to review the test process once more. The practice of the non-contact video system is shown as Figure 3.

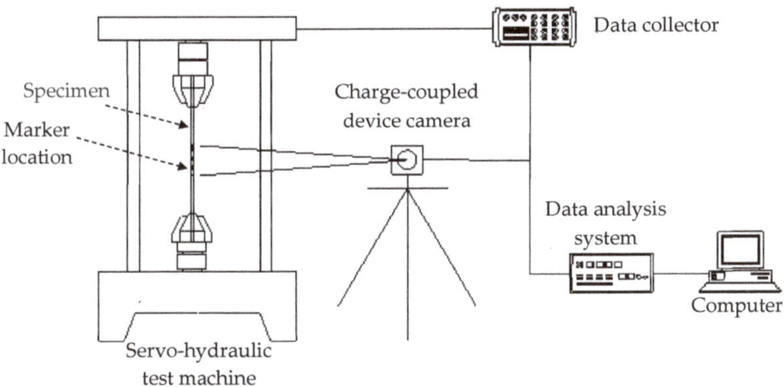

Figure 2. Charge-coupled device camera system.

Figure 3. Standard tensile test set-up for S355 structural steel at ambient temperature.

In present study, the CCDC pictured with 2452 pixels × 2056 pixels and a constant aperture of 1:1.8. The object distance of the special lens was 174 mm for material tests, and the magnification ratio of the lens was 0.338. The maximum frequency of data collection was 15 Hz. The resolution ratio of displacement was 0.012 μm and that of the strain was 5 με.

3. Calibration Tensile Test at Ambient Temperature

3.1. Uniaxial Tensile Tests on S355 Structural Steel

Both of the strain gauge and CCDC were used to investigate the stress-strain relationship of S355 structural steel. The dimension of the specimen for a standard uniaxial tensile test is shown as

Figure 4. The standard uniaxial tensile tests were conducted on a servo-hydraulic testing machine, with a maximum stroke displacement of 75 mm and a capacity of 500 kN, shown in Figure 5.

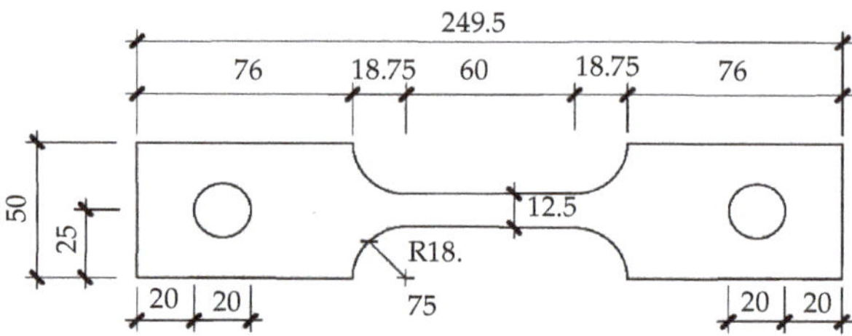

Figure 4. Dimension of steel specimens for standard tensile test (unit = mm).

Figure 5. Servo–hydraulic testing machine.

There are two groups of specimen measuring by the strain gauge and CCDC system, respectively. For CCDC system measuring, the first step is to sprinkle black paint on the surface of each specimen to reduce the metallic brightness, and then, markers with white spots painting on a black color background were prepared within the middle part of a specimen surface of about 30 mm length. On the other side of the same specimen, a strain gauge with the measuring range of 100,000 $\mu\varepsilon$ was pasted. The specimen was lightly tensioned to about 1 kN to eliminate any mechanical relaxation. Then, the tensile specimen was tested until failure at a straining rate of 0.0005 /s, in keeping with the strain rates set out in the EN 6892-1 [17]. As shown in Figure 6, the S355 steel specimen experienced a ductile model with necking phase. The broken cross section in the middle of length formed a cone-shape with cross section area gradually reduced towards the broken surface.

Figure 6. View of tensile failure of S335 steel at ambient temperature.

The stress-strain curves were obtained by a tensile test with strain gauge and CCDC system, respectively, as shown in Figure 7a. The elastic modulus can be determined by the tangent modulus of the initial linear phase of the stress-strain relationship. The effective yield strength is the intersection point of the stress-strain curve and the proportional line offset by 0.15% strain. The ultimate strength is the maximum stress in the stress-strain curve, and the fracture strain is dependent on the specimen broken. The mechanical properties of S355 steel are listed in Table 1.

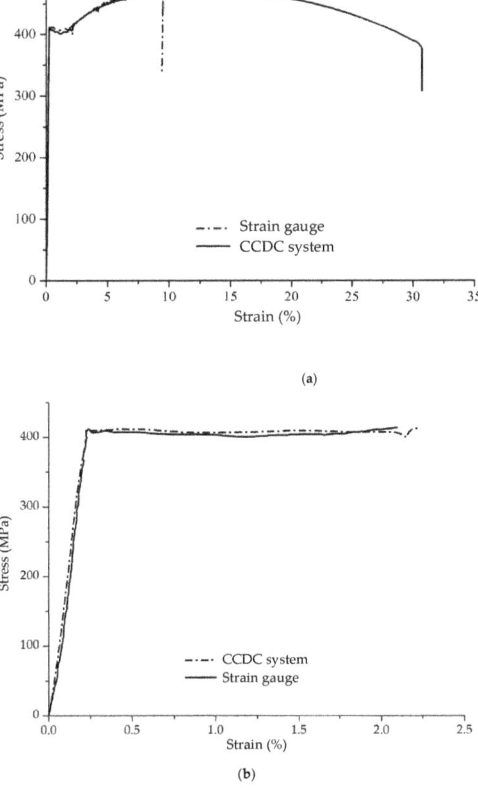

Figure 7. Comparison of stress-strain curves of S355 steel measured by different methods at ambient temperature. (**a**) Full range of stress-strain curve. (**b**) Partial range of stress-strain curve.

Table 1. Mechanical properties of S355 steel at ambient temperature.

Mechanical Properties	Measurement		Deviation
	Strain Gage	CCDC	
Elastic Modulus (E/GPa)	201.5	204.2	1.34%
Yield Strength (f_y/MPa)	400.6	406.8	1.55%
Ultimate Strength (f_u/MPa)	472.6	472.6	0%
Fracture Strain (ϵ_u)	0.095	0.308	225%

Test results indicate that the ultimate strength obtained by different measurements was the same. As shown in Figure 7b, the elastic modulus and yield strength, which obtained by strain gauge were smaller than those obtained by CCDC, and the deviation was 1.34% and 1.55%, respectively, as listed in Table 1. The test process displayed that the tensile specimen also experienced a large strain from the ultimate strength till fracture. As the limited of measuring range, the strain gauge can only capture a partial range of the stress-strain curve before ultimate strength. Comparing the strain gauge, the CCDC system can obtain a full range of stress-strain curve until broken as shown in Figure 7a.

Therefore, it is more efficient to use the CCDC system than strain gauge to measure the specimens with large deformation through the test process.

3.2. Uniaxial Tensile Tests on Steel Cable

The dimension of a steel cable specimen with ultimate strength of 1860 MPa is shown in Figure 8. The nominal diameter is 15.2 mm and cross section area is 139 mm^2. The process of marker painted is the same as mentioned in Section 3.1. Shown in Figure 9, the CCD camera was settled right ahead of the specimen in the distance of 1.5 m to form a measure plan and supplement brightness using a white lamp. Shown in Figure 10, an extensometer with the gauge length of 50 mm was settled in the mid-length of the same specimen. A couple of strain gauge was pasted on the surface of the mid-length. Thus, the stress-strain relationship of the steel cable was measured by three methods, i.e., strain gauge, extensometer and CCDC system, at ambient temperature. The standard tensile test was carried out by a servo-hydraulic test machine with a maximum stroke displacement of 75 mm and a capacity of 500 kN. There were four specimens repeatedly tested in a tensile test to get the average stress-strain curve as shown in Figure 10.

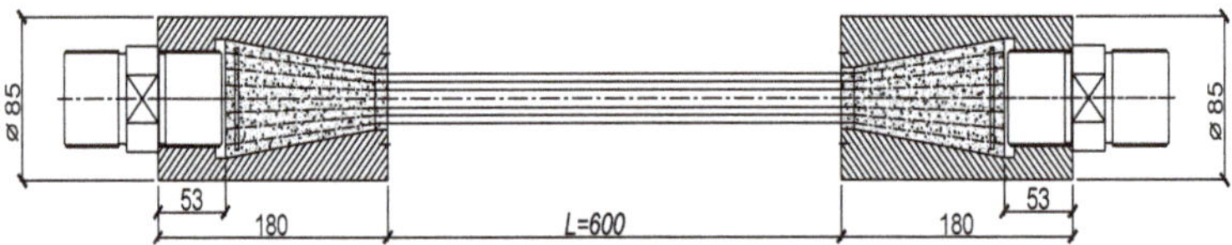

Figure 8. Dimension of a pre-stressed steel cable (Unit = mm).

Figure 9. Tensile test set-up for pre-stressed steel cables at ambient temperature.

Figure 10. Strain gauge and extensometer installation.

Shown in Figure 11, the stress-strain curves were obtained by different measuring methods. The effective yield strength, elastic modulus, ultimate strength and rupture strength were derived from each stress-strain curve with the same method as mentioned in Section 3.1. The values of the mechanical properties of steel cables obtained by different measuring methods are listed in Tables 2–4. There is a blank in Table 2 as a strain gauge is difficult to paste on the surface of a wire with 5 mm diameter, one of strain gauges split away off the surface of a wire and induced a blank in Table 2. As the arm of an extensometer cannot grab the steel cable tightly enough, there was the slipping displacement between the surface of the arm and steel cable. Thus, there are a few of blanks in Table 3. Comparison of the values of elastic modulus and effective yield strength, which were obtained by an extensometer and CCDC system, showed that they matched well with only 1.13% and 0.6% deviation, respectively. However, the values of elastic modulus and effective yield strength measured by a strain gauge were larger than those obtained by the extensometer and CCDC system. For an elastic modulus obtained by strain, the deviation was 2.83% and 3.99% compared to that by an extensometer and CCDC system. For the effective yield strength, the deviation was 16.6% and 17.1%, respectively.

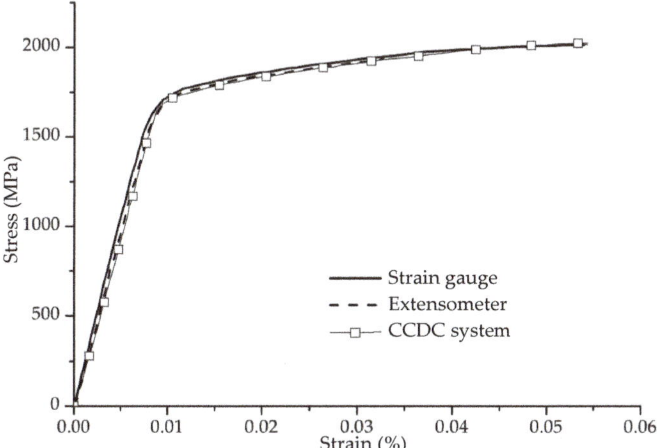

Figure 11. Comparison of stress-strain curves of pre-stressed steel cables measuring by different measurement methods.

As the strain gauge was pasted along twist wires, the strain measured by a strain gauge is off the longitudinal axis, which is smaller than the axis strain in the longitudinal direction. Thus, the smaller strain would induce the larger elastic modulus and yield strength. It also indicated that strain gauge cannot be suitable to measure the stress-strain curve of steel cables, which were twisted by a group of wires.

As listed in Table 3, the extensometer cannot captured the ultimate strength and rupture strain as its arm must be removed before the ultimate strength due to limited travel distance. Thus, an extensometer cannot capture the full range of stress-strain curves in a material properties test, especially for steel cables. The rupture strain of steel cables measured by the CCDC system was smaller than that by a strain gauge with 2.55% deviation. It also indicated that the strain gauge could only obtain stress-strain curve of one wire in a steel cable. Thus, the CCDC system was more efficient and accurate to measure the mechanical properties of steel cables than other methods.

Table 2. Mechanical properties of pre-tension steel cable obtained by strain gauge.

Mechanical Properties	Specimen 1	Specimen 2	Specimen 3	Specimen 4	Average Value
Elastic Modulus (E/GPa)	204.6	214.7	198.7	199.4	200.9
Yield Strength (f_y/MPa)	1778	1751	1729	1724	1746
Ultimate Strength (f_u/MPa)	2017.2	2008.2	2022.2	2015.9	2015.9
Fracture Strain (ϵ_u)	-	0.045	0.056	0.051	0.051

Table 3. Mechanical properties of pre-tension steel cables obtained by extensometer.

Mechanical Properties	Specimen 1	Specimen 2	Specimen 3	Specimen 4	Average Value
Elastic Modulus (E/GPa)	196.1	195.5	194.8	195.1	195.4
Yield Strength (f_y/MPa)	1732	1729	1725	-	1729
Ultimate Strength (f_u/MPa)	-	-	-	-	-
Fracture Strain (ϵ_u)	-	-	-	-	-

Table 4. Mechanical properties of pre-tension steel cables obtained by CCDC system.

Mechanical Properties	Specimen 1	Specimen 2	Specimen 3	Specimen 4	Average Value
Elastic Modulus (E/GPa)	195.1	191.2	193.7	192.9	193.2
Yield Strength (f_y/MPa)	1729	1725	1710	1709	1718.
Ultimate Strength (f_u/MPa)	2017.2	2008.2	2014.3	2015.9	2015.9
Fracture Strain (ϵ_u)	0.0558	0.0548	0.0462	0.0527	0.0524

4. Calibration Tensile Test at Elevated Temperature

4.1. Uniaxial Tensile Tests on S690 Structural Steel

The dimension of a S690 structural steel specimen for a standard uniaxial tensile test at an elevated temperature is the same as shown in Figure 4. The standard uniaxial tensile tests at an elevated temperature were also conducted on a servo-hydraulic testing machine, with a maximum stroke displacement of 75 mm and a capacity of 500 kN as shown in Figure 12. A specimen of S690 high-strength steel was installed and heated by a tubular high-temperature furnace as shown in Figure 13. The heat part was a split-tube furnace with a three-zone configuration and a side view window. A type K thermocouple was mounted at the centre of each zone to measure the heating temperature. The specimen was heated to 400 °C at the heat rate of 10 °C/min and subject to tension [18]. Both the extensometer and CCDC system were used to measure the strain in the specimen under the applied tension force.

Figure 12. Uniaxial tensile test set-up of S690 high strength steel under the high temperature.

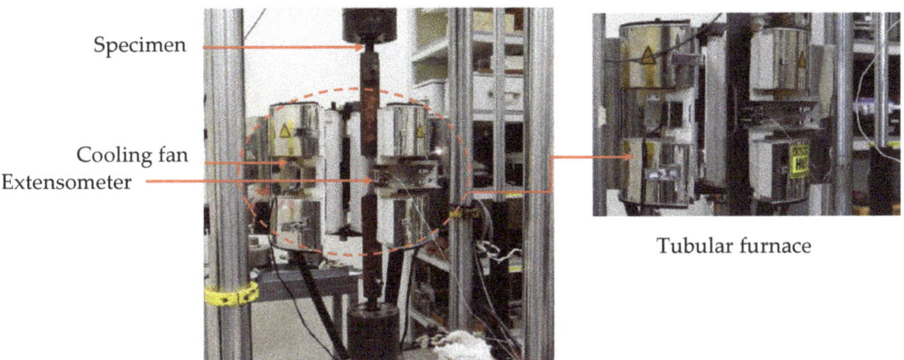

Figure 13. Tubular furnace and extensometer.

As shown in Figure 13, the arms of the extensometer are ceramic rods with quartz chisel ends, which passed through the side entry port and contacted the test coupons. It is specified for a gauge length of 25 mm and could travel 2.5 mm. Thus, the maximum measurable strain was 10%. During testing, the displacements of the test coupons were directly detected and transmitted by the rods to the strain gauges placed within the extensometer body. A cooling fan was mounted close to the extensometer to maintain the temperature of the extensometer lower than 150 °C. There were four specimens repeatedly tested in a tensile test to measure the average high temperature stress-strain curve. Two specimens measured by an extensometer, and the other by the CCDC system.

The same as for the marker paint process in Section 3.1. For the high temperature test, the high temperature paint was used to make markers on the surface of test specimens.

Both the tests measured by the extensometer and CCDC system were conducted on the S690 steel. The target temperature was up to 400 °C at the heating rate of 10 °C/min and held for 30 min to obtain uniform temperature distributions in both the furnace and the test specimens. The strain rate was 0.003 /min [18]. For the CCDC system, the white light lamp was used to supply the brightness of the specimen, which was mounted in the furnace. The function of the exposure compensation was used to help the CCD camera eliminate the extra exposure due to a high temperature.

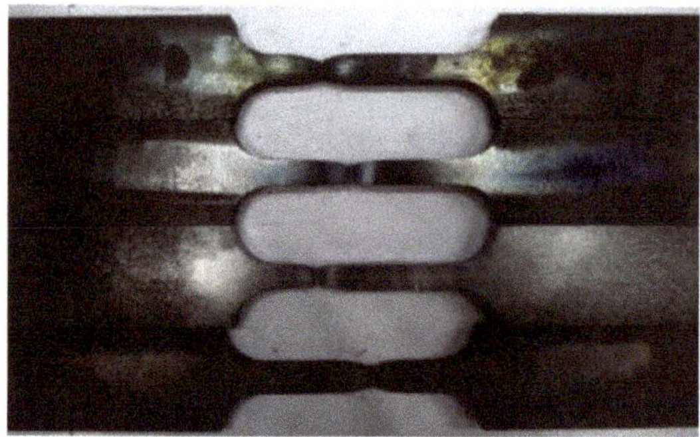

Figure 14. Post-high temperature tensile test high strength steel specimens.

The tests observed that the high strength steel S690 experienced a necking at 400 °C. As shown in Figure 14, the broken cross section area gradually reduced towards the broken surface.

The stress-strain curves of S690 at 400 °C was shown in Figure 15, and the mechanical properties was listed in Table 5.

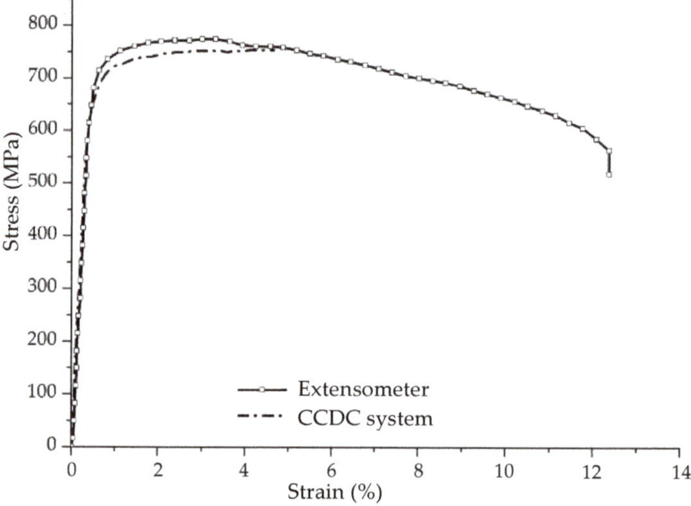

Figure 15. Stress-strain curves of high strength steel at 400 °C obtained by different measurement methods.

Table 5. Mechanical properties of S690 high strength steel obtained by different measurement methods at 400 °C.

Mechanical Properties	Extensometer			CCDC System		
	Specimen 1	Specimen 2	Average Value	Specimen 1	Specimen 2	Average Value
Elastic Modulus (E/GPa)	181.6	175.9	178.7	170.7	175.4	173
Yield Strength (f_y/MPa)	775	734	755	786	751	768.5
Ultimate Strength (f_u/MPa)	767.4	782.6	775	803.4	754.3	778.8
Fracture Strain (ϵ_u)	-	-	-	-	0.1240	0.1240

The mechanical properties listed in Table 5, i.e., elastic modulus, yield strength and ultimate strength of S690 steel at an elevated temperature obtained by the CCDC system was slightly larger than those by the extensometer. The deviation between them was 3.3%, 1.8% and 0.49%, respectively.

Since the ceramic arm of the extensometer has to be removed away from the test specimen before the rupture of the test specimen due to its limited travel distance as mentioned above; the extensometer reading can only cover the partial range of the stress-strain curve at an elevated temperature. The CCDC system can obtain the full range of the curve. A good fit between the extensometer and CCDC readings is generally observed as shown in Figure 15 and listed in Table 5. It is worth noting that the S690 steel experienced a long strain history after the ultimate strength at an elevated temperature. This ductility behavior of the S690 high strength steel at an elevated temperature is important to evaluate the mechanical response of structural members exposed to fire.

The test results showed that both the CCDC system and extensometer could capture the stress-strain curves accurately. However, the extensometer could only capture the partial range of the stress-strain curve due to its limited travel distance. The CCDC system was more efficient than the extensometer to obtain the full range of stress-strain curves, which is a benefit for analysis on the mechanical response of structural members exposed to fire.

4.2. Uniaxial Tensile Tests on 1860 MPa Pre-Tensioned Steel Cable

As the calibration tests have conducted as above, the CCDC system was used to measure the strain in the steel cables at different temperatures. The uniaxial tensile test process is the same as those illustrated in the Section 3.1. Three specimens were repeatedly tested at each temperature and the average of stress-strain relationships was obtained by the CCDC system as shown in Figure 16. Comparison of test curves and those resulted by EN1992-1-2 (2005) [19] indicated that the strain after the ultimate strength was ignored by EN1992-1-2 at an elevated temperature. The full range of stress-strain curve showed that steel cable specimens experienced a large strain after the ultimate strength, which would induce the large deformation of a pre-tensioned steel cable at an elevated temperature. Thus, the used of CCDC system could measure the stress-strain curves accurately and efficiently.

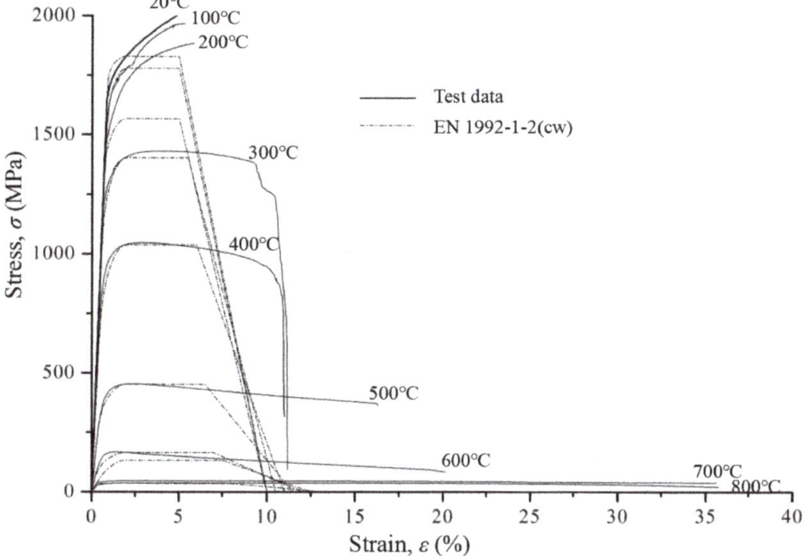

Figure 16. Stress-strain relationships of high tensile strength steel cables at elevated temperature.

5. Cautions

As the CCDC system is an optical measurement technology based on the digital image correlation method, the speckles distribution, random vibration of servo-hydraulic testing machine can induce the deviation of test data in a limit range. The use of a high-temperature filter, exposure compensation system and rigid tripod can decrease the deviation. Meanwhile, the data collection frequent should be high enough to get more numbers of test data in each group. The more high data collection frequent reached, the more accurately the average value can be gained to improve the smoothness of data curves.

6. Conclusions

This paper investigated a series of calibrate standard uniaxial tension test at ambient temperature and an elevated temperature, respectively. The following conclusions can be drawn:

- At ambient temperature, the data measured with the strain gauge and CCDC system fit well before the ultimate strength in a uniaxial tensile test of structural steel. The strain gauge could not capture the full range of stress-strain curves, but the CCDC system could.
- Compared to the strain gauge and extensometer method, the CCDC system not only could measure the test data accurately as other measurement methods, but also was able to capture the full range of the stress-strain relationship curves for steel cables with a complex shape.
- The use of high temperature paint could make speckle to form on the gauge length, which could be captured by a CCD camera. The high temperature filter also could improve the efficient workability of the CCDC system for mechanical properties tested at an elevated temperature.

Author Contributions: Conceptualization, Funding acquisition, Investigation, Project administration, Resources, Supervision, Writing—review & editing, Y.D.; Data curation, Investigation, Validation, Visualization, Writing—original draft, Z.-m.G.

Funding: The Civil Engineering Disaster Prevention National Key Laboratory in China (SLDRCE14-05), the National Natural Science Foundation of China (51878348).

Acknowledgments: The authors gratefully acknowledge the International Structural Fire Research Laboratory (ISFRL) in Nanjing Tech University, China.

Conflicts of Interest: The authors declare no conflict of interest. The founding sponsors had no role in the design of the study; in the collection, analyses, or interpretation of data; in the writing of the manuscript, and in the decision to publish the results.

References

1. Xiong, M.X.; Liew, J.R. Mechanical properties of heat-treated high tensile structural steel at elevated temperature. *Thin-Walled Struct.* **2016**, *98 Pt A*, 169–176. [CrossRef]
2. Mellegard, K.D.; Pfeifle, T.W.; Fossum, A.F.; Senseny, P.E. Pressure and flexible membrane effects on direct-contact extensometer measurements in axisymmetric compression tests. *J. Test. Eval.* **1993**, *21*, 530–538.
3. Winkler, J.; Fischer, G.; Georgakis, C.T. Measurement of local deformations in steel monostrands using digital image correlation. *J. Bridge. Eng.* **2014**, *19*, 04014042. [CrossRef]
4. Lyons, J.S.; Liu, J.; Sutton, M.A. Sutton High-temperature deformation measurements using digital-image correlation. *Exp. Mech.* **1996**, *36*, 64–70. [CrossRef]
5. Hu, Y.; Wang, Y.; Chen, J.; Zhu, J. A new method of creating high-temperature speckle patterns and its application in the determination of the high-temperature mechanical properties of metals. *Exp. Tech.* **2018**, *42*, 523–532. [CrossRef]
6. Pan, B.; Wu, D.; Wang, Z.; Xia, Y. High-temperature digital image correlation method for full-field deformation measurement at 1200 C. *Meas. Sci. Technol.* **2010**, *22*, 015701. [CrossRef]
7. Gales, J.A.; Bisby, L.A.; Stratford, T. New parameters to describe high-temperature deformation of prestressing steel determined using digital image correlation. *Struct. Eng. Int.* **2012**, *22*, 476–486. [CrossRef]

8. Yang, Y.; Li, X.; Xiao, R.; Zhang, H. Digital image correlation and complex biaxial loading tests on thermal environment as a method to determine the mechanical properties of Gh738 using warm hydroforming. *High. Temp. Mater. Process.* **2015**, *19*, 47–69. [CrossRef]

9. Konopík, P.; Džugan, J.; Rund, M.; Procházka, R. Determination of local mechanical properties of metal components by hot micro-tensile test. In Proceedings of the 25th Anniversary International Conference on Metallurgy and Materials, METAL 2016, Brno, Czech Republic, 25–27 May 2016.

10. Le, D.B.; Tran, S.D.; Dao, V.T.; Torero, J. Deformation capturing of concrete structures at elevated temperatures. *Procedia Eng.* **2017**, *210*, 613–621.

11. Barba, D.; Reed, R.; Alabort, E. Ultrafast miniaturised assessment of high-temperature creep properties of metals. *Mater. Lett.* **2019**, *240*, 287–290. [CrossRef]

12. Sakanashi, Y.; Gungor, S.; Forsey, A.; Bouchard, P. Measurement of creep deformation across welds in 316H stainless steel using digital image correlation. *Exp. Mech.* **2017**, *57*, 231–244. [CrossRef]

13. Chen, Y.; Lei, Z.; Bai, R.; Wei, Y.; Tao, W. Study on Elastoplastic Crack Propagation Behavior of Laser-welded 6061 Aluminum Alloy Using Digital Image Correlation Method. *IOP Conf. Ser. Mater. Sci. Eng.* **2017**, *280*, 012040. [CrossRef]

14. Bai, R.; Wei, Y.; Lei, Z.; Jiang, H.; Tao, W.; Yan, C.; Li, X. Local zone-wise elastic-plastic constitutive parameters of Laser-welded aluminium alloy 6061 using digital image correlation. *Opt. Lasers Eng.* **2018**, *101*, 28–34. [CrossRef]

15. Pan, B.; Qian, K.; Xie, H.; Asundi, A. Two-dimensional digital image correlation for in-plane displacement and strain measurement: A review. *Meas. Sci. Technol.* **2009**, *20*, 062001. [CrossRef]

16. Hung, P.C.; Voloshin, A.S. In-plane strain measurement by digital image correlation. *J. Braz. Soc. Mech. Sci. Eng.* **2003**, *25*, 215–221. [CrossRef]

17. *EN 6892-1: Metallic Materials-Tensile Testing–Part 1: Method of Test at Room Temperature*; European Committee for Standardization: Brussels, Belgium, 2009.

18. *BS EN 10002-5: Tensile Testing of Metallic Materials – Part 5: Method of Test at Elevated Temperatures*; European Committee for Standardization: Brussels, Belgium, 1992.

19. *EN 1992-1-2, Design of Concrete Structures. Part 1. 2 General Rules—Structural Fire Design*; CEN: Brussels, Belgium, 2004.

 *applied sciences*

Article

New Four Points Initialization for Digital Image Correlation in Metal-Sheet Strain Measurements

Alejandro-Israel Barranco-Gutiérrez [1,*], José-Alfredo Padilla-Medina [2], Francisco J. Perez-Pinal [2], Juan Prado-Olivares [2], Saúl Martínez-Díaz [3] and Oscar-Octavio Gutiérrez-Frías [4]

[1] Cátedras CONACyT—TecNM-Instituto Tecnológico de Celaya, Celaya 38010, Mexico
[2] TecNM-Instituto Tecnológico de Celaya, Celaya 38010, Mexico; alfredo.padilla@itcelaya.edu.mx (J.-A.P.-M.); francisco.perez@itcelaya.edu.mx (F.J.P.-P.); juan.prado@itcelaya.edu.mx (J.P.-O.)
[3] TecNM-Instituto Tecnológico de la Paz, La Paz 28080, Mexico; saulmd@itlp.edu.mx
[4] UPIITA-Instituto Politécnico Nacional, Ciudad de México 07340, Mexico; ogutierrezf@ipn.mx
*   Correspondence: israel.barranco@itcelaya.edu.mx; Tel.: +52-612-183-7484

Received: 1 March 2019; Accepted: 14 April 2019; Published: 24 April 2019

Abstract: Nowadays, the deformation measurement in metal sheets is important for industries such as the automotive and aerospace industries during its mechanical stamping processes. In this sense, Digital Image Correlation (DIC) has become the most relevant measurement technique in the field of experimental mechanics. This is mainly due to its versatility and low-cost compared with other techniques. However, traditionally, DIC global image registration implemented in software, such as MATLAB 2018, did not find the complete perspective transformation needed successfully and with high precision, because those algorithms use an image registration of the type "afine" or "similarity", based on a 2D information. Therefore, in this paper, a DIC initialization method is presented to estimate the surface deformation of metal sheets used in the bodywork automotive industry. The method starts with the 3D points reconstruction from a stereoscopic digital camera system. Due to the problem complexity, it is first proposed that the user indicates four points, belonging to reference marks of a "Circle grid". Following this, an automatic search is performed among the nearby marks, as far as one desires to reconstruct it. After this, the local DIC is used to verify that those are the correct marks. The results show reliability by reason of the high coincidence of marks in experimental cases. We also consider that the quality of mark stamping, lighting, and the initial conditions also contribute to trustworthy effects.

Keywords: DIC; initial condition; image registration; strain measurement

1. Introduction

The quality control of mechanical stamping is a need in modern industries. Engineers are always looking to improve the mechanical work optimization. Specifically, in the mechanical industry of metal-sheets manufacture, it is necessary to know their mechanical properties to use them in the most suitable way [1,2]. In this industry, the material properties such as Young's modulus, Poisson's ratio, anisotropic plastic ratio parameters and others are required for improving the design and manufacture processes by finite element analyses. On the other hand, the most commonly used method for finding the material properties is a tensile test with a strain-gauge type extensometer. However, the results from the extensometer are not applicable to measure the strain at a local point and the onset of diffuse necking. In contrast, the Digital Image Correlation (DIC) method is a state-of-the-art technique that can be used for an accurate strain measurement of material properties [3]. DIC has become a very popular technique in mechanics, particularly for measurement materials deformation without contact. It is an economic and versatile method which also offers a big amount of experimental data [4]. As digital cameras technology constantly improves in terms of resolution, lenses and frames per second (FPS), this technique allows covering a wide range of scale in space and time. DIC in stereo systems even reach non-planar cases, very common in the industry, meaning the three-dimensional reconstruction and access to volumetric data about the material and structure behavior through time. Nonetheless, the use of DIC data to validate models in a quantitative way or to identify with precision several constitutive parameters, remains an open problem. One of the reasons for this is the complicate compromise between the measuring resolution and the large space for the measuring to be done. A second reason is the state of the frontiers. A third reason is that the measured displacements are not directly compared with common simulations. Of course, there are several efforts presenting promising results, as well as [5], a work analyzing the damage in a super pression aerostatic balloon (SPB) used by the Japanese Aerospace Exploration Agency (JAXA) during the insufflate prove. The expansion, when it is not uniform, generates a concentration of stress and makes the balloon explode. They proved a new method called the Simplified Digital Image Correlation Method (SiDIC), in a rubber balloon, to verify if the SPB could be measured. The SiDIC identified in a correct way the non-deformed region even though the deformation was not precise.

Below, a brief historical background of DIC is given. An open code 3D-DIC Toolbox for MATLAB, the one implemented algorithm for multiple cameras, inspired by biomechanics, was reported in [6]. They were verified using the Multiple Digital Image Correlation Method (MultiDIC) and low-cost hardware. In [7], a quick method to correlate digital images (3D-DIC) was proposed to implement real-time measuring. Two improvements came up: the development of an efficient algorithm of the inverse composition of Gauss-Newton (IC-GN) with parallel computing to avoid redundant calculation, and an efficient IC-GN algorithm which was used to reach a speed of 10 frames/second with a resolution of 5000 points per frame. For validation, the displacement field of a four-point bending beam was determined by the real-time 3D-DIC. The experimental results were verified in traditional Chinese medicine. In the publication of [8], a procedure was developed to generate a sequence of intermediate synthetic images for a gradual following of the transformation of the pixel's subset between the two external configurations. An adequate distortion function of image was defined in the whole image through adopting an algorithm based in characteristics and followed by a scheme of interpolation based in non-uniform rational B-spline (NURBS). This allowed a fast and trustworthy estimation of the initial deformation parameters for the refinement phase after the DIC analysis, and tests were made on aluminum plates. Document [9] proposed a method of deconvolution to recover the real fields of displacement and tension of their counterparts provided by local DIC. The proposed algorithm could be considered an extension of the Van Cittert deconvolution, based on the small tension assumption. The authors claimed an improvement on the fine details in displacement and tension maps, as well as in space resolution. The guidelines to evaluate the precision and resolution of the 2D analysis under the patronage of the Society of Experimental Mechanics were reported in [10].

On the other hand, algorithms to map relative displacements of the deformed material points in opposition to non-deformed material by using DIC were reported in [11,12]. The current possibilities for the DIC scale allows the study of deformation on different levels, from meters to the nanoscale [13,14], with the condition of recording it correctly, following a known pattern like a circle grid or square grid. Additionally, studies have looked at how different systems behave, like biological materials [15–18], metallic alloys [19–21], memory-shape alloys [22,23], porous metals [24–26], polymers [27] and polymeric foams [28]. A key step in the process of DIC image tracking is the definition of the initial assumption, for the non-linear optimization routine aimed at finding the parameters that describe the transformation of the subset of pixels. This initialization can be very challenging and possibly fail when it comes to pairs of highly deformed images, such as those obtained from two angled views of non-planar objects. For example, in Figure 1, it can be seen how the global DIC cannot exactly match the circle mark to measure the strain. This was performed with the MATLAB function called "imregcorr", which was reported in [29].

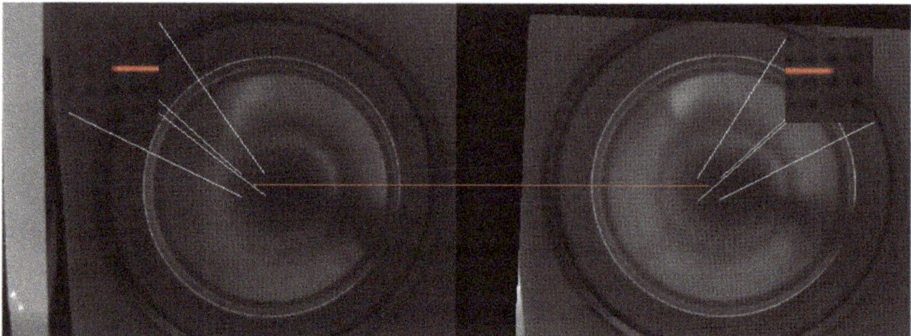

Figure 1. Global correlation with an error of image registration transform.

As can be observed from the preceding discussion, there still is the question of whether it is possible to improve the DIC accurateness. In brief, these previous works highlight the need for developing a systematic approach for achieving this goal, which to the author's awareness has not yet been stated. Therefore, in this paper we propose the use of a mark-search as an initial condition and of DIC as the system to verify the point-pairing. We propose the use of a mark-search as an initial condition and the use of DIC as the system to verify the point-pairing, because the number of mechanical part images are plentiful, the search for features to measure is very broad, as a result of which providing the initial conditions is necessary to fully automate the process gradually. The lack of a correct initial condition generates slowness and inaccurateness in current implementations of DIC. An appropriate initial condition is especially useful to know the material properties, specifically, its deformation curve and characterization zones such as: elastic, inelastic and rupture [30]. The proposed method in this work is applied to galvanized steel sheet, used in the outside bodywork of a pickup truck at 620 μm-thick and 96.5 μ hardness. Figure 2 shows the surface view of the material obtained from a metallographic microscope model:AX @ 200X, Company: Carl Zeiss AG, Jena, Turinga, Germany. It is necessary to mention that the experiments presented in this paper were all made for the automotive industry sector.

Figure 2. The surface view of a metal sheet from a metallographic microscope AX ZEISS @ 200X.

2. Theoretical Bases

The dimensional estimation using stereoscopic vision systems was described in detail by Tsai in [31], who quantitatively concluded that the accuracy and precision relied on the distance between the camera and the object to be measured, as well as the resolution of the cameras used. Stereoscopic vision is a technique frequently used to locate points in three dimensions (3D) based on points in two or more 2D images [6,32]. To achieve a stereoscopic point triangulation, it is necessary to calibrate both cameras. In this paper, we used the PinHole model, which is described by Equation (1) [33]:

$$s\widetilde{m} = A[R|t]\widetilde{M} = \begin{bmatrix} f_x & c & u_0 \\ 0 & f_y & v_0 \\ 0 & 0 & 1 \end{bmatrix} \begin{bmatrix} r_{11} & r_{12} & r_{13} & t_x \\ r_{21} & r_{22} & r_{23} & t_y \\ r_{31} & r_{32} & r_{33} & t_z \end{bmatrix} \widetilde{M} = \begin{bmatrix} m_{11} & m_{12} & m_{13} & m_{14} \\ m_{21} & m_{22} & m_{23} & m_{24} \\ m_{31} & m_{32} & m_{33} & m_{34} \end{bmatrix} \begin{bmatrix} x \\ y \\ z \\ 1 \end{bmatrix} \tag{1}$$

since s is the number that defines the scale of the objects with respect to their real size in the image, and $[R|t]$ is the matrix of extrinsic parameters of the camera that describe the rigid transformation (rotation and translation) between the coordinate system of the camera and the coordinate system of an object outside the camera. A is the matrix of intrinsic parameters that describes the position of the center of the image in pixels (u_0, v_0), and the ratio of the size of the pixel (f_x, f_y) has units expressed in $\frac{pixels}{meters}$ in the axes x and y regarding the focal distance between the entrance hole of the light and the matrix of the light sensors of the camera. Parameter c describes the asymmetry of the two axes of a pixel where a zero expresses an angle of 90 degrees. On the other hand, the entry $M = \widetilde{M} = \begin{bmatrix} x \\ y \\ z \\ 1 \end{bmatrix}$ is a 3D point (x, y, z) in homogeneous coordinates of the scene or object expressed in meters in its own coordinate system, and $\widetilde{m} = \begin{bmatrix} u \\ v \\ 1 \end{bmatrix}$ is the corresponding 2D point in the image expressed in pixels (u, v). Finally, the correction of the lens distortion is made with a polynomial based on the idea that the distortion changes as a circumference, as in Equations (2) and (3):

$$\breve{u} = u + (u - u_0)\left[k_1\left(x^2 + y^2\right) + k_2\left(x^2 + y^2\right)^2\right] \tag{2}$$

$$\breve{v} = v + (v - v_0)\left[k_1\left(x^2 + y^2\right) + k_2\left(x^2 + y^2\right)^2\right] \tag{3}$$

Subsequently, to achieve a stereoscopic calibration, the translation vector that joins each camera reference system is calculated [34,35]. The point position in three dimensions can be estimated from the coordinates in two dimensions and the Equation (4) [33,36–38]:

$$\begin{bmatrix} um_{31} - m_{11} & um_{32} - m_{12} & um_{33} - m_{13} \\ vm_{31} - m_{21} & vm_{32} - m_{22} & vm_{33} - m_{23} \\ u'm'_{31} - m'_{11} & u'm'_{32} - m'_{12} & u'm'_{33} - m'_{13} \\ v'm'_{31} - m'_{21} & v'm'_{32} - m'_{22} & v'm'_{33} - m'_{23} \end{bmatrix} \begin{bmatrix} \hat{x} \\ \hat{y} \\ \hat{z} \end{bmatrix} = \begin{bmatrix} m_{14} - um_{34} \\ m_{24} - vm_{34} \\ m'_{14} - u'm'_{34} \\ m'_{24} - v'm'_{34} \end{bmatrix} \tag{4}$$

where (u, v) and (u', v') are the coordinates of the paired points of the left and right cameras that correspond to the 3D point to be reconstructed. The scalars m_{ij} are obtained by multiplying the intrinsic and extrinsic parameters of the left camera $A[R|t]$; in an analogous way, the scalars m'_{ij} are obtained by multiplying the intrinsic and extrinsic parameters of the right chamber $A'[R'|t']$.

2.1. Normalized Cross Correlation (NCC)

Points matching, which consists of locating a point (photographed from different positions) in the left photograph and in the right image, is necessary to carry out the triangulation process. 2D-DIC is used to correlate a given set of points in the two stereo views of the reference configuration and match these points along the sequence of images. The correlated image points that are set are used to reconstruct and track the 3D position of the material points of the ROI (region of interest) through time [6]. In the literature, there are several works that report using 2D-DIC to accomplish this process, because this technique is so meaningful for 3D point reconstruction based on 2D images. The first advantage is that cross correlation is equally simple to calculate. Once a coincidence for a patch in a typical position inside an image is achieved, the Fourier methods can be used to calculate the correlation in a fast way. The second advantage is that cross correlation is independent from translations and invariant in the scaling dominance of the intensity. Equation (5) presents the definition of a normalized cross correlation:

$$NCC = \frac{\sum_{(i,j)\in S}\left[f(u,v) - \overline{f}\right]\left[g(u,v) - \overline{g}\right]}{\sqrt{\sum_{(i,j)\in S}\left[f(u,v) - \overline{f}\right]^2 \sum_{(i,j)\in S}\left[g(u,v) - \overline{g}\right]^2}} \tag{5}$$

Here, f and g are each the grayscale functions of the windows of the current image at a specific location (x, y). The functions $\overline{f}$ and $\overline{g}$ correspond to the gray scale mean of the reference image and the current subset.

2.2. Superficial Strain Estimation

To estimate the level of superficial strain, the engravings of speckle, circular or square marks on the sheet are used, which involves an electrochemical process in which an electrolyte is applied as a reagent on the surface of the sheet to engrave a thin pattern so that when the sheet is painted, the marking is imperceptible. The use of a circles pattern was chosen because it is considered a non-uniform mark deformation. In this proposed method, the facet sizes (window sizes) for the correlation function are defined by the distance between the circles' centroids. The advantages and disadvantages of a circle pattern in contrast with a speckle pattern are show in Table 1.

Table 1. The advantages and disadvantages of circle pattern versus speckle pattern.

Characteristics	Speckle Pattern	Circle Pattern
The normalized correlation matches well with the patterns.	YES	NO
Mark deformation is assumed as irregular.	NO	YES
A pixel can represent a mark.	Some cases	YES
Independence of the distance between the cameras and the specimen.	Some cases	YES

After the sheet is subjected to the sausage process, the circles are distorted in the form of ellipses and the distances between their centroids are modified; the latter are used in the digital images to determine the deformation states in the testing metal sheet. Figure 3a,b shows the changes of the circular grid with the deformation of the metal sheet. A simple way to estimate the deformation in the centroid of each ellipse is by averaging the distance with its four neighboring centroids $\hat{C}_{k-1,l}$, $\hat{C}_{k,l-1}$, $\hat{C}_{k+1,l}$ and $\hat{C}_{k,l+1}$, as expressed by Equation (6):

$$\Delta f_{k,l} = \frac{d\left(\hat{C}_{i,j}, \hat{C}_{i-1,l}\right) + d\left(\hat{C}_{i,j}, \hat{C}_{i,j-1}\right) + d\left(\hat{C}_{i,j}, \hat{C}_{i+1,j}\right) + d\left(\hat{C}_{i,j}, \hat{C}_{i,j+1}\right)}{4} \tag{6}$$

$$\text{where } d\left(\begin{bmatrix} x_1 \\ y_1 \\ z_1 \end{bmatrix}, \begin{bmatrix} x_2 \\ y_2 \\ z_2 \end{bmatrix}\right) = \sqrt{(x_1 - x_2)^2 + (y_1 - y_2)^2 + (z_1 - z_2)^2}.$$

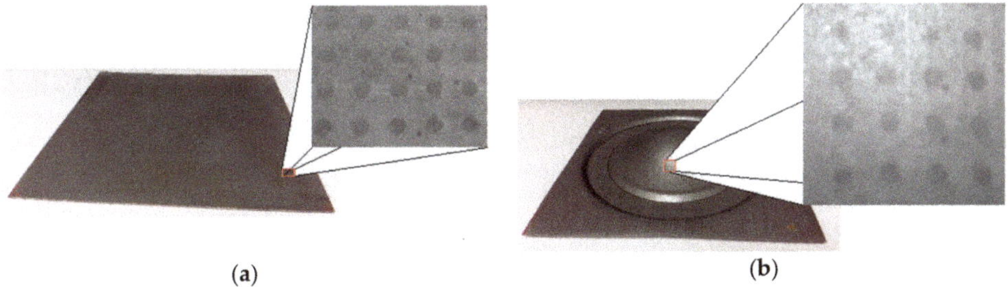

(a)   (b)

Figure 3. (**a**) Circles on metal sheet without deformation; and (**b**) circles on formed metal sheet.

3. Materials and Methods

3.1. Materials

To capture the images, a pair of Posilica GT 2750 (Allied vision, Exton, PA, USA) cameras with a sensitivity in the range of 400 nm to 670 nm as used. It has ethernet communication, a maximum capture rate of 33 fps, and resolution of 2750 to 2200 pixels at 230–250 lux. They were mounted on a metal structure that allows lighting control. Figure 4 shows the arrangement of the cameras, the light source and the metal test piece used to test the proposed method. LED lighting was chosen due to the contrast with the circles marked on the sheet: a blue LED of 640 nm, which coincides with the work [39]. To compare the dimensional measurements, a digital microscope (Jiusion 6-06814-24289-8) with a 100 μm scale was used.

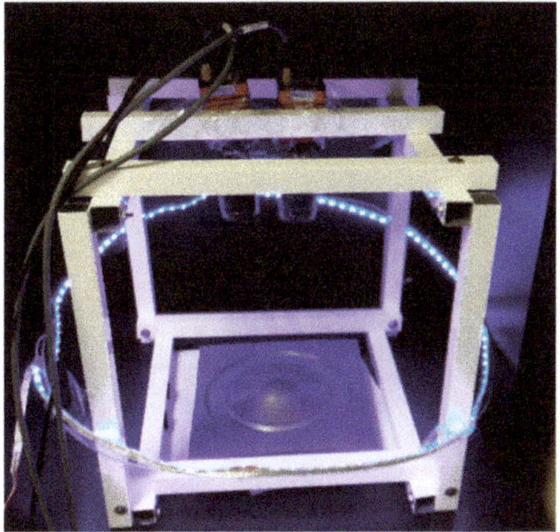

Figure 4. Setting of stereo cameras with illumination and metal sheet.

3.2. Method

The proposed method has the purpose of measuring the surface deformation in a metal sheet specimen used for truck exteriors, and it was divided into ten stages from (a) to (j). The peculiarity of this procedure consists of the incorporation of an initial condition based on points normalization to perform the matching using 2D-DIC. The general procedure is as follows:

(a) Stamping of known circle grid on the unformed metal sheet.
(b) Deformation of the metal-sheet through the mechanical stamping process.
(c) Calibration of cameras.
(d) Illumination of the piece with LED blue light for measuring.

(e)   Capture of stereo images.

(f)   Selection of four landmarks.

(g)   Search for neighbor's centroid.

(h)   Calculation of the NCC in the proposed neighborhood.

(i)   Triangulation of the points to obtain their position in 3D space.

(j)   Strain estimation from averaging the centroids' differences with the four neighbors using Equation (6).

The process includes the stamping of the known circle grid on the unformed metal sheet and the deformation of the sheet through the mechanical stamping process, because it is important to delimit the application of the proposed method.

The individual camera calibration can be performed with the "calib" function, while the stereoscopic calibration can be performed with the function "stereo_gui". Both are from the library "Camera Calibration Toolbox for Matlab" published by Jean-Yves Bouguet [40]. The illumination can be performed through a blue LED set. It is important that the image capture is done with the calibrated camera system.

3.3. Four-Points Initialization

The construction of the 3D centroid mesh $CM = C_{i,j}$ of the deformed circles pattern begins with the process of obtaining the binarized image, which is detailed in Appendix A. After this, two clicks given on two consecutive ellipses of the left binarized image and two on the same circles of the right binarized image (Figure 5), allows for a knowledge of the distance and slope between two centroids of pseudo-ellipses.

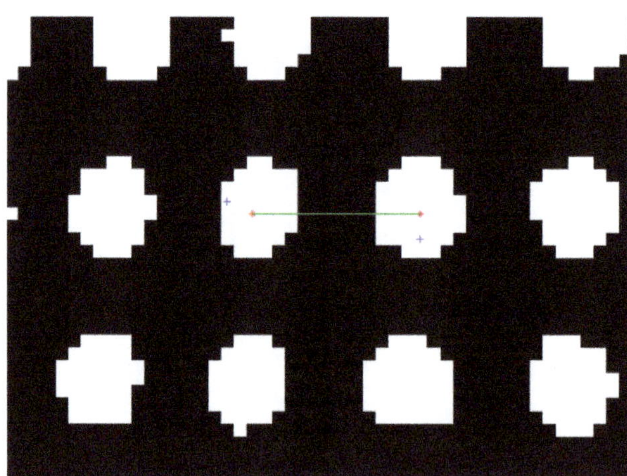

Figure 5. The user with blue (+) and the centroids of each blob in red (*).

The clicks do not necessarily have to be done on the centroids; it is only necessary that they are done on the two consecutives pseudo-ellipses. The selected points are referred to as $\left(x_{0,0}^{left},\ y_{0,0}^{left}\right)$ and $\left(x_{1,0}^{left},\ y_{1,0}^{left}\right)$, while in the right image the points are $\left(x_{0,0}^{right},\ y_{0,0}^{right}\right)$ and $\left(x_{1,0}^{right},\ y_{1,0}^{right}\right)$. To work with discrete values, the values of each point, in each image, are rounded up because the computational functions give fractions of the pixels where the mouse is clicked via the computer. From the rounded coordinates provided by the user, the blob number nb_i^s (considering a blob as a group of pixels with a connected connection equal to four) is retrieved to provide the coordinates of the centroid, also rounded; as an index of the tagged image (7).

$$nb_i^s = L^s\left(\widetilde{x}_{i,j}^s, \widetilde{y}_{i,j}^s\right) \tag{7}$$

where s can be the left or right side, and $L^s(x, y)$ is the labeled image as explained in Appendix A.

Next, the centroid of each blob $C_{i,j}^s$ is calculated using the statistical moments for 2D (8) with the help of Equations (9) and (10):

$$C_{i,j}^s = \left(xc_{i,j}^s, yc_{i,j}^s\right) = \left(\frac{m_{10}}{m_{00}}, \frac{m_{01}}{m_{00}}\right) \tag{8}$$

$$m_{pq} = \sum_{x=1}^{k}\sum_{y=1}^{l} x^p y^q BO^s(x, y) \tag{9}$$

$$BO^s(x, y) = \begin{cases} 1 & si \ L^s(x,y) = nb_i^s \\ 0 & otherwise \end{cases} \tag{10}$$

The distance and the slope between the two centroids in Equations (11) and (12) are calculated to initiate an iterative automatic process of estimation of the 2D position of blobs. The estimation of the next blob position from the information of the two clicks on each image $yc_{i+1,j}^s, yc_{i,j}^s, xc_{i+1,j}^s$ and $xc_{i,j}^s$:

$$\widetilde{d}_{g,h}^s = \sqrt{\left(yc_{i+1,j}^s - yc_{i,j}^s\right)^2 + \left(xc_{i+1,j}^s - xc_{i,j}^s\right)^2} \tag{11}$$

$$\widetilde{m}_{g,h}^s = \frac{\left(yc_{i+1,j}^s - yc_{i,j}^s\right)}{\left(xc_{i+1,j}^s - xc_{i,j}^s\right)} \tag{12}$$

where g and h are the index of the distances and slopes matrixes.

From Equations (11) and (12), the pseudo-row of marks is estimated using Equations (13) and (14):

$$\widetilde{xc}_{i+1,j}^s = xc_{i,j}^s + \widetilde{d}_{g,h}^s \cos\left(\widetilde{m}_{g,h}^s\right) \tag{13}$$

$$\widetilde{yc}_{i+1,j}^s = yc_{i,j}^s + \widetilde{d}_{g,h}^s \cos\left(\widetilde{m}_{g,h}^s\right) \tag{14}$$

Meanwhile, to find the next row, Equations (15) and (16) are used:

$$\widetilde{xc}_{i,j+1}^s = xc_{i,j}^s + \widetilde{d}_{g,h}^s \cos\left(\widetilde{m}_{t-1}^s + \pi/2\right) \tag{15}$$

$$\widetilde{yc}_{i,j+1}^s = xc_{i,j}^s + \widetilde{d}_{g,h}^s \cos\left(\widetilde{m}_{t-1}^s + \pi/2\right) \tag{16}$$

The normalized cross-correlation (5) is used to ensure a good pairing of centroids. This drastically decreases the number of operations performed for the matching between the interest points and at the same time takes advantage of the reliability granted by the NCC. The Matlab code can be downloaded from https://sites.google.com/itcelaya.edu.mx/dic/.

4. Results

The results of a typical initialization of the DIC are displayed in Figure 6 to show an initialization comparison. The working conditions were: one facet of 21×21, the search for the maximum normalized correlation, one-pixel step, and 100 marks that were pre-selected in the left image.

To observe the behavior of the normalized correlation internally, the correlation between a left image ellipse against all the possible 21×21 facets of the right image was plotted to qualitatively locate the maximum peak of the correlation (Figure 7).

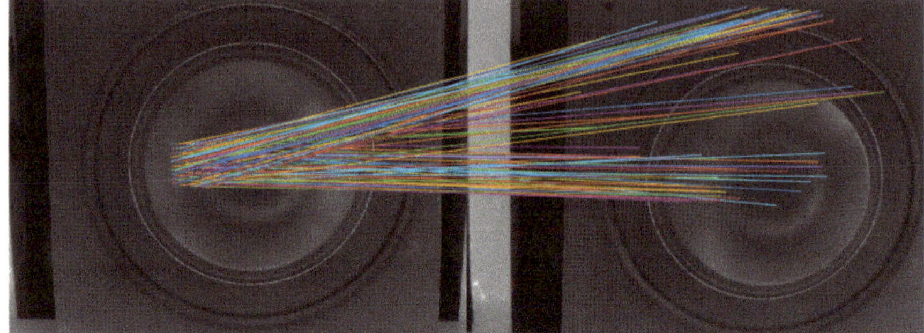

Figure 6. Pairing of points with typical initialization of the Digital Image Correlation.

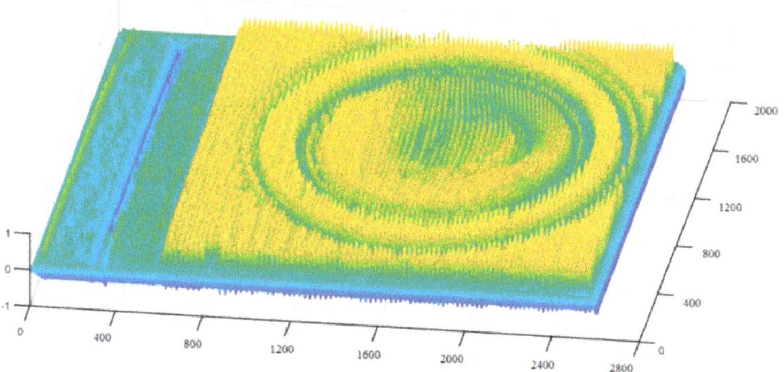

Figure 7. Correlation between a left image ellipse versus the overall possible 21 × 21 facets of the right image.

The engraving of light reference circles on the metal sheet resulted from an equidistance of 1.55 mm between the centroids when the metal sheet had not been deformed. The measurement made with a microscope is shown in Figure 8, where the before (Figure 8a) and after (Figure 8b) deformation is appreciated.

(a)   (b)

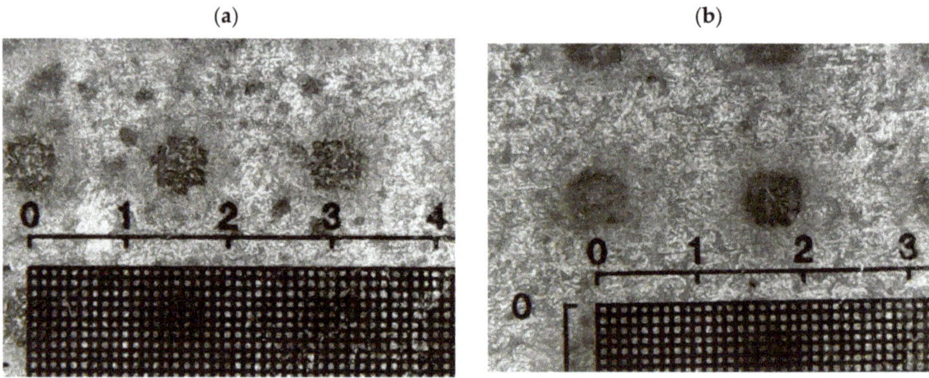

Figure 8. (**a**) Image captured before the forming mechanical process of the metal sheet; and (**b**) image captured after the forming mechanical process of the metal sheet.

On the other hand, the calibration parameters of the right and left cameras, as well as the translation vector between both cameras are the following:

Intrinsic parameters of the left camera:

Focal length: fc_left = [4139.77293 4125.08567] ± [262.92862 287.85915]

Principal point: cc_left = [1386.44233 1290.81589] ± [145.82339 108.75101]

Skew: alpha_c_left = [0.00000] ± [0.00000] => pixel angle = 90.00000 ± 0.00000 degrees

Distortion: kc_left = [−0.03826 −0.81032 −0.00495 −0.00231 0.00000] ± [0.15686 2.09125 0.00925 0.01010 0.00000]

Intrinsic parameters of the right camera:

Focal length: fc_right = [4009.97898 4233.47842] ± [271.16800 339.84438]

Principal point: cc_right = [1149.40996 962.27029] ± [198.63016 96.89436]

Skew: alpha_c_right = [0.00000] ± [0.00000] => pixel angle = 90.00000 ± 0.00000 degrees

Distortion: kc_right = [−0.12949 0.25429 0.00496 0.02934 0.00000] ± [0.13731 0.93124 0.00825 0.01424 0.00000]

Extrinsic parameters (position of the right chamber with respect to the one on the left):

Rotation (In Rodrigues format): om = [−0.00856 0.37660 0.00093] ± [0.03213 0.04287 0.00594]

Translation vector: T = [−85.14492 33.53035 24.13032] ± [2.97387 1.09323 11.19968]

The results of the image processing step are shown in Figure 9. In the top row, the pair of stereo images before the area filter is presented, while in the row below, the images after the area filter application are reported.

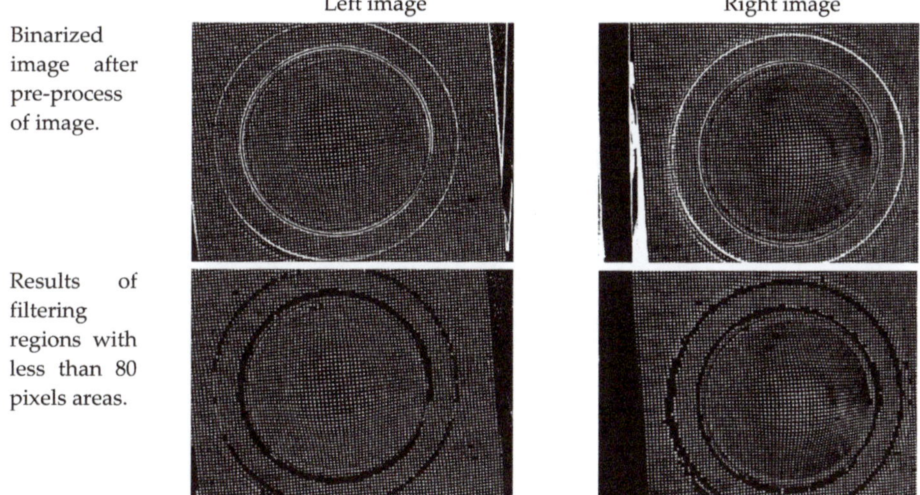

<table>
<tr><td></td><td style="text-align:center">Left image</td><td style="text-align:center">Right image</td></tr>
<tr><td>Binarized image after pre-process of image.</td><td></td><td></td></tr>
<tr><td>Results of filtering regions with less than 80 pixels areas.</td><td></td><td></td></tr>
</table>

Figure 9. Image segmentation to obtain referential ellipses for the deformation measurement.

We selected one hundred points to show the method's accuracy in a space of 100×100 marks. To verify if each point matching was right, the local DIC was calculated for the proposed points estimated in the initialization stage. The best correspondence was in the same position projected by the four-point method. Figure 10 shows a case of the correlation using a window of 31×31 pixels on a zone of 21×21 pixels. The MATLAB function used for this aim was "corr2".

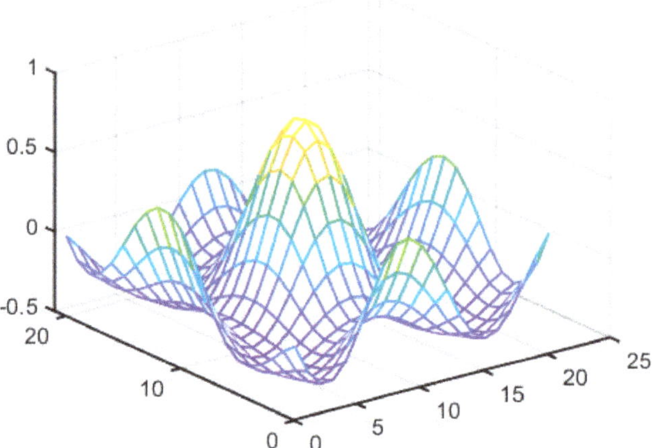

Figure 10. Correlation on one matching projected by a four-point initialization using a window of 31×31 pixels on a zone of 21×21 pixels.

Figure 11 shows the points matching of an area of 10×10 deformed circular marks. Those are the same taken to perform the 3D reconstruction. On one side, the detail of the points matching is shown with a zoom.

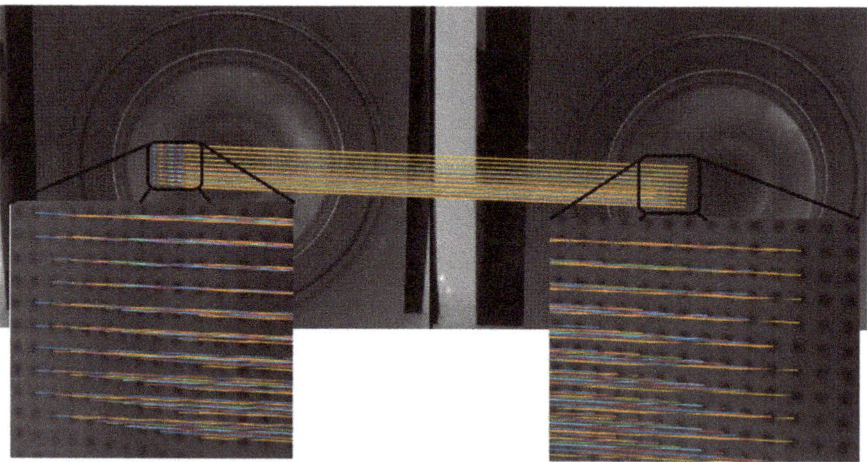

Figure 11. Points matching using the initial condition proposed and verified by Normalized Cross Correlation.

Figure 12 shows the three-dimensional reconstruction of the marks found and verified for the point matching stage. From the 3D reconstruction the strain was calculated using Equation (6), averaging the distances between the four neighboring marks of the mesh of circles of a radius of 0.5 mm and a separation between the centers of 2.55 mm. The strain estimated by the proposed method is shown in Figure 13.

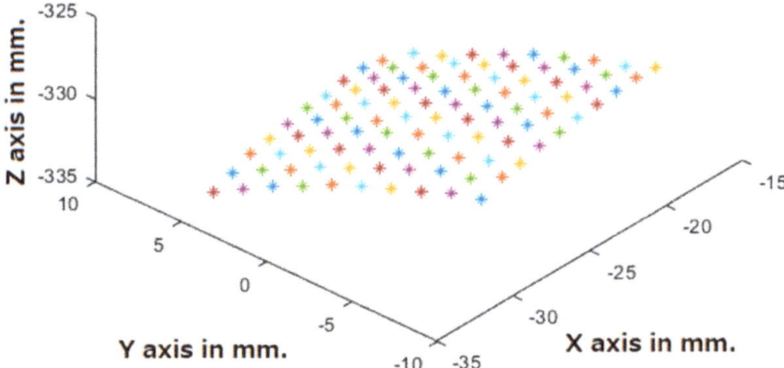

Figure 12. Three-dimensional reconstruction of marks on the metal-sheet surface from point matching.

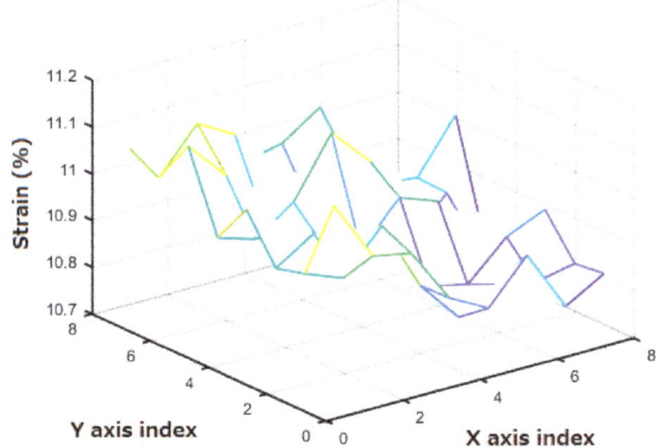

Figure 13. Estimated strain in the selected zone.

The strains estimated in points having their four neighbors from a matrix of 100×100 were reduced to a matrix of 8×8. The values in Tables 2 and 3 were obtained with the stereoscopic vision system. From these results, it can be calculated that the average error in the experiment was, in this case, $\epsilon = 1.290878\%$ (0.019995 mm.). This takes into account a distance between the centroids of the marks of 1.55 mm, without deformation.

Table 2. Estimated strain (%) using the proposed method.

Position Indexes	j = 1	j = 2	j = 3	j = 4	j = 5	j = 6	j = 7	j = 8
i = 1	10.4623	10.334	10.2988	10.1945	10.2066	10.2369	10.1932	10.1931
i = 2	10.2814	10.2721	10.3268	10.2013	10.1416	10.2001	10.1461	10.209
i = 3	10.2739	10.3202	10.2906	10.2571	10.1665	10.1584	10.2454	10.2757
i = 4	10.3615	10.2886	10.2542	10.2351	10.3094	10.2866	10.2113	10.217
i = 5	10.2692	10.282	10.2885	10.4412	10.3556	10.2272	10.2494	10.158
i = 6	10.4365	10.3514	10.2451	10.4047	10.361	10.2592	10.3007	10.3751
i = 7	10.3587	10.4371	10.3885	10.1398	10.215	10.2778	10.1877	10.3029
i = 8	10.3664	10.3276	10.3599	10.2999	10.314	10.3569	10.2394	10.172

Table 3. Estimated strain (%) using a digital microscope.

Position Indexes	$j = 1$	$j = 2$	$j = 3$	$j = 4$	$j = 5$	$j = 6$	$j = 7$	$j = 8$
$i = 1$	11.2903	9.6774	9.3548	9.6774	8.871	8.0645	8.0645	7.2581
$i = 2$	10.4839	8.871	7.2581	8.0645	8.0645	5.6452	4.8387	5.6452
$i = 3$	6.4516	8.0645	8.5484	7.7419	6.4516	6.4516	6.4516	6.4516
$i = 4$	8.0645	9.6774	9.6774	7.2581	6.4516	6.4516	5.6452	6.4516
$i = 5$	8.0645	10.1613	10.1613	5.6452	5.6452	6.4516	5.6452	7.2581
$i = 6$	11.2903	11.2903	8.0645	7.2581	8.0645	10.4839	8.0645	10.9677
$i = 7$	13.7097	10.4839	8.871	8.871	8.871	9.6774	8.871	12.9032
$i = 8$	9.6774	6.4516	7.2581	8.0645	10.4839	7.2581	6.4516	8.871

5. Discussion

The principal contribution of this work consists in the correct points-pairing between images (100%), in the specific case of almost-identical marks, to measure the deformation of a metal sheet with a small error. The method is semi-automatic due to the high complexity of the shown problem. This is evident when using the MATLAB function called "imregcorr" [41], and it is solved when the user indicates only two pairs of coinciding points. This works as an initial condition to begin a search of correspondent marks among the images. Basically, we noticed that the initial conditions of the search increased, in a notable way, the efficiency of the point-pairing system (with *efficiency* referring to the amount of paired marks, divided between the total of marks in the space of analysis). We assumed that the quality of the mark stamping, lighting, and the initial conditions, also contribute to trustworthy effects. According to the results, it can be observed that the correlation applied to the detection of deformed circular patterns (ellipses) is not ordered. This is because the objective of the construction of the pattern is to make the marks identical and that, in opposition to this, the correlation uses the differences between marks to emit a factor of similarity. The system was also compared with a manual deformation measurement using a microscope, which achieved a low sensitivity of the camera system and a displacement. This was due to the resolution of the camera system. In terms of processing time, it is well known that the correlation has a computational complexity of m^2n^2 for an nxn image and an mxm template. Meanwhile, the proposed method depends on the number of points on the analyzed sheet; which is many times less than the number of pixels of an image, and it also depends on the distance between the camera and the printed sheet that is to be studied.

6. Conclusions

The proposed methodology solves the corresponding points-pairing for cases where multiple marks tend to be identical. This is one of the most complex and difficult problems DIC experiences in stereoscopic vision. The philosophical basis where most of points-pairing algorithms reside is the search for differences between the objects inside images. This goes against the goal of the *Circle-grid* of obtaining identical circles. The proposal in this work takes advantage of the characteristics of the problem to reduce its fallibility and to reach very important results, as well as the knowledge of an equal-points grid and a superficial deformation of less than 50%; which is of interest. The results are compared to the predetermined MATLAB function to show the advantages of this work´s proposal, which may be of use with a high degree of trustworthiness, thanks to the large amount of accurate correspondences. The processing time for the search in the initial conditions is relatively low because only distances, slopes and cosine functions are calculated. This way of solving the problem could be used in systems requiring patterns or points-grids, such as squares, hexagons and others. As a future work, we suggest to automatize in a total way the points´ correspondence without losing the gained benefits.

Author Contributions: Conceptualization, A.-I.B.-G. and J.-A.P.-M.; formal analysis, J.P.-O.; methodology F.J.P.-P.; software, S.M.-D.; validation, O.-O.G.-F.

Funding: The authors wish to thank the Instituto Poliécnico Nacional for its support provided through the project SIP-20195901. In addition, the authors would like to express their gratitude to the COFAA for its financial support and to the CONACYT for Cátedra ID 1641, TecNM and PRODEP.

Acknowledgments: We would also like to thank Karla Cárcamo Pérez for her valuable editorial help.

Conflicts of Interest: The authors declare no conflict of interests. The funders had no role in the design of the study, in the collection, analyses, or interpretation of data; nor in the writing of the manuscript, or in the decision to publish the results.

Appendix A

In this section, an image process is made to binarize the ellipses engraved on the metal sheets. With this, the initial condition for 2D-DIC is found. The first step is to situate the threshold in the image by using the Otsu method; that way, absolute-white circular marks will be obtained. Immediately after this, a number must be assigned to the ellipses (blobs) so they can be indexed onto the process of labelling the connected components. By the end, these connected components are discharged from areas bigger than 80 pixels. Considering that the pair of images is represented in monochrome, with a size of $N \times M$ and a pixel resolution of b bits, the pair of images can be represented as:

$$I^s(x, y, z) \in \left\{ 0 \leq \mathbb{Z} \leq 2^b \right\} \forall \{1 \leq x \leq M\}, \{1 \leq y \leq N\} \tag{A1}$$

where s represents the right or the left camera.

We receive an image in grayscale. Following this, we make a complement of this image to enhance the importance of the circles, as indicated in Equation (A2), and to obtain fair marks:

$$C^s(x, y) = 2^b - I_g^s(x, y) \tag{A2}$$

where s refers to left or right image.

To obtain a wide image-operation range, these images are normalized, as indicated in Equation (A3):

$$NI^s(x, y) = 2^b \frac{C^s - min(C^s)}{max(C^s) - min(C^s)} \tag{A3}$$

A difficulty of the non-uniform-illumination is presented on the metal-sheet surface by the nature of the problem. To decrease it, an adaptive filter of average window (A3) is applied to the normalization process on the image, where the size of the window is three times the average of the circular mark's ratios, in pixels. The filter is defined by Equations (A4) and (A5):

$$SM^s(x, y) = \sum_{k=1}^{M} \sum_{l=1}^{N} NI^s(k, l) \frac{1}{(3r)^2} J^s_{3r,3r}(x - k, y - l) \tag{A4}$$

$$SI^s(x, y) = NI^s(x, y) - \alpha * SM^s(x, y) \tag{A5}$$

Afterwards, another normalization is applied, through Equation (A3) on Equation (A5), and a thresholding with the Otsu method is applied to obtain a binarized version $B^s(x, y)$ of $SI^s(x, y)$. In such a process, we receive as a result the marks in white and the other regions in black. The next step is to label the binarized image elements $O_i^s(x, y) = labeling(B^s(x, y))$. Here, every binarized segment of the image (blob) is independent: $O_i^s \cap O_j^s \in \varnothing \ \forall \ j \neq i$ for $i, j = 1, 2, \ldots, LN$. Nonetheless, it is common to have small non-representative stains in that set of blobs $O_i^s(x, y)$, and this is why an interconnected pixels selection is applied with areas less than 80 pixels (A6):

$$L_i^s(x, y) = \left\{ O_i^s(x, y) \ \forall \ area\left[O_i^s(x, y) \right] < a^s \right\} \tag{A6}$$

The overall previous process is shown in Figure A1 and dot matrix position indexes Figure A2.

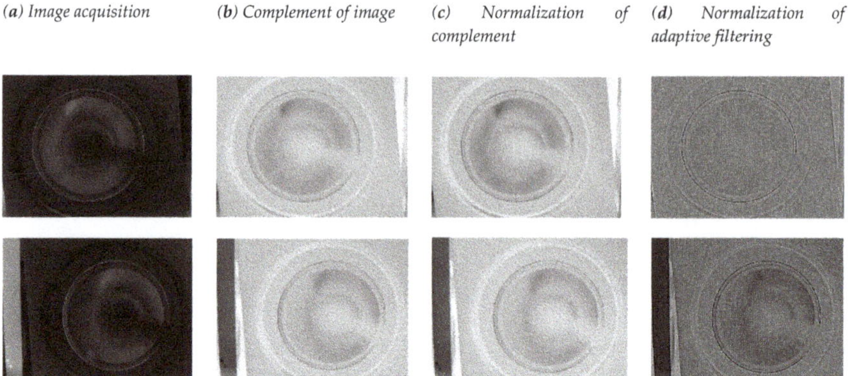

Figure A1. Pair of images of the stereoscopic system showing the result of each stage of the pre-processing of images. The upper row presents the left-side images, and the lower row presents those from the right-side.

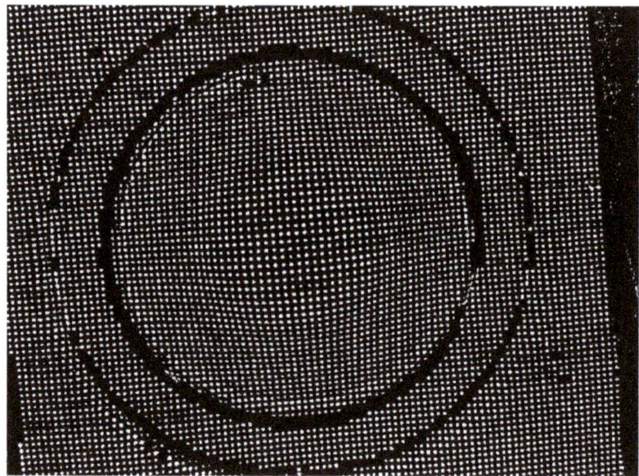

Figure A2. Binarized left-side photography to obtain a candidate's marks.

References

1. Choi, J.; Choi, B.; Heo, S.; Oh, Y.; Shin, S. Numerical modeling of the thermal deformation during stamping process of an automotive body part. *Appl. Therm. Eng.* **2018**, *118*, 159–172. [CrossRef]
2. Lee, H.H.; Yoon, J.I.; Kim, H.S. Single-roll angular-rolling: A new continuous severe plastic deformation process for metal sheets. *Scr. Mater.* **2018**, *146*, 204–207. [CrossRef]
3. Nguyen, V.T.; Kwon, S.J.; Kwon, O.H.; Kim, Y.S. Mechanical Properties Identification of Sheet Metals by 2D-Digital Image Correlation Method. *Procedia Eng.* **2017**, *184*, 381–389. [CrossRef]
4. Blaber, J.; Adair, B.; Antoniou, A. Ncorr: Open-Source 2D Digital Image Correlation Matlab Software. *Exp. Mech.* **2015**, *55*, 1105–1122. [CrossRef]
5. Kazuki, K.; Takuma, M.; Shuichi, A. Measurement of Super-Pressure Balloon Deformation with Simplified Digital Image Correlation. *Appl. Sci.* **2018**, *8*, 9. [CrossRef]
6. Solav, D.; Moerman, K.M.; Jaeger, A.M.; Genovese, K.; Herr, H.M. MultiDIC: An Open-Source Toolbox for Multi-View 3D Digital Image Correlation. *IEEE Access* **2018**, *16*, 30520–30535. [CrossRef]
7. Xinxing, S.; Xiangjun, D.; Chen, Z.; He, X. Real-time 3D digital image correlation method and its application in human pulse monitoring. *Appl. Opt.* **2016**, *55*, 696–704. [CrossRef]

8. Genovese, K.; Sorgente, D. A morphing-based scheme for large deformation analysis with stereo-DIC. *Opt. Lasers Eng.* **2018**, *104*, 159–172. [CrossRef]

9. Reu, P.L.; Toussaint, E.; Jones, E.; Bruck, H.A.; Iadicola, M.; Balcaen, R.; Turner, D.Z.; Siebert, T.; Lava, P.; Simonsen, M. DIC Challenge: Developing Images and Guidelines for Evaluating Accuracy and Resolution of 2D Analyses. *Exp. Mech.* **2017**, *58*, 1067–1099. [CrossRef]

10. Grédiac, M.; Blaysat, B.; Sur, F. A Robust-to-Noise Deconvolution Algorithm to Enhance Displacement and Strain Maps Obtained with Local DIC and LSA. *Exp. Mech.* **2018**, *1*, 1–25. [CrossRef]

11. Bruck, H.A.; McNeill, S.R.; Sutton, M.A.; Peters, W.H. Digital image correlation using Newton–Raphson method of partial differential correction. *Exp. Mech.* **1989**, *29*, 261–267. [CrossRef]

12. Cheng, P.; Sutton, M.A.; Schreier, H.W.; McNeill, S.R. Full-field speckle pattern image correlation with B-spline deformation function. *Exp. Mech.* **2002**, *42*, 344–352. [CrossRef]

13. Kammers, A.D.; Daly, S. Small-scale patterning methods for digital image correlation under scanning electron microscopy. *Meas. Sci. Technol.* **2011**, *22*, 125501. [CrossRef]

14. Rubino, V.; Lapusta, N.; Rosakis, A.; Leprince, S.; Avouac, J. Static laboratory earthquake measurements with the digital image correlation method. *Exp. Mech.* **2015**, *55*, 77–94. [CrossRef]

15. Dickinson, A.S.; Taylor, A.C.; Ozturk, H.; Browne, M. Experimental validation of a finite element model of the proximal femur using digital image correlation and a composite bone model. *J. Biomech. Eng.* **2011**, *133*, 014504. [CrossRef]

16. Zhang, D.; Eggleton, C.; Arola, D. Evaluating the mechanical behavior of arterial tissue using digital image correlation. *Exp. Mech.* **2002**, *42*, 409–416. [CrossRef]

17. Franck, C.; Maskarinec, S.A.; Tirrell, D.A.; Ravichandran, G. Three-dimensional traction force microscopy: A new tool for quantifying cell-matrix interactions. *PLoS ONE* **2011**, *6*, e17833. [CrossRef]

18. Wang, H.; Lai, W.; Antoniou, A.; Bastawros, A. Application of digital image correlation for multiscale biomechanics. In *Handbook of Imaging in Biological Mechanics*; CRC Press: Oxfords, UK, 2014; pp. 141–151.

19. Carroll, J.D.; Abuzaid, W.; Lambros, J.; Sehitoglu, H. High resolution digital image correlation measurements of strain accumulation in fatigue crack growth. *Int. J. Fatigue* **2013**, *57*, 140–150. [CrossRef]

20. Tong, W. Detection of plastic deformation patterns in a binary aluminum alloy. *Exp. Mech.* **1997**, *37*, 452–459. [CrossRef]

21. Rehrl, C.; Kleber, S.; Antretter, T.; Pippan, R. A methodology to study crystal plasticity inside a compression test sample based on image correlation and EBSD. *Mater. Charact.* **2011**, *62*, 793–800. [CrossRef]

22. Daly, S.; Ravichandran, G.; Bhattacharya, K. Stress-induced martensitic phase transformation in thin sheets of Nitinol. *Acta Mater.* **2007**, *55*, 3593–3600. [CrossRef]

23. Reedlunn, B.; Daly, S.; Hector, L.; Zavattieri, P.; Shaw, J. Tips and tricks for characterizing shape memory wire part 5: Full-field strain measurement by digital image correlation. *Exp. Technol.* **2013**, *37*, 62–78. [CrossRef]

24. Bastawros, A.; Bart-Smith, H.; Evans, A. Experimental analysis of deformation mechanisms in a closed-cell aluminum alloy foam. *J. Mech. Phys. Solids* **2000**, *48*, 301–322. [CrossRef]

25. Bart-Smith, H.; Bastawros, A.-F.; Mumm, D.; Evans, A.; Sypeck, D.; Wadley, H. Compressive deformation and yielding mechanisms in cellular Al alloys determined using X-ray tomography and surface strain mapping. *Acta Mater.* **1998**, *46*, 3583–3592. [CrossRef]

26. Antoniou, A.; Onck, P.; Bastawros, A.F. Experimental analysis of compressive notch strengthening in closed-cell aluminum alloy foam. *Acta Mater.* **2004**, *52*, 2377–2386. [CrossRef]

27. Jerabek, M.; Major, Z.; Lang, R. Strain determination of polymeric materials using digital image correlation. *Polym. Test.* **2010**, *29*, 407–416. [CrossRef]

28. Wang, Y.; Cuitiño, A.M. Full-field measurements of heterogeneous deformation patterns on polymeric foams using digital image correlation. *Int. J. Solids. Struct.* **2002**, *39*, 3777–3796. [CrossRef]

29. Reddy, B.S.; Chatterji, B.N. An FFT-Based Technique for Translation, Rotation, and Scale-Invariant Image Registration. *IEEE Trans. Image Proc.* **1996**, *5*, 1266–1271. [CrossRef]

30. Min, J.; Stoughton, T.B.; Carsley, J.E.; Lin, J. Comparison of DIC methods of determining forming limit strains. *Procedia Manuf.* **2017**, *7*, 668–674. [CrossRef]

31. Tsai, T.R. A versatile camera calibration technique for high-accuracy 3D machine vision metrology using off-the-shelf TV cameras and lenses. *IEEE J. Robot. Autom.* **1987**, *RA-3*, 323–344. [CrossRef]

32. Barranco-Gutiérrez, A.I.; Martínez-Díaz, S.; Gómez-Torres, J.L. *Visión estereoscópica con Matlab y OpenCV*, 1st ed.; Pearson Education: Mexico City, Mexico, 2018; pp. 10–53.

33. Zhang, Z. A flexible new technique for camera calibration. *IEEE Trans. Pattern Anal. Mach. Int.* **2000**, *22*, 1330–1334. [CrossRef]

34. Cofaru, C.; Philips, W.; Paepegem, W.V. A novel speckle pattern Adaptive digital image correlation approach with robust strain calculation. *Opt. Laser Eng.* **2012**, *50*, 187–198. [CrossRef]

35. Garcia, D.; Orteu, J.J.; Penazzi, L. A combined temporal tracking and stereo-correlation technique for accurate measurement of 3D displacements: Application to sheet metal forming. *J. Mater. Proc. Technol.* **2002**, *125–126*, 736–742. [CrossRef]

36. Shi, J.; Chen, F.; Lu, J.; Chen, G. An evolutionary image matching approach. *Appl. Soft Comput.* **2013**, *13*, 3060–3065. [CrossRef]

37. Cyrille, B.; Philippe, D. Automatic Camera Calibration. U.S. Patents US20160350921 A1, 2 May 2017.

38. Jia, Z.; Yang, J.; Liu, W.; Wang, F.; Liu, Y.; Wang, L.; Fan, C.; Zhao, K. Improved camera calibration method based on perpendicularity compensation for binocular stereo vision measurement system. *Opt. Expres* **2015**, *23*, 15205–15223. [CrossRef] [PubMed]

39. García-Rodenas, L.A.; Araujo, P.; Bruyère, V.I.E.; Morando, P.J.; Regazzoni, A.E.; Blesa, M.A. A Model for the Dissolution of Metal Oxides Mediated by Heterogeneous Charge Transfer. Anales de la Asociación Química Argentina, 2004, Volume 92, n.1-3. Available online: http://www.scielo.org.ar/scielo.php?script=sci_arttext&pid=S0365-03752004000100007 (accessed on 16 April 2019).

40. Computational Vision at Caltech. Camera Calibration Toolbox for Matlab. Available online: http://www.vision.caltech.edu/bouguetj/calib_doc/ (accessed on 24 February 2019).

41. Matlab Company. *Image Processing Toolbox™ User's Guide*; Mathworks: Sherborn, MA, USA, 2018; pp. 6-18–6-60. Available online: https://www.mathworks.com/help/images/ (accessed on 16 April 2019).

Article

Dynamic Response of Copper Plates Subjected to Underwater Impulsive Loading

Kaida Dai [1,*], **Han Liu** [2], **Pengwan Chen** [1,*], **Baoqiao Guo** [1], **Dalin Xiang** [3] and **Jili Rong** [3]

[1] State Key Laboratory of Explosion Science and Technology, Beijing Institute of Technology, Beijing 100081, China; baoqiao_guo@bit.edu.cn
[2] Beijing Institute of Structure and Environment Engineering, Beijing 100076, China; 20042577@bit.edu.cn
[3] School of Aerospace Engineering, Beijing Institute of Technology, Beijing 100081, China; xiangdalin1985@sina.com (D.X.); rongjili@bit.edu.cn (J.R.)
*   Correspondence: daikaida@bit.edu.cn (K.D.); pwchen@bit.edu.cn (P.C.)

Received: 26 March 2019; Accepted: 6 May 2019; Published: 10 May 2019

Abstract: Understanding the mechanical response and failure behaviors of thin plates under impact loading is helpful for the design and improvement of thin plate structures in practical applications. The response of a copper plate subjected to underwater impulsive loading has been studied in fluid-structure interaction (FSI) experiments. Three typical copper plates, (a) without a pre-notch, (b) with a cross-shaped pre-notch (+), and (c) with a ring-shaped pre-notch (○) were selected. A high-speed photography system recorded the full-field shape and displacement profiles of the specimens in real time. The 3D transient deformation fields' measurements were obtained using a 3D digital image correlation (DIC) technique. Strain results from DIC and the strain gauges technique were in good agreement. A dimensionless deflection was used to analyze the effect of plate thickness and loading intensity on the deformation of the copper plates. The typical failure modes of different copper plates were identified. The test plates exhibited large ductile deformation (mode I) for copper plates without a pre-notch, and large ductile deformation with local necking (mode Ic), splitting (mode II), splitting and tearing (mode IIc), and fragment (mode III) for the copper plate with a pre-notch.

Keywords: copper plate; underwater impulsive loading; non-liner dynamic deformation; 3D digital image correlation

1. Introduction

Flat-panel structures that are widely used in naval assets and warships are sometimes affected by underwater explosions (UNDEX) because of, for example, torpedo attacks [1]. In the design and application of ships, small curvature hull panels with welded stiffeners can be considered flat-panels [2]. Currently, experimental and computational methods have been used to investigate the dynamic response of plate structures with different geometric dimensions and materials. However, the response of these structures to dynamic loading is not well understood. Due to high deformation rates and their corresponding short loading times and material nonlinearity, the structural dynamic response caused by an UNDEX is complex. Hence, it is difficult to accurately measure deformation fields in the dynamic response process of a metal plate subjected to underwater impulsive loading.

The material response to different loading rates has been examined through shock loading experiments. Ahmed et al. experimentally studied the large deformation behavior of hull panels subjected to an UNDEX [3]. Ramajeyathilagam et al. investigated the deformation and failure modes of thin rectangular plates and cylindrical shell panels through experimental and numerical methods [4–6]. The results schowed that the pressure on the plate is approximated by Cole's empirical formula and Taylor's plate theory [7,8]. Hung et al. carried out studies on aluminum alloy plate underwater blast loading, and further measured the underwater pressure, acceleration, and strain histories on the plate [9].

Experimental measurements are affected by some uncertainties (pressure reflections, vibrations, etc.) because of the complex and hazardous nature of UNDEX, and therefore many experimental apparatuses incorporating fluid-structure interaction (FSI) effects have been developed. In recent years, a cylinder-shaped underwater shock simulator, in lieu of explosive detonation, was utilized by Espinosa et al. to produce shock loading on stainless steel similar to the loading generated in an UNDEX [10]. According to the study, the shock loading pressure generated in the water chamber can be independently controlled by changing the velocity and mass of the flyer plate. A combination of the shadow moiré technique and high speed photography was used to record the full deformation fields of steel plates. Mori et al. used this experiment setup to study the deformation and energy absorption performance of I Core sandwich structures [11]. Avachat et al. presented a similar experimental setup to investigate the effect of panel thickness on the dynamic response of composite sandwich plates [12]. A similar equivalent device was also designed by McShane et al. to load polymer-metal bilayer plates. They analyzed the influence of polymer coating and found four typical failure models [13].

With the current rapid development of the digital image correlation (DIC) method, the measurement of complex and inhomogeneous deformation fields has become relatively easy [14]. Pan et al. experimentally investigated the deformation and failure mechanisms of an aluminum panel and a stationary carbon fiber reinforced polymer (CFRP) panel under transient ballistic impact. Full-field 3D deformation has been measured by a single-camera high-speed stereo-DIC technique [15]. The development of high-speed photography technology has allowed 3D full-field deformation measurements using two high-speed digital cameras. Spranghers et al. applied a 3D DIC technique to measure the dynamic response of aluminum plates subjected to free air blast loading conditions and an explosively driven shock tube (EDST) [16,17]. These studies showed a different structural response, a linear elastic-plastic deformation, and elastic vibration for the free air blast loading and the EDST loading, respectively. Chen et al. studied the dynamic response of thin metal plates under confined air blasting loading, and measured 3D full deformation fields using the 3D DIC technique [18]. Aune et al. reported the structural response of thin steel and aluminum plates with different stand-off distances under air blasting [19]. These experimental investigations showed that thin plates experienced larger plastic deformation and the failure of supports. Tiwari et al. investigated full-field transient plate deformation measurements during buried blasting loading using the 3D DIC method. They well-defined yield boundaries on the plate surface were based on the Cowper-Symonds constitutive relation, with full-field strain and strain rate measurements [20]. Gagliardi et al. obtained blast-driven displacement measurements of an aluminum 6061-T6 plate as one side of an aquarium-like structures using a DIC system [21]. Arora et al. used high-speed photography and the DIC method to monitor the deformation and reveal the failure mechanism of glass-fiber reinforced polymer (GFRP) sandwich panels and laminate tubes during underwater shocks [22]. LeBlanc and Shukla carried out experimental and numerical investigations on the transient response of e-glass/vinyl ester curved composite panels subjected to underwater explosive loading using a 3D DIC system, along with high-speed photography [23,24]. Similar investigations on aluminum honeycomb sandwich panels in FSI experiments were reported by Xiang et al. [25]. Huang et al. performed an experimental study on dynamic deformation and failure modes of circular composite laminates, based on the processing of the 3D DIC method, and observed how the impulse intensity, thickness, and failure of panels clearly ly affect the response of laminate plates [26]. Shukla et al. presented a comprehensive summary review of recent underwater implosion studies on metallic cylindrical structures. These studies demonstrated that the 3D DIC technique can be used for accurate dynamic deformation measurements during underwater implosions [27]. Recently, there is increasing interest in the dynamic response and failure at the supports of materials subject to shock loading, but only limited studies have reported on metal panel and composite materials' response to underwater impulses generated by FSI equipment. The microscopic damage mechanism of these materials under impact loading conditions is not well understood. The quantification of the response of these materials as a function of loading intensity has not yet been fully investigated. Studies on the effect of initial damage on failure modes of these materials to underwater impulses have been rarely reported.

This paper presents the transient responses and failure modes of clamped copper plates subjected to underwater shock loading. The shock pressure histories generated in FSI apparatus' are well monitored. Two typical pre-notches are selected to evaluate the effect of pre-notches on the failure modes. A high-speed photography system with two cameras is applied to record real-time images of the copper plate. The time-resolved 3D full-field deformations are measured using the DIC technique. Visual observations and scanning electron microscope are carried out to identify the failure modes. A dimensionless deflection is performed to explain the effects of experimental conditions on the dynamic responses of circular plates subjected to underwater impulsive loading.

2. Experimental Configuration

2.1. DIC Method

The principle of DIC is based on the corresponding relationship between the speckled gray values in a rectangular area (subset) on an undeformed image (reference image) and a deformed image (target image) [28]. The deformation field of the subsets can be calculated by comparing a subset from a reference image with another subset from a target image [29,30]. The 3D DIC technique is based on the principles of stereo triangulation, whereby two imaging sensors can be used to reconstruct the stereo profile of a specimen, see Figure 1.

The principle diagram of the 3D stereo matching and displacement field is shown in Figure 1. Camera calibration is necessary to determine the relative position and internal distortion of two cameras before testing [21,31]. On the basis of the calibration results of both cameras, the left and right images are stereos matched to reconstruct the three-dimensional surface contour. Then, the deformed image is correlated to match with the corresponding reference image to obtain the 3D full-field displacement profile of the sample. Schmidt et al. believe that the sub-pixel accuracy of the out-of-plane measurement is approximately 0.03 pixels [32], while the in-plane measurement is more accurate, with a 0.01 pixel accuracy.

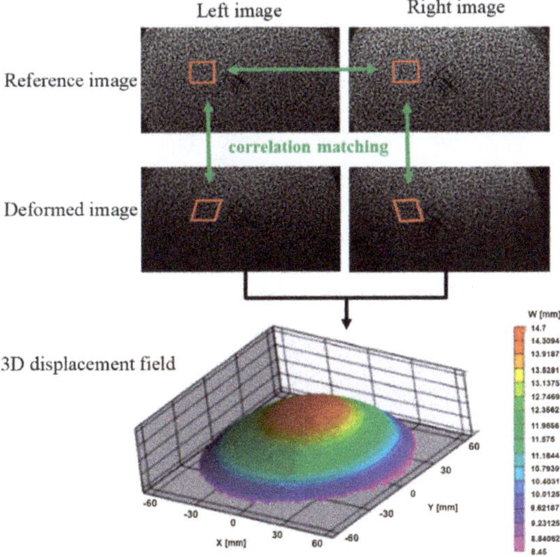

Figure 1. Principle diagram of 3D stereo matching.

2.2. Experimental Details

A simplified FSI experimental setup was developed in a laboratory setting for resembling underwater explosive loading conditions by Xiang and Chen based on Espionosa's work [10,24,33]. Figure 2 shows the schematic diagram of the FSI experimental setup. In the FSI setup, a steel water chamber is incorporated into a gas gun apparatus. The test copper plate and a steel piston are installed at the rear (left) end and the front end with an O-ring, respectively. A gas gun is used to drive a 5 mm thick flyer plate, to prompt the piston to produce the exponentially decaying pressure history. The impulsive pressure histories at A and B positions (the center and end) are measured by dynamic high-pressure transducers.

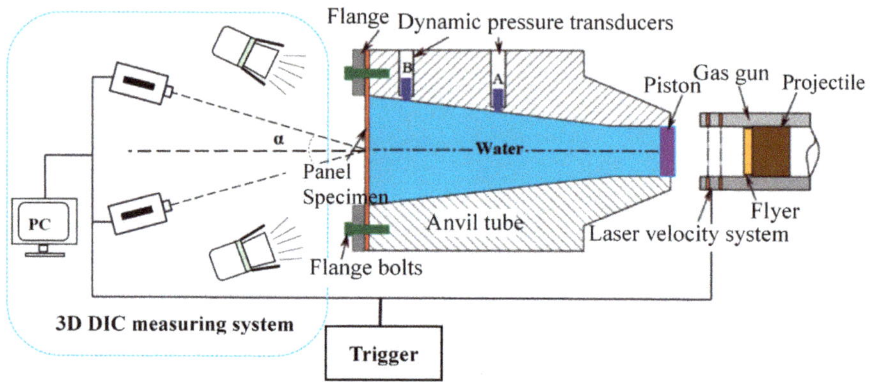

Figure 2. Schematic diagram of the experimental setup.

The material used in this study is ASTM C11000 copper. The mechanical properties of the material are obtained through open literature [34]. The density ρ is 8930 kg/m^3 and the static tensile yield stress σ_y is 205 MPa.

The circular specimen plates with a diameter of 200 mm, as shown in Figure 3a, were clamped at the end of the anvil tube by a steel ring and 12 bolts. The plate is only restricted by a steel ring because the bolts do not pass through the plate. The diameter (D) of the specimen exposed to the water shock loading is 152.4 mm. The strain gauge is used to measure the in-plane components of strain, and is placed at a location 30 mm from the center O, as shown in Figure 3b. The three different thicknesses (T) of the copper plates are 1 mm, 2 mm, and 3 mm. In order to study the effect of initial damage of the plate on failure modes, except on a copper plate without pre-notch, two kinds of shapes of pre-notches, include cross (+) and ring (O), are prepared in the center of the specimens, see Figure 3c,d. The width (W) of the pre-notch is 1 mm, the lengths (L) of the cross-shaped pre-notches are 30 mm and 50 mm, the depths (H) are 0.5 mm and 1.5 mm, the depth and diameter (D_c) of ring-shaped pre-notches are 0.5 mm and 30 mm.

In the present study, a 3D DIC technique is used to capture the dynamic response of materials under shock loading. The deformation process of the outside surface of the copper plates was measured by 3D DIC measuring system, see Figure 2. Figure 4a shows the pictures of a 3D DIC measurement system consisting of two halogen lamps and two synchronized high-speed cameras in a stereoscopic setup. Two Photron Fastcam SA5 high-speed digital cameras (Photron Inc., Tokyo, Japan) were positioned behind the water chamber to record the speckles on the outside surface of the copper plate. The safety distance was about 0.8 m and each camera was angled at approximately 10° with respect to the symmetry plane. The power of the two halogen lamps as the lighting source is 1 kilowatt. The photography system is synchronically triggered by a laser trigger while the flyer impacting the piston. A framing rate of 50,000 frames per second (fps) is chosen in tests with an image resolution of

512 × 272 pixels. In order to study the fracture process of cross-shaped pre-notches, a higher framing rate of 75,000 fps is used with an image resolution of 320 × 264 pixels.

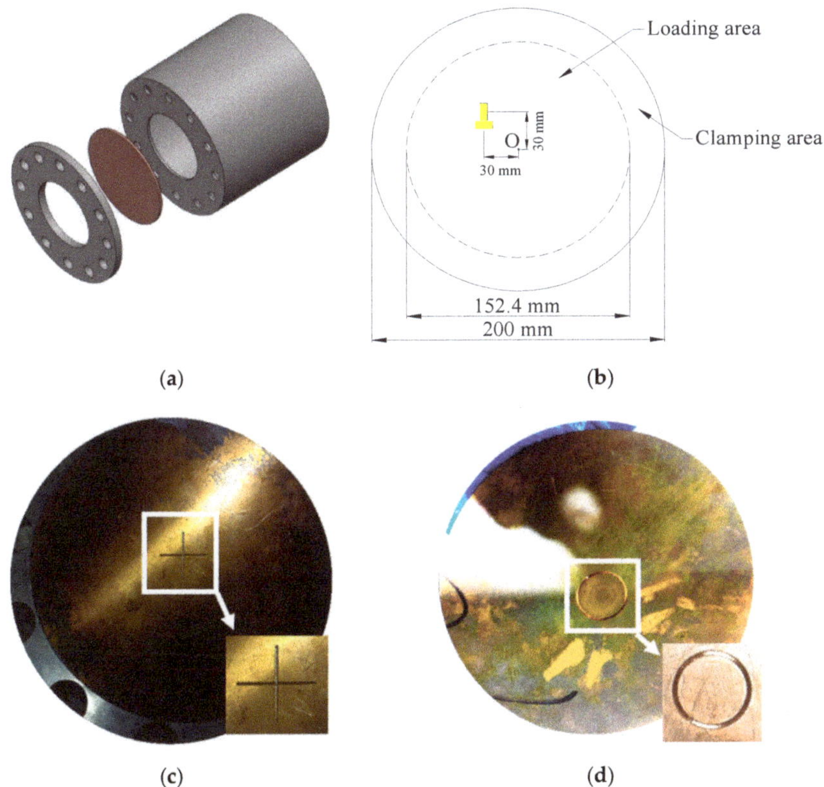

Figure 3. Outline of different specimens: (**a**) schematic of insufficient clamping condition; (**b**) schematic of a specimen with strain gauges; (**c**) a copper plate with a cross-shaped pre-notch; (**d**) a copper plate with a ring-shaped pre-notch.

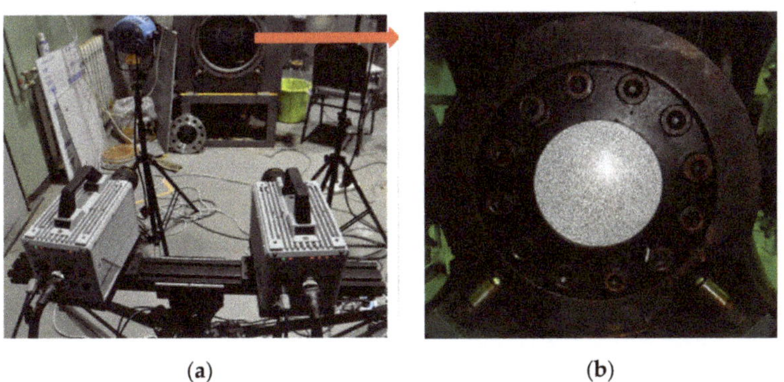

Figure 4. Diagrams of the fluid-structure interaction (FSI) experimental setup: (**a**) the 3D digital imafge correlation (DIC) measuring system with two high-speed digital cameras and two light spots; (**b**) the interest area of the specimen surface with a speckle pattern.

Figure 4b shows the specimen coated with a randomized speckle pattern. It is necessary to clean and polish the specimen surface before preparing a randomized speckle pattern. The speckle quality directly affects the accuracy of the measurement results. In order to gain a high contrast speckle pattern, the outside surface of the copper plate is painted white and then marked with random black points. A randomized speckle pattern can be considered perfect when the black speckle dots have a diameter of approximately 5 pixels as seen by the two cameras [35]. The post-processing is performed with the VIC-3D software package (Correlated Solutions, Inc., Columbia, SC, USA) to obtain the full-field measurement.

2.3. Underwater Pressure Peak and Impulse Estimation

The free-field incident UNDEX pulse in the fluid can be idealized as an exponential pressure attenuation:

$$p = p_0 e^{-t/t_0} \tag{1}$$

where p_0 is the initial peak pressure and t_0 is a characteristic decay time [10]. The free-field momentum (impulse/area) can be given by

$$I_0 = \int_0^{\infty} p\,dt = p_0 t_0 \tag{2}$$

In the FSI setup, the peak pressure p_{mx} and impulse of the shock wave I_{mx} is given by

$$p_{mx} = \frac{sf}{s+f} V_0 \left(\frac{D}{D_x} \right)^2, \tag{3}$$

$$I_{mx} = \sum_0^{\infty} p_{mx} \left[\frac{s-f}{s+f} \right]^n \Delta t, \tag{4}$$

where f and s are the acoustic impedance values of the fluid and solid, V_0 is the flyer impact velocity, and D and D_x are the diameters of the tube at the impact and pressure prediction locations, respectively. n is the number of wave reverberations in the flyer and Δt is the time required for the elastic longitudinal wave to twice traverse the flyer plate.

In this investigation, the diameters of impact locations, the tube at location A and location B are 66 mm, 106 mm, and 142 mm, respectively. The acoustic impedance values of solids and fluids are 40.82×10^6 kg/(s·m^2) and 1.46×10^6 kg/(s·m^2). Therefore, the peak pressure values at the two pressure transducer locations A and B can be simplified as follows:

$$p_A = 0.5464\, V_0, \tag{5}$$

$$p_B = 0.3027\, V_0, \tag{6}$$

3. Results and Discussion

3.1. Pressure Results

The typical pressure profiles of the A1# test at positions A and B are shown in Figure 5. The profiles were obtained using a flyer plate with a thickness of 5.0 mm and a launching velocity of 158.3 m/s. At position A, the recorded pressure history showed similar characteristics to that of the free-field incident UNDEX pulse, i.e., short rise time, a peak value of about 84.9 MPa, and subsequent exponential pressure decay. On the pressure profile at position B, the peak pressure is about 50.4 MPa (Figure 5b) and the reflective wave is observed as a second peak at about 250 μs.

The peak pressures at position A and B are calculated using Equations (5) and (6), respectively. The results are summarized in Table 1. It is clearly shown that the errors of peak pressure at positions A and B are all below 9%. Due to energy attenuation, the experimental peak pressures at position A are lower than the predicted value, but the experimental peak pressures at position B are higher than the predicted value, because of the stack of reflection waves.

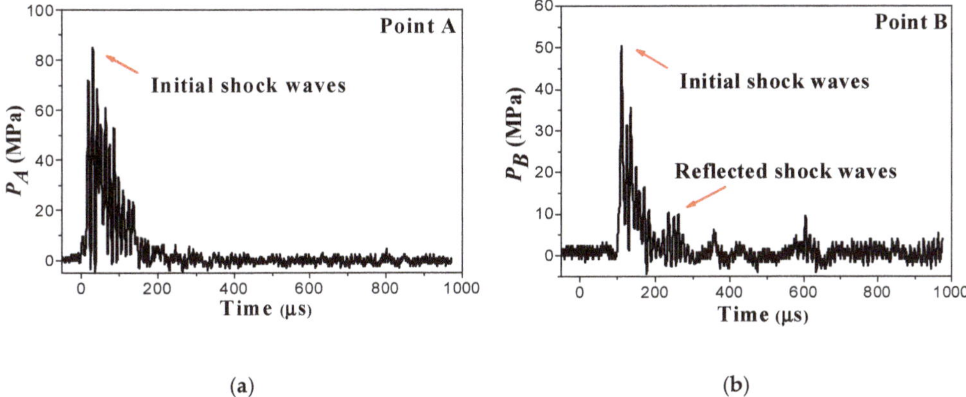

(a)   (b)

Figure 5. Typical pressure-time history at positions A and B of the A1# test; (a) point A (b) point B.

Table 1. Characteristic parameters of underwater impulsive loads.

Sample Number	V_0 (m/s)	p_A (MPa)		p_B (MPa)	
		Exp.	Equation (5)	Exp.	Equation (6)
A1#	158.3	84.9	86.5	50.4	47.8
A2#	124.3	62.8	67.9	40.3	37.6
A3#	129.5	65.6	70.8	42.0	39.2
A4#	130.2	66.3	71.1	42.0	39.4
A5#	46.6	23.4	25.5	15.5	14.1
A6#	121	64.5	66.1	39.2	36.6
A7#	128.9	65.7	70.4	41.8	39.0

3.2. Specimens without Pre-Notches

As shown in Figure 6, a reduced area surrounded by the red line at the specimen center was selected as the AOI (area of interest) to obtain full-field deformation. The green region corresponding to the strain gauges cannot be correlation calculated, because of the speckle lack.

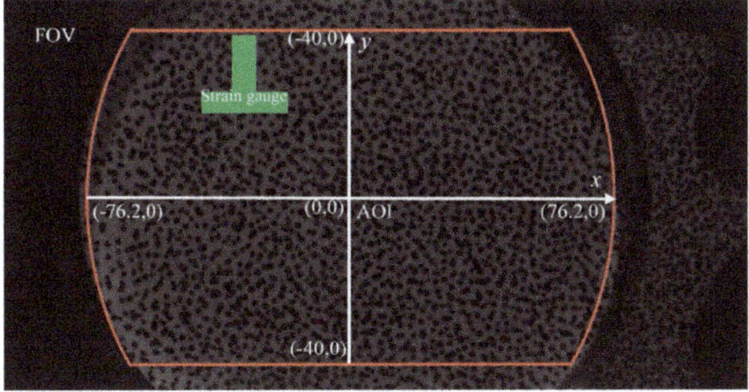

Figure 6. Diagram of selected area of interest (AOI).

Figure 7 shows the out-of-plane displacement fields of the A5# specimen ($T = 1$ mm, $V_0 = 46.6$ m/s) at different time steps. The major component is the out-of-plane displacement (δ). The shock wave acts on the internal surface of the copper plate as a plane shock wave during the early period. More shock waves act on the plate boundary, bringing the clear displacement ring up to 0.08 ms, then the displacement increases from the boundary to the center of the plate until 0.4 ms. Note that the deformable contour exhibits an approximately planar circular shape in the first 0.4 ms, and evolves into a symmetric dome shape after 0.48 ms.

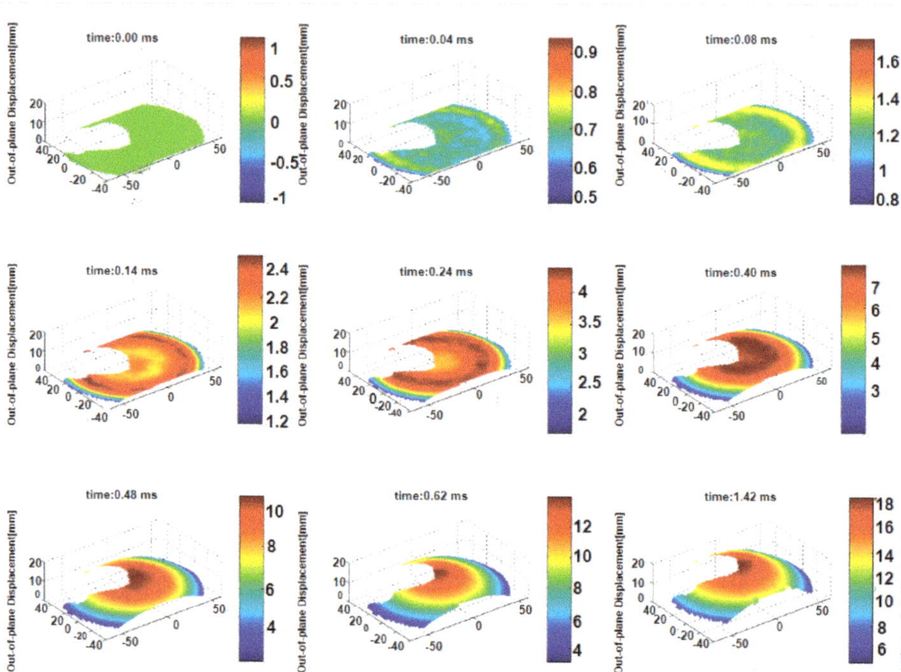

Figure 7. Out-of-plane displacement fields of the A5# specimen ($T = 1$ mm, $V_0 = 46.6$ m/s).

The out-of-plane displacement and velocity profiles along the x-axis are shown in Figure 8. Notice that the profiles are plotted from −60 to 70 mm, while the boundary is located between −76.2 and 76.2 mm. As indicated in Figure 8a, the plate is forced to move out of the plane underwater shock wave. The displacement is first produced at the boundary region (see lines from 0.02 ms to 0.24 ms), the shape after deforming of the plate is approximately symmetrical and is almost constant until 0.14 ms. The plate continues to deform when the pressure of the shock wave has vanished. At 1.52 ms, the displacement reaches its maximum at the center first, and then rebounds at 2.24 ms, due to elastic recovery.

Particles close to the boundary are constrained and have a very limited initial velocity. The plate has the same given velocity at 0.02 ms. From this moment on, the velocity increases from the boundary to the center, because of the centripetal radial flow of water. The outside surface obtains a maximum deformation velocity of 34.1 m/s at 0.42 ms (see Figure 8b), and then presents attenuation of velocity repeatedly. The inertia forces cause the plate to be further deformed.

Figure 9 presents the evolution of the in-plane maximum principal strain fields of the A5# specimen at different time steps. It is found that higher principal strain appeared close to the boundaries until 0.30 ms, because the plate's borders are constrained by steel flange. Further, due to the interaction of the inertia forces and refection waves, strain develops towards the center of the plate, and reaches a maximum value at 0.66 ms.

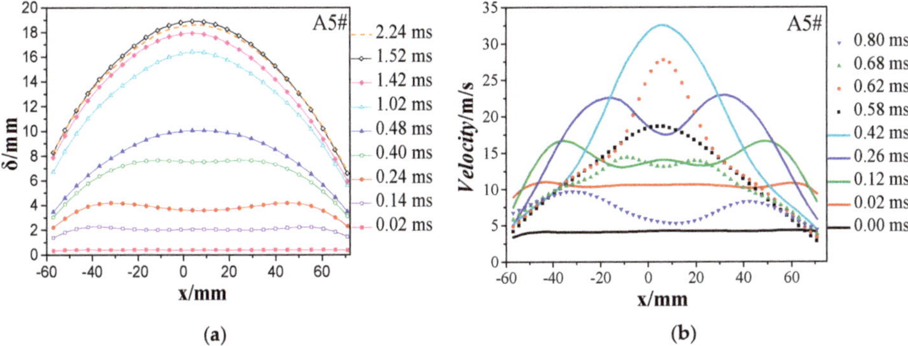

(a)   **(b)**

Figure 8. The out-of-plane displacement and velocity profiles: (**a**) profile of displacement along the x-axis; (**b**) profile of velocity along the x-axis.

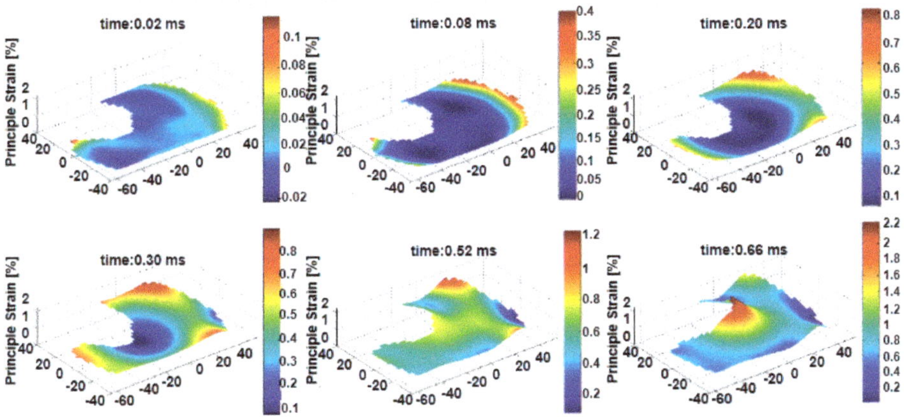

Figure 9. In-plane maximum principal strain fields of the A5# specimen at different time steps.

The comparison of strain histories measured by the DIC and strain gauges (see Figure 3b) are shown in Figure 10. The DIC results are obtained from strain data of the corresponding symmetric points. The evolution trends of the two history curves are similar. It can be seen that results from DIC and strain gauges show a good agreement in the initial growth stage until 0.6ms. From this moment on, strain gauges falling off from the sample cause strain curves to stop rising and start falling.

Figure 11 shows photographs of the recovered A5# specimens after the test. It can be seen the main failure mode of the specimen is only a large ductile deformation (mode I), as defined by Smith and Nurick [36]. Because the test plate slides out from the clamped location, an annular sliding trace can be clearly observed on the boundary (see Figure 11b).

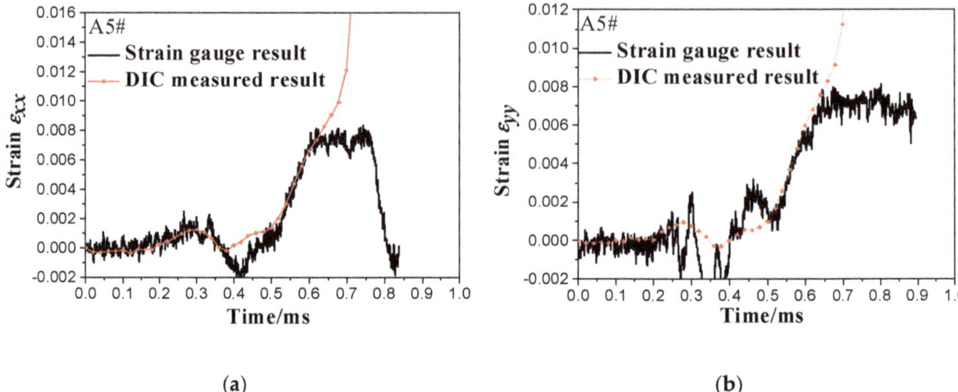

(a)   **(b)**

Figure 10. Comparison of the normal strain results from DIC and strain gauges; (**a**) strain of x-direction and (**b**) strain of y-direction.

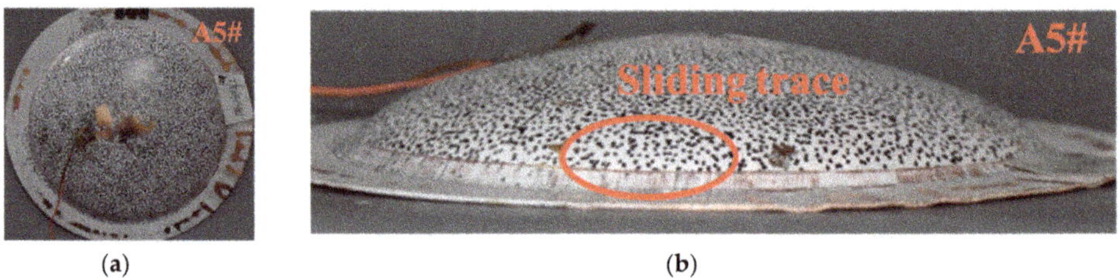

(a)   **(b)**

Figure 11. Photographs of recovered specimens after shock experiment, (**a**) shows the outside face; (**b**) shows a side view.

The results of the deformation and failure mode for all seven specimens without pre-notches are summarized in Table 2. The peak pressure p_{ms} and impulse I_{ms} are predicted values by Equations (3) and (4), the dimensionless impulse is given by $\bar{I} = I_{ms}/\overline{M}\sqrt{\sigma_y/\rho}$, where $\overline{M}$ is the dimensionless mass, the final deflection δ_{t1} and δ_{t2} are obtained by DIC and the altimeter, the dimensionless final deflection δ_{t1}/L is also given, where $L = 152.4$ mm is the specimen span. The relationship between the dimensionless impulse and normalized final deflection is shown in Figure 12. The result shows that the normalized final deflection of the plate increases linearly with the dimensionless impulse. The failure modes are always large ductile deformations (mode I). The relative out-of-plane measurement errors of DIC for seven tests are from 0.84%–6.83%, which can be estimated by $|\delta_{t1} - \delta_{t2}|/\delta_{t1}$.

Table 2. Experimental conditions of specimen plates without pre-notches.

Number	Thickness T (mm)	V_0 (m/s)	p_{ms} (MPa)	I_{ms} (Pa·s)	δ_{t1} (mm)	δ_{t2} (mm)	$\bar{I}$	δ_{t1}/L	Measurement Error (%)	Failure Mode
A1#	1	158.3	42.1	2769	27.9	27.3	2.05	0.183	2.15	Mode I
A2#	1	124.3	33.0	2174	23.9	24.1	1.61	0.157	0.84	Mode I
A3#	2	129.5	34.4	2264	22.2	22.8	0.84	0.146	2.7	Mode I
A4#	3	130.2	34.6	2277	16.1	17.2	0.56	0.106	6.83	Mode I
A5#	1	46.6	12.4	815	18.6	17.8	0.60	0.122	4.3	Mode I
A6#	1	121	32.2	2116	23.6	22.2	1.57	0.155	5.93	Mode I
A7#	1	128.9	34.3	2254	24	23.7	1.67	0.158	1.25	Mode I

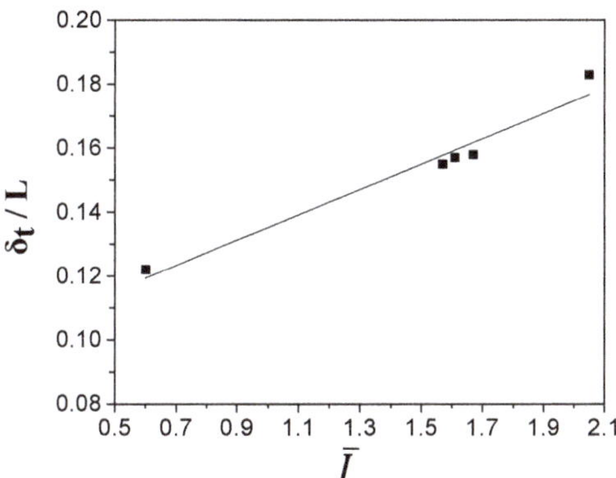

Figure 12. Normalized final deflection of specimens as a function of the dimensionless impulse.

3.3. Specimens with Pre-Notches

Table 3 lists the experimental conditions and the corresponding failure mode of the specimens with pre-notches. The specimens with pre-notches have been investigated to understand the effect of pre-notches on the dynamic deformations and failure modes. Figures 13 and 14 show the evolution of the in-plane maximum principal strain of C1# and C2# specimen ($T = 1$ mm, $L = 30$ mm, and $H = 0.5$ mm). The selected frame rate and image resolution are 75,000 fps and 320×264 pixels for the C1# test. It is very clear from Figure 13 that weak strain concentration has appeared at the center of the outside surface at 559 μs. From this moment on, because the pre-notch causes the strength of the central region of the plate to decrease, higher principal strain localizes and concentrates around the pre-notch. The shape of strain concentration is close to the cross shape with a 45 degree angle, the position and direction of strain concentration correlate well with the pre-notch.

Compared with the test for C1#, the impact velocity has been increased to 118.9 m/s in the C2# test. A framing rate of 50,000 fps is used with an image resolution of 512×272 pixels. Similarly to the result of the test without a pre-notch, the principal strain increases from the boundary to center regions. Similarly to the test for C1#, a cross-shaped strain concentration appears in the correlated position of the pre-notch at 520 μs. At 600 μs, a mismatched region is also found on the displacement profile, demonstrating that a crack rupture occurs in the plate center (labeled "1"). After 620 μs, the water escapes through the crack from the anvil, and the crack continues to expand in the direction of the cross pre-notch.

Table 3. Experimental conditions of specimen panels with pre-notches.

| Sample Number | Thickness T (mm) | Pre-Notch | | | V_0 (m/s) | Failure Mode |
		Shape	Depth H(mm)	Length L(mm)	Diameter D_c(mm)		
C1#	1	Cross	0.5	30	—	101.9	Mode Ic
C2#	1	Cross	0.5	30	—	118.9	Mode IIc
C3#	2	Cross	1.5	30	—	119.2	Mode II
C4#	1	Cross	0.5	50	—	140.8	Mode IIc
R1#	1	Ring	0.5	—	30	125.3	Mode III
R2#	1	Ring	0.5	—	50	121.0	Mode III

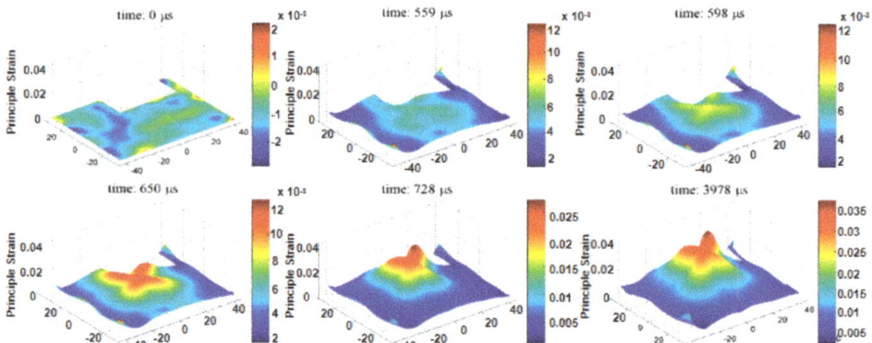

Figure 13. In-plane maximum principle strain of the C1# specimen at different time steps.

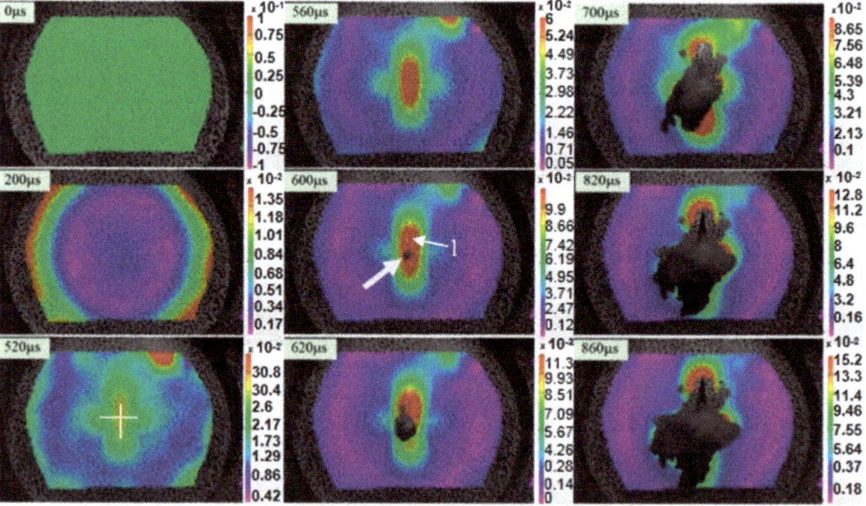

Figure 14. In-plane maximum principal strain of the C2# specimen at different time steps.

Figure 15 shows the out-of-plane displacement fields on the surface of the R1# specimen with a ring-shaped pre-notch ($T = 1$ mm, $D_c = 30$ mm). The camera framing rate and image resolutions are the same as the test for C2#. The pre-notch is indicated by a white line. At 520 µs, a crack forms in the center area indicated by white ring A. After this moment, the crack extends along the ring-shaped pre-notch, and then the circular flyer falls off from the plate (see 840 µs in Figure 15). Figure 16 shows the in-plane maximum principal strain profile along the x-axis. Notice that strain concentration is located at ±15 mm from the center, where is the same location as the pre-notch. This explains how strain concentration also occurs around ring pre-notch.

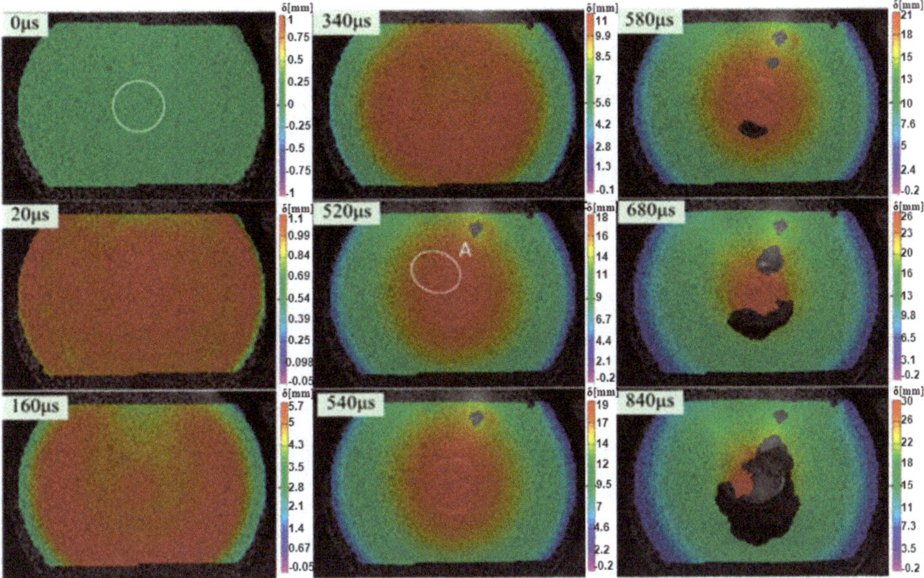

Figure 15. Out-of-plane displacement fields of the R1# specimen at different time steps.

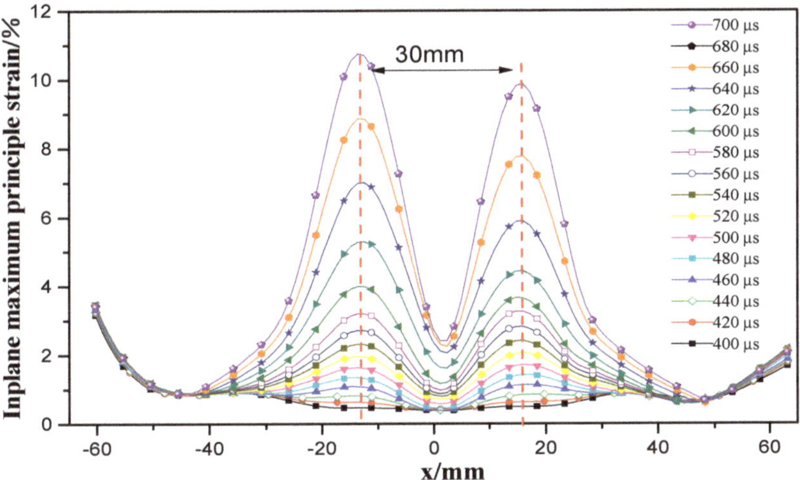

Figure 16. The in-plane maximum principal strain profile along the x-axis of the R1# specimen.

After underwater loading tests, the specimens with the cross-shaped and ring-shaped pre-notches have been recovered for further analysis. The four typical failure modes observed in these experiments are reported in Figure 17. The 1 mm-thick plate with the depth cross pre-notch of 0.5 mm has been tested using a flyer with a velocity of 101.9 m/s. The larger ductile deformation and a cross-shaped local necking (mode Ic) appear in the center area of the C1# specimen, which is only presented in the pre-notch position marked with a white circle in Figure 17a.

Figure 17. Failure modes of copper plates with pre-notches: (**a**) large ductile deformation and local necking (mode Ic failure); (**b**) splitting (mode II failure); (**c**) splitting and tearing (mode IIc failure); (**d**) fragment (mode III failure).

For the same case, while the velocity of the flyer is increased to 118.9 m/s and 140.8 m/s, the failure modes of the C2# and C4# specimens evolve to splitting of pre-notch and tearing (mode IIc), see Figure 17c, the new cracks grow from the pre-notch tips. Scanning electron microscopy (SEM) was used to measure the microstructure of the fracture surfaces of the C2# specimen. Figure 18 shows the microstructures of different fracture surfaces of the C2# specimen, including the pre-notched part labeled by the white lines and the new crack part labeled by the red line. In the pre-notched part, due to the fact that the residual thickness of the plate is only 0.5 mm, a plug-like shear failure is easily caused by the underwater shock wave, therefore, considerable parallel striations are found on the fracture surface. In contrast, the fracture surface of the new crack part has many dimple structures, which are considered a typical feature of tensile tearing failure. This can be explained by the fact that after fracture of the pre-notch, the sectional tension by shock wave causes the tearing failure and a radial new crack.

For the 2 mm-thick plate with the depth of pre-notch increased to 1.5 mm, only a splitting of pre-notch (mode II) is observed on the outside face of the C3# specimen, labeled by a white ring line in Figure 17b. Furthermore, the shape of the pre-notch also influences the failure mode of the circular plate under clamped condition subjected to underwater impulsive loading. For the R1# and R2# specimens, the specimens occur failure along the pre-notch, and the fragment (mode III) produced as the circular flyer is shown in Figure 17d.

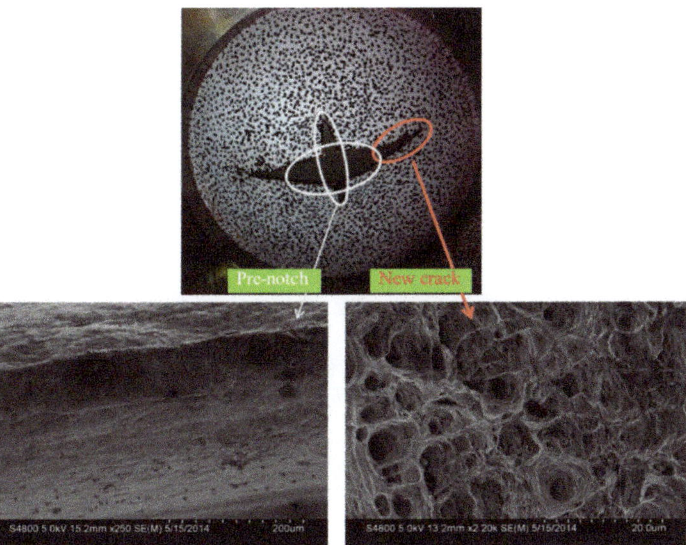

Figure 18. Microstructures of different fracture surfaces of the C2# specimen.

4. Conclusions

Experimental investigations on the non-linear dynamic responses of circular copper plates subjected to underwater impulsive loading using a fluid-structure interaction experimental setup were presented. The different types of copper plates, with and without pre-notches, were studied and analyzed. The shock pressure histories were successfully measured by the dynamic pressure transducers at locations A and B. A stereoscopic camera system combined with a 3D DIC technique was utilized to monitor the out-of-plane deformation, velocity, and in-plane maximum principal strain of the copper plates. Results showed that dynamic deformation is the interaction effect of the initial underwater shock wave and water medium flow.

For the clamped circular specimens without a pre-notch, a dimensionless deflection was performed to further investigate the relationship between the impulsive wave intensity, thickness of the specimens, and the final deflection. The linear relation between the normalized final deflection and the dimensionless impulse was observed.

The effects of the impulsive wave intensity, geometric dimension of the specimens, and the dimension of the pre-notch on the failure modes of the copper plates were also studied. Based on the observation of the current experiments, a large ductile deformation (mode I) was observed for the plate without a pre-notch, and other four typical failure modes, including large ductile deformation with local necking (mode Ic), splitting (mode II), splitting and tearing(mode IIc), and fragment (mode III) were found for the plates with different pre-notches.

This study provides insight into the relationship between the impulsive wave intensity, geometric dimensions of specimens and pre-notches, and the non-liner dynamic responses of copper plates subjected to underwater impulsive loading. This discussion will be helpful in understanding failure processes and predicting failure modes.

Author Contributions: Conceptualization, P.C.; experimental design and measurement, K.D., H.L., B.G. and D.X.; data analysis, K.D., H.L. and B.G.; writing-original draft manuscript, K.D.; Writing-review and editing, P.C. and J.R.

Funding: The present work was supported by the National Natural Science Foundation of China under Grants 11472047, 11472054 and 11521062.

Conflicts of Interest: The authors declare no conflict of interest.

References

1. Rajendran, R.L.J. Blast loaded plates. *Mar. Struct.* **2009**, *22*, 99–127. [CrossRef]
2. Fox, E.N. A review of underwater explosion phenomena. *Comp. Underwater Explos. Res.* **1947**, *1*, 1–83.
3. Ahmed, W.K.P.J. Non-linear dynamic analysis assessment of explosively loaded submarine hull panels. *Shock Vib. Bull.* **1990**, *1*, 139–171.
4. Ramajeyathilagam, K.; Vendhan, C.P. Deformation and rupture of thin rectangular plates subjected to underwater shock. *Int. J. Impact Eng.* **2004**, *30*, 699–719. [CrossRef]
5. Ramajeyathilagam, K.; Vendhan, C.P.; Rao, V.B. Non-linear transient dynamic response of rectangular plates under shock loading. *Int. J. Impact Eng.* **2000**, *24*, 999–1015. [CrossRef]
6. Ramajeyathilagam, K.; Vendhan, C.P.; Rao, V.B. Experimental and numerical investigations on deformation of cylindrical shell panels to underwater explosion. *Shock Vib.* **2001**, *8*, 253–268. [CrossRef]
7. Cole, R.H. *Underwater Explosions*, 2nd ed.; Princeton University Press: Princeton, NJ, USA, 1948.
8. Taylor, G.I. The pressure and impulse of submarine explosion waves on plates. In *Aerodynamics and the Mechanics of Projectiles and Explosions*; Batchelor, G.K., Ed.; Cambridge University Press: Cambridge, UK, 1963; Volume III, pp. 287–303.
9. Hung, C.F.; Hsu, P.Y.; Hwang-Fuu, J.J. Elastic shock response of an air-backed plate to underwater explosion. *Int. J. Impact Eng.* **2005**, *31*, 151–168. [CrossRef]
10. Espinosa, H.D.; Lee, S.; Moldovan, N. A novel fluid structure interaction experiment to investigate deformation of structural elements subjected to impulsive loading. *Exp. Mech.* **2006**, *46*, 805–824. [CrossRef]
11. Mori, L.F.; Queheillalt, D.T.; Wadley, H.N.G.; Espinosa, H.D. Deformation and failure modes of I-core sandwich structures subjected to underwater impulsive loads. *Exp. Mech.* **2009**, *49*, 257–275. [CrossRef]
12. Avachat, S.; Zhou, M. Effect of facesheet thickness on dynamic response of composite sandwich plates to underwater impulsive loading. *Exp. Mech.* **2012**, *52*, 83–93. [CrossRef]
13. McShane, G.J.; Stewart, C.; Aronson, M.T.; Wadley, H.N.G.; Fleck, N.A.; Deshpande, V.S. Dynamic rupture of polymer-metal bilayer plates. *Int. J. Solids Stuct.* **2008**, *45*, 4407–4426. [CrossRef]
14. Pan, B. Recent progress in digital image correlation. *Exp. Mech.* **2011**, *51*, 1223–1235. [CrossRef]
15. Pan, B.; Yu, L.P.; Yang, Y.Q.; Song, W.D.; Guo, L.C. Full-field transient 3D deformation measurement of 3D braided composite panels during ballistic impact using single-carmera high-speed stereo-digital image correlation. *Compos. Struct.* **2016**, *157*, 25–32. [CrossRef]
16. Sprangher, K.; Vasilakos, I.; Lecompte, D.; Sol, H.; Vantomme, J. Full-field deformation measurements of aluminum plates under free air blast loading. *Exp. Mech.* **2012**, *52*, 1371–1384. [CrossRef]
17. Louar, M.A.; Belkassem, B.; Ousji, H.; Spranghers, K.; Kakogiannis, D.; Pyl, L.; Vantomme, J. Explosive driven shock tube loading of aluminium plates: Experimental study. *Int. J. Impact Eng.* **2015**, *86*, 111–123. [CrossRef]
18. Chen, P.W.; Liu, H.; Ding, Y.S.; Guo, B.Q.; Chen, J.J.; Liu, H.B. Dynamic deformation of clamped circular plates subjected to confined blast loading. *Strain* **2016**, *52*, 478–491. [CrossRef]
19. Aune, V.; Fagerholt, E.; Hauge, K.O.; Langseth, M.; Borvik, T. Experimental study on the response of thin aluminum and steel plates subjected to airblast loading. *Int. J. Impact Eng.* **2016**, *90*, 106–121. [CrossRef]
20. Tiwari, V.; Sutton, M.A.; McNeill, S.R.; Xu, S.W.; Deng, X.M.; Fourney, W.L.; Bretall, D. Application of 3D image correlation for full-field transient plate deformation measurements during blast loading. *Int. J. Impact Eng.* **2009**, *36*, 862–874. [CrossRef]
21. Gagliardi, F.J.; Cunningham, B.J. The use of digital image correlation in explosive experiments. In Proceedings of the 14th international detonation symposium, Coeur d'Alene, ID, USA, 11–16 April 2010.
22. Arora, H.; Hooper, P.A.; Dear, J.P. The effects of air and underwater blast on composite sandwich panels and tubular laminate structures. *Exp. Mech.* **2012**, *52*, 59–81. [CrossRef]
23. LeBlanc, J.; Shukla, A. Dynamic response of curved composite panels to underwater explosive loading: Experimental and computational comparisons. *Compos. Struct.* **2011**, *93*, 3072–3081. [CrossRef]
24. LeBlanc, J.; Shukla, A. Response of E-glass/vinyl ester composite panels to underwater explosive loading: Effects of laminate modifications. *Int. J. Impact Eng.* **2011**, *38*, 796–803. [CrossRef]
25. Xiang, D.L.; Rong, J.L.; He, X. Experimental investigation of dynamic response and deformation of aluminium honeycomb sandwich panels subjected to underwater impulsive loads. *Shock Vib.* **2015**, *2015*, 650167. [CrossRef]

26. Huang, W.; Zhang, W.; Chen, T.; Jiang, X.W.; Liu, J.Y. Dynamic response of circular composite laminates subjected to underwater impulsive loading. *Compos. Part. A* **2018**, *109*, 63–74. [CrossRef]

27. Shukla, A.; Gupta, S.; Matos, H.; LeBlanc, J.M. Dynamic collapse of underwater metallic structures-recent investigations: Contributions after the 2011 Murray Lecture. *Exp. Mech.* **2018**, *58*, 387–405. [CrossRef]

28. Siebert, T.; Becker, T.; Spiltthof, K.; Neumann, I. High-speed digital image correlation: Error estimations and applications. *Opt. Eng.* **2007**, *46*, 0510045. [CrossRef]

29. Pan, B.; Xie, H.M.; Guo, Z.Q.; Hua, T. Full-field strain measurement using a two-dimensional Savitzky-Golay digital differentiator in digital image correlation. *Opt. Eng.* **2007**, *46*, 0336013. [CrossRef]

30. Pan, B.; Xie, H.M. Full-field strain measurement based on least-square fitting of local displacement for digital image correlation method. *Acta Opt. Sin.* **2007**, *27*, 1980–1986.

31. Yu, L.; Pan, B. Color stereo-digital image correlation method using a single 3CCD color camera. *Exp. Mech.* **2017**, *57*, 649–657. [CrossRef]

32. Schmidt, T.; Tyson, J.; Galanulis, K. Full-field dynamic displacement and strain measurement using advanced 3D image correlation photogrammetry: Part I. *Exp. Tech.* **2006**, *27*, 47–50. [CrossRef]

33. Chen, P.W.; Liu, H.; Zhang, S.L.; Chen, A.; Guo, B.Q. Full-field 3D deformation measurement of thin metal plates subjected to underwater shock loading. In *Advancement of Optical Methods in Experimental Mechanics*, 1st ed.; Jin, H., Yoshida, S., Lamberti, L., Lin, M.T., Eds.; Springer: Cham, Switzerland, 2016; Volume 3; pp. 211–223.

34. Jones, N. *Structural Impact*, 2nd ed.; Cambridge University Press: Cambridge, UK, 1997; pp. 348–349.

35. Lecompte, D.; Smits, A.; Bossuyt, S.; Sol, H.; Vantomme, J.; Hemelrijck, D.V.; Habraken, A.M. Quality assessment of speckle patterns for digital image correlation. *Opt. Laser Eng.* **2006**, *44*, 1132–1145. [CrossRef]

36. Teeling-Smith, R.G.; Nurick, G.N. The deformation and tearing of circular plates subjected to impulsive loads. *Int. J. Impact Eng.* **1991**, *11*, 77–91. [CrossRef]

 *applied sciences*

Article

Gradient Correlation Functions in Digital Image Correlation

Mikael Sjödahl

Department of Engineering Sciences and Mathematics, Luleå University of Technology, SE-971 87 Luleå, Sweden; mikael.sjodahl@ltu.se

Received: 18 March 2019; Accepted: 17 May 2019; Published: 24 May 2019

Abstract: The performance of seven different correlation functions applied in Digital Image Correlation has been investigated using simulated and experimentally acquired laser speckle patterns. The correlation functions were constructed as combinations of the pure intensity correlation function, the gradient correlation function and the Hessian correlation function, respectively. It was found that the correlation function that was constructed as the product of all three pure correlation functions performed best for the small speckle sizes and large correlation values, respectively. The difference between the different functions disappeared as the speckle size increased and the correlation value dropped. On average, the random error of the combined correlation function was half that of the traditional intensity correlation function within the optimum region.

Keywords: image correlation; gradient correlation functions; laser speckles

---

1. Introduction

Digital image correlation (DIC), or Particle Image Velocimetry (PIV), has since its introduction in the 1980s evolved into one of the most versatile and widespread techniques in experimental mechanics [1–6]. Recent examples are found in diverse scientific fields such as biomechanics [7], infrastructure [8], material science [9], composite structures [10], and microfluidistics [11], but the technique is not restricted to these scientific fields. A quick search on a popular search engine lists over 50,000 contributions out of which 16,000 are published throughout the last four years. The great versatility of the technique comes from its simplicity and flexibility, scalability in both space and time, the fact that it is non-intrusive and that it produces deformation fields. Furthermore, the deformation fields can be generated as Lagrangian fields (typically used in DIC) or Eulerian fields (common in PIV), depending on which images that are compared. In general, the technique requires a unique feature to be present in the plane considered. In the absence of natural features, a pattern needs to be added, usually using spray (DIC) or by adding small particles to a flow (PIV). The general approach is then to follow the features in between successive frames using a model of the deformation field, which most often is performed locally involving a limited number of pixels, but global approaches have been demonstrated [6]. At the core of this calculation is a numerical optimization routine whose underlying function most often is defined as an intensity cross-covariance or as a sum of squared intensity differences. The performance of the two approaches differs only in details. It has been shown that the random error in the deformation calculation roughly scales with average feature size, subimage width and correlation value [12]. The quotient between subimage width and feature size defines in principle the number of independent contributions to the correlation and relates also to the reliability of the calculation. In addition, the feature size defines the curvature of the correlation peak and hence its susceptibility to random noise. The amount of random noise is specified by the correlation value. For a laser speckle pattern, the decay of the correlation value is almost completely dominated by speckle decorrelation, while for a white-light pattern algorithm dependent features such as precision in interpolation becomes important. This is the reason feature sizes slightly larger than the sampling limit often are preferred with these patterns.

Calculated image gradients are used frequently in DIC algorithms as a means of interpolation and as an aid to improve the performance of the optimization routine. Neggers et al. has recently published an extensive review of current algorithms using image gradients for DIC calculations [13]. The conclusions were that image gradients speed up search algorithms considerably and the most effective gradient vector is formulated from a weighted blend of image gradients from both images. However, these image gradients are used to speed up the search for the most probable set of correlation parameters. The underlying function is still an intensity correlation. On the other hand, feature detection based on image gradients are frequently used in computer-vision applications. For example O'Callaghan and Haga has published a paper on the use of a normalized gradient correlation function for detection of changes in a video stream [14]. The motivation for the use of gradient correlation in their application was mainly that the detection becomes more robust against varying background intensity. Such correlation functions have to my knowledge not yet been explored in connection with DIC.

The purpose of this paper is to investigate the performance of correlation functions based on derivatives of intensity images as a function of feature size and image degradation. Three fundamental normalized correlation functions are formulated based on intensity, intensity gradients, and intensity Hessian, respectively. From these an additional four correlation functions can be constructed as a combination of the three fundamental ones. These functions are then evaluated using simulations and experiments using laser speckles. The different correlation functions are introduced and discussed in Section 2. The set of evaluations are introduced and presented in Section 3 where the simulations are detailed in Section 3.1 and the experiments in Section 3.2, respectively. The paper ends with a discussion and some concluding remarks.

2. Theory

Consider two images $I_1(\mathbf{x}_1)$ and $I_2(\mathbf{x}_2)$ registered at the two time instances t_1 and t_2, respectively, containing approximately the same features. The images are assumed sampled on an $[M, N]$ sized grid with pixel pitch $[p_y, p_x]$ in row and column directions, respectively. It is assumed that the motions of the features between the two images are small wherefore local information can be used to estimate local motion. Apart from the intensities it is assumed that the gradient vector $\mathbf{G}_i(\mathbf{x}_i)$ and the Hessian matrix $H_i(\mathbf{x}_i)$ may be generated in each pixel, where $i = 1, 2$ for each of the images respectively. These additional fields are generated from application of the in-plane gradient column vector $\nabla_\perp = (\partial/\partial x, \partial/\partial y)^t$ as $\mathbf{G}_i(\mathbf{x}_i) = \nabla_\perp I(\mathbf{x}_i)$ and $H_i(\mathbf{x}_i) = \nabla_\perp I(\mathbf{x}_i)\nabla_\perp^t$, respectively. Associated with each pixel are therefore an intensity value, an intensity gradient vector, and an intensity Hessian matrix, respectively.

Figure 1 shows as a cropped example the intensity, the magnitude of the gradient vector and the determinant of the Hessian, respectively, of a laser speckle image. It is obvious that these three images contain different information and that they will perform differently in a correlation calculation. While the intensity image presents the distribution in intensity, the gradient image shows where the intensity changes most rapid. In regions with zero image gradient, the intensity correlation is essentially insensitive. The gradient image therefore dictates with what precision the intensity correlation function can be positioned. In addition, the Hessian matrix shows regions with a large intensity curvature. As the determinant can take on negative values, the background appears grayish, but regions close to speckle peaks light up. It is obvious that these peaks appear at the same positions as speckle maxima, but that they are locally more confined. Given these images, two continuous approximations of the intensity distributions can be formed using a nine-node quadratic Hermitian Finite Element. The intensity information in each node is taken from the intensity images and the gradient information required along the edges of the element is taken from the gradient and Hessian matrices, respectively. This approximation allows for a continuous description of image values, image gradients, and image curvatures, respectively, that is used throughout the remaining part of this paper.

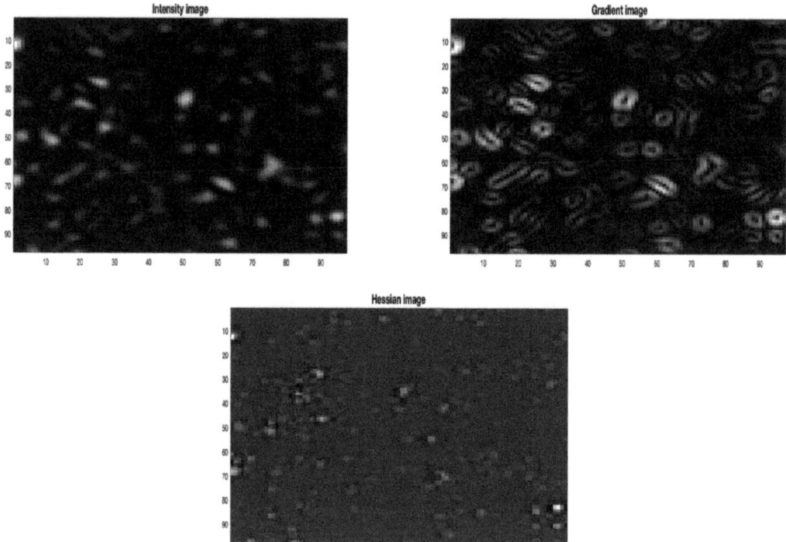

Figure 1. Images generated from the acquired set of intensity values. The upper left image shows the intensity distribution, the upper right image shows the magnitude of the calculated gradient vector, and the lower left image shows the determinant of the calculated Hessian matrix.

An intensity correlation is usually expressed as,

$$CI(\Delta \mathbf{x}) = \frac{\sum i_1(\mathbf{x}_1 + \Delta \mathbf{x}) i_2(\mathbf{x}_2)}{\sqrt{\sum i_1^2(\mathbf{x}_1 + \Delta \mathbf{x}) \sum i_2^2(\mathbf{x}_2)}}, \tag{1}$$

where $i = I - <I>$ is the zero-mean intensity image, $\mathbf{x}_1$ and $\mathbf{x}_2$ are coordinates associated with the first and second image, respectively, and the summation is taken over all image points considered. The correlation variable, $\Delta \mathbf{x}$, contains all translations and translation gradients considered in the correlation. The general procedure in image correlation is to pick out a subimage of size $m \times n$ pixels from I_1 and search in I_2 for the set $\Delta \mathbf{x}_{max}$ of $\Delta \mathbf{x}$ that maximizes the correlation value $CI(\Delta \mathbf{x})$ in Equation (1). The two translation components $[u, v]$ of $\Delta \mathbf{x}_{max}$ are then taken as an estimate of the local displacement vector in the region of the chosen subimage. A displacement field is generated from repetition of the procedure for a multitude of different subimages. The random error, e, with which $[u, v]$ can be determined has been shown to vary as [12],

$$e = k \frac{S^2}{M} \sqrt{\frac{1 - \gamma}{\gamma}}, \tag{2}$$

where k is an algorithm dependent constant of order unity, S is the average feature size, $M = [m, n]$ is the correlation window width and γ is the intensity correlation value. The quotient M/S defines in principle the number of independent contributions to the correlation and relates also to the reliability of the calculation while the additional S defines the curvature of the correlation peak and hence its susceptibility to random noise. The amount of random noise is specified by the correlation value γ.

As with Equation (1), a gradient correlation may be expressed as,

$$CG(\Delta \mathbf{x}) = \frac{\sum \mathbf{G}_1(\mathbf{x}_1 + \Delta \mathbf{x}) \cdot \mathbf{G}_2(\mathbf{x}_2)}{\sqrt{\sum |\mathbf{G}_1(\mathbf{x}_1 + \Delta \mathbf{x})|^2 \sum |\mathbf{G}_2(\mathbf{x}_2)|^2}}, \tag{3}$$

where $\mathbf{G} = \partial I / \partial x \hat{\mathbf{x}} + \partial I / \partial y \hat{\mathbf{y}}$ is the local image gradient vector, and $\hat{\mathbf{x}}$ and $\hat{\mathbf{y}}$ are orthogonal coordinate axes, respectively. Finally, a Hessian correlation function may be formulated as,

$$CH(\Delta\mathbf{x}) = \frac{\sum\left[\mathbf{v}_{11}(\mathbf{x}_1 + \Delta\mathbf{x}) \cdot \mathbf{v}_{21}(\mathbf{x}_2) + \mathbf{v}_{12}(\mathbf{x}_1 + \Delta\mathbf{x}) \cdot \mathbf{v}_{22}(\mathbf{x}_2)\right]}{\sqrt{\sum\left[|\mathbf{v}_{11}|^2 + |\mathbf{v}_{12}|^2\right]\sum\left[|\mathbf{v}_{21}|^2 + |\mathbf{v}_{22}|^2\right]}}, \tag{4}$$

where in the denominator special reference to $\mathbf{x}_1 + \Delta\mathbf{x}$ and $\mathbf{x}_2$ are omitted for ease of reading. In Equation (4), $\mathbf{v}_{11}$ and $\mathbf{v}_{12}$ are the two orthogonal principal vectors associated with the local intensity Hessian of image I_1. The corresponding vectors for I_2 are expressed as $\mathbf{v}_{21}$ and $\mathbf{v}_{22}$, respectively.

The three correlation functions in Equations (1), (3) and (4) are all normalized between $[-1, 1]$; however, their correlation features differ significantly. For example, as both the gradient correlation and the Hessian correlation involves vectors they are expected to drop off more quickly to a change in feature structure. Because of the normalization, the three correlation functions in Equations (1), (3) and (4) can be combined to produce the additional four correlation functions:

$$CIG(\Delta\mathbf{x}) = CI(\Delta\mathbf{x})CG(\Delta\mathbf{x}), \tag{5}$$

$$CIH(\Delta\mathbf{x}) = CI(\Delta\mathbf{x})CH(\Delta\mathbf{x}), \tag{6}$$

$$CGH(\Delta\mathbf{x}) = CG(\Delta\mathbf{x})CH(\Delta\mathbf{x}), \tag{7}$$

$$CIGH(\Delta\mathbf{x}) = CI(\Delta\mathbf{x})CG(\Delta\mathbf{x})CH(\Delta\mathbf{x}). \tag{8}$$

Hence, in total seven different correlation functions can be used to estimate the local deformation between the two set of images. Under what conditions either of them is preferable is investigated in the coming sections.

Figure 2 shows a comparison of the correlation properties between the seven correlation functions. The left image shows the width of the auto-correlation functions produced by setting $I_2 = I_1$ in Equations (1) and (3)–(8), respectively, as a function of displacement $\Delta\mathbf{x} = u$. It is seen that all additional correlation functions are narrower and more well-defined than the intensity correlation. In particular all correlation functions that include the Hessian matrix have a significantly sharper peak. One may also notice that all mixed correlation functions are essentially free from ringing, which indicates that the background fluctuations generated from the three images are uncorrelated. The right image shows the drop-off in correlation as a response to an intensity decorrelation between the two images considered. These results were produced from simulated images with speckle diameter of five pixels and a 64×64 pixels correlation window was used. Details are found in Section 3.1. It is seen in the right image of Figure 2 that the drop-off for the additional correlation functions are significantly quicker as compared to the intensity correlation. In particular, the functions that includes the Hessian correlation drops off fast. Whether this is a positive feature will be investigated in coming sections. In one respect, it is this feature that provides the sharp correlation peaks in the left image. However, a high sensitivity to small changes may also make the function unreliable.

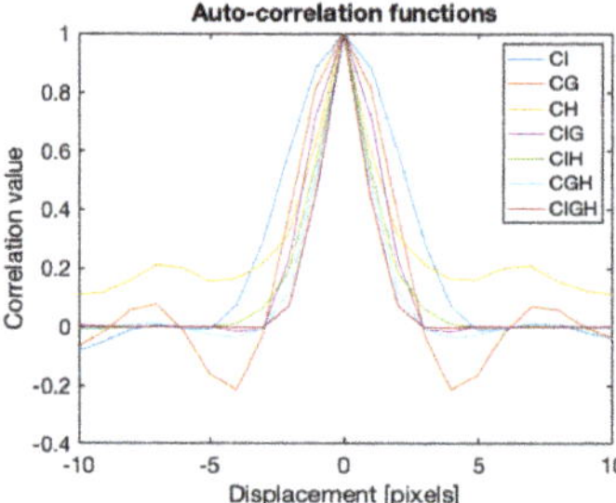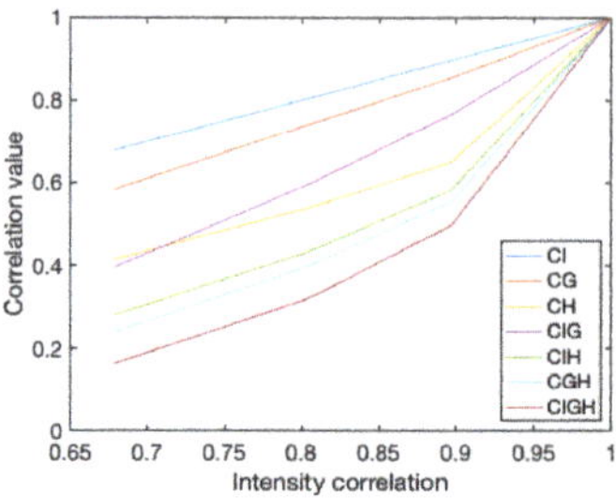

Figure 2. Sensitivity of the different correlation functions. The left image shows the width of the different auto-correlation functions as a function of displacement. The right image shows the decrease in maximum correlation value as a function of intensity correlation value. These results are generated according to the theory in Sections 2 and 3.1, respectively.

3. Evaluation of Correlation Bases

The different correlation functions described in Section 2 are evaluated using simulations and are demonstrated on real images using laser speckles. In the simulations, the average speckle size is varied between three and seven pixels, and the speckle motion can vary randomly between -5 and 5 in both orthogonal directions, respectively. Simultaneously, the speckle correlation is varied between unity and 0.7. In all simulations the correlation window was chosen to be 32×32 pixels and 225 independent windows are evaluated for each set of parameters. The reason for ignoring the effect of the correlation window size on the performance of the different correlation functions is that according to Equation (2) the important parameter to consider is the quotient M/S. Hence, it is sufficient to vary only the speckle size to capture the general behavior of the different correlation functions. The experiments were performed using laser speckles whose sizes were controlled by the objective aperture. In contrast to painted speckles, laser speckles are generated from random interference on the detector and do not exist on the object surface. Their extension thus depends on the numerical aperture of the imaging and on the wavelength of the laser [15]. In addition, for a well-focused system they will appear to follow the movement of the surface [16]. A set of ten rigid body translations were performed for each setting and the motion between the acquired images were analyzed in 225 independent regions with a correlation window size of 32×32 pixels using each of the described correlation functions, respectively. Details are provided in the subsections below.

3.1. Simulations

The simulations are performed on computer generated laser speckle image pairs in accordance with the procedure described by Sjödahl and Benckert [17]. Consider an $N \times N$ matrix A filled with random complex numbers were the real and imaginary parts are independently picked from a normal distribution. In all simulations $N = 1024$. The matrix A is taken to represent the exit pupil plane of a general imaging system. A quadratic aperture W of width $w = \lfloor N/S \rfloor$ pixels is placed centrally in the matrix, were S is the speckle size in pixels on the detector and $\lfloor \cdot \rfloor$ rounds down to the nearest integer. The reference speckle pattern I_1 is then generated as,

$$u_1(s_x, s_y) = W(s_x, s_y)A(s_x, s_y) \longrightarrow I_1(x, y) = |FT[u_1(s_x, s_y)]|^2, \tag{9}$$

where FT performs a 2D Fourier transform and (s_x, s_y) are spatial frequency components spanning the domain $[-1/2, 1/2 - 1/N]$ in both orthogonal directions, respectively. The deformed pattern I_2 is generated as,

$$u_2(s_x, s_y) = W(s_x, s_y)A(s_x - d, s_y)\exp[2\pi i(us_x + vs_y)] \longrightarrow I_2(x, y) = |FT[u_2(s_x, s_y)]|^2, \tag{10}$$

where $d = \lfloor w(1 - \sqrt{\gamma}) \rfloor$ represents a shift of the exit pupil coherence cells between the two recordings, γ is the intensity correlation value generated and $\mathbf{u} = [u, v]$ is the image plane speckle movement. In this way, speckle pattern pairs with a defined speckle size, S, relative motion, $\mathbf{u}$, and a defined intensity correlation, γ, can be generated. Mean deformation, standard deviation of the deformation magnitude, mean correlation value, and reliability are calculated for each of the speckle image pairs generated and for each of the seven correlation base functions. A single deformation estimate is in this case considered reliable if the estimate is within ± 1 pixel from the correct value. The results from the simulations are shown in Figure 3.

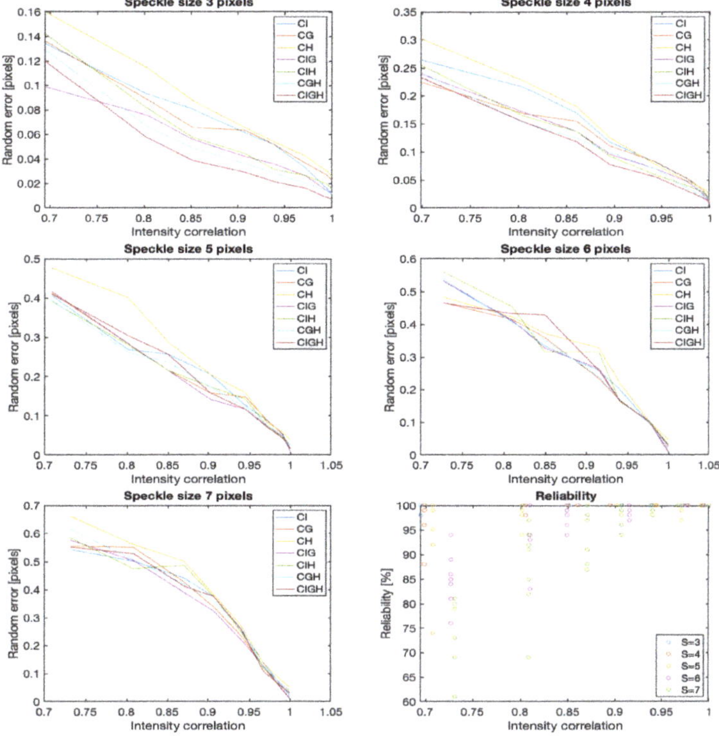

Figure 3. Results from the simulation. The plots show, row-wise, random errors from the different types of correlation functions defined by Equations (1) and (3)–(8) as a function of intensity correlation value for speckle sizes $S = 3, 4, 5, 6$, and 7 pixels, respectively. The lower right plot shows the reliability of the evaluations as a function of intensity correlation.

3.2. Experiments

The experimental set-up is sketched in Figure 4. The set-up consists of a 10 mW continuous wave He-Ne laser (632.8 nm wavelength) as illumination source, a white-painted aluminum plate, an $f = 55$ mm Mikro-Nikkor objective and a monochrome Dalsa nano camera (3.5 µm pixel size, resolution 2056×2464 pixels). The magnification was $m = 0.9$ which translates into 3.9 µm/pixel in object coordinates. One acquired speckle image with aperture setting $f/32$ is seen to the right in Figure 4. The experiments were performed with aperture settings $[f/22, f/16, f/11, f/8]$, which resulted in average speckle sizes $[7.6, 5.5, 3.8, 2.7]$ pixels, respectively. Ten consecutive in-plane translations are performed using a fine-pitch micrometer. Each incremental translation was approximately 2.5 µm, which translates into approximately 0.65 pixels on the detector. The total translation between the first and last image is therefore approximately 25 µm, which translates into 6.5 pixels in detector coordinates. The result from the experiment is summarized in Figure 5. The reliability was unity for each of the analyzed image pairs, and is not presented separately.

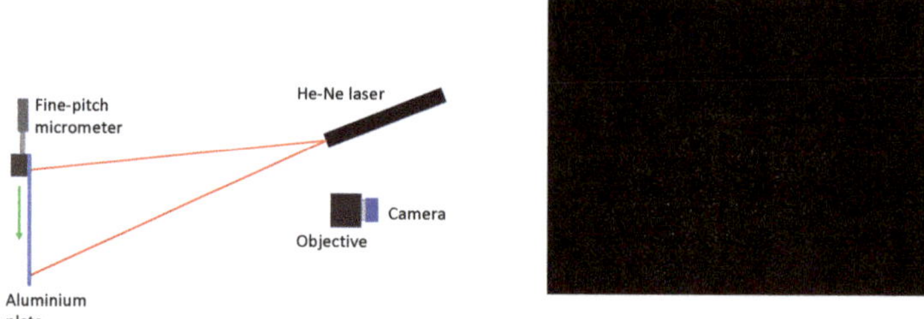

Figure 4. Sketch of the experimental set-up. The set-up consists of a 10 mW He-Ne laser, a monochrome camera from Dalsa equipped with a mikro-Nikkor camera objective with focal length $f = 55$ mm, and a white aluminum plate that can be translated in-plane using a fine-pitch micrometer. The right part of the figure displays one of the acquired speckle images. This image is acquired with an aperture setting of $f/32$.

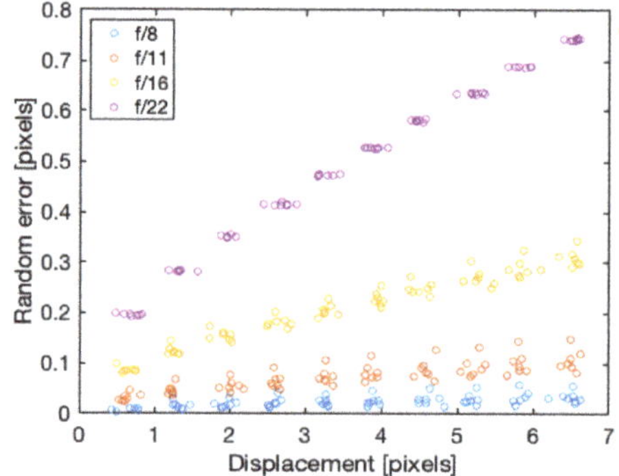

Figure 5. Random error in displacement as a function of speckle motion using different aperture numbers and analyzed using the seven different correlation functions. All results for the same aperture setting are lumped together.

4. Discussion and Conclusions

The structure highlighted in the three images in Figure 1 shows the different features that contributes to the different types of correlation functions defined by Equations (1), (3) and (4), respectively. These features are also responsible for the different shapes and drop-offs of the correlation functions shown in Figure 2. It is seen that out of the fundamental correlation functions, both the gradient correlation and to a greater extent the Hessian correlation produces a narrower and more well-defined peak and drops off more rapidly in response to feature degradations. With reference to the general behavior of correlation functions, these two effects should have contradictory effects on the accuracy of the deformation calculation [12]. The sharper peak should make the peak position more well-defined. The decrease in correlation value will on the other hand increase noise. The combined correlation functions show up a similar behavior with the most well-defined correlation peak produced by the multiplication of all three fundamental correlation functions. This is also the correlation function that drops off most rapid. One can also notice that this combined correlation function is basically free from higher order ringing, which indicates that the three fundamental correlation functions basically are uncorrelated.

A few interesting observations emerge from the results shown in Figure 3. For the high correlation values, the random errors always show up in the same order with the mixed correlation values the lowest. The difference between the best performing correlation function, which is the combination of all three fundamental functions, and the pure intensity correlation function is roughly a factor of two. The fundamental Hessian correlation performs constantly the worst, which is not surprising as the intensity Hessian is most susceptible to image noise. This disadvantage seems to be cancelled by multiplication with the other, more noise tolerant, correlation functions enhancing its advantage of being sharp. In fact, all mixed correlations involving the Hessian correlation performs well. One can also notice that the intensity correlation follows the general behavior $\sqrt{(1-\gamma)/\gamma}$, where γ is the intensity correlation value for all speckle sizes, while the other correlation functions do not. In fact, for the smallest speckle sizes the mixed correlation functions involving the Hessian correlation seem to be more robust to decorrelation and show up an opposite curvature as compared with the pure intensity correlation. This effect seems to be valid up to a correlation of roughly 0.85 where these functions turns up and approach the other functions. At this point the combined correlation function is approximately three times more accurate as compared to the pure intensity correlation. This effect is not as pronounced for larger speckle sizes and for speckle sizes in the range 5–7 pixels all correlation functions perform approximately the same. By this one can notice two things. Firstly, for small speckle sizes both the gradient image and the Hessian image becomes more pronounced and will dominate in areas where they generate large numbers. Hence, their positive feature of being sharp is pronounced. As the features grow larger their relative weight decrease. An additional positive effect of this is for images that contain sharp edges. Such images are notoriously tricky to analyze using traditional intensity correlation, but with the combined correlations the gradient and Hessian correlations will act as filters, which actually is the motivation for the gradient correlation introduced by O'Callaghan and Haga [14]. Secondly, as the relative weight of the gradient and Hessian correlations decrease all three correlation functions behave approximately Gaussian and their combined effect will follow the same general trend. In addition, as they are uncorrelated their combined effect will also be Gaussian, and one gains very little to combine them. In conclusions therefore, combinations that include gradient and Hessian correlations contribute positively for sharp and dense patterns, but their positive effect decrease rapidly for larger features.

The reliability presented in the lower right corner of Figure 3 shows a dramatic behavior on speckle size and correlation value. It is seen that the reliability is always unity for highly correlated patterns for all speckle sizes and for all correlation functions. As the correlation drops the reliability starts to drop, in particular for the larger speckle sizes. As a matter of fact, the reliability drops to as low as 0.6 for a speckle size of 7 pixels and a correlation value of 0.7. The significant drop-off in reliability for the larger speckle sizes shown in Figure 3 is to a large extent associated with the pure Hessian correlation and to some extent the combined gradient and Hessian correlation. All other correlation functions are unaffected. The susceptibility of these two correlation functions comes from the magnification of noise caused by numerical differentiation in combination with a small sample size characterized by the quotient N/S, where in this case $N = 32$ and S is the speckle size. A larger correlation window would significantly improve the reliability. As a matter of fact, it is recommended to keep the quotient N/S above ten for reliable results in low-correlation images [12]. The reliability is therefore considered manageable for all relevant correlation functions considered.

The speckle image shown in Figure 4 shows a typical feature often encountered in practice, that of an uneven illumination. Uneven illumination is often difficult to circumvent as the intensity profile of most illumination sources is uneven. A TEM_{00} laser, for example, has a Gaussian beam profile. Unless compensated for, uneven illumination will bias the displacement estimate towards the brighter regions [12]. However not explicitly tested, one can speculate that the gradient and Hessian correlation functions would help to prevent this unwanted effect. In this investigation, however, no significant bias was noted.

The pattern shown in Figure 4 is a laser speckle pattern. A laser speckle pattern has the unique quality of providing a random pattern defined by the full spatial bandwidth of the field with unit contrast. Hence, the feature size of the pattern can be controlled by choosing an aperture size optimum for the resolution of the detector. The great disadvantage with laser speckles is that they decorrelate in response to any change in the optical set-up [16], an effect that has prevented widespread use of laser speckles in experimental mechanics. This effect is even more pronounced for the small numerical apertures used with former digital detectors. Modern lines of digital detectors can however be purchased with as small pixels as 1 μm, which considerably opens up applications with laser speckles in experimental mechanics because of their ease of use. Figure 5 shows the results from application of the seven different correlation functions using the lens f−numbers $f/22$, $f/16$, $f/11$, and $f/8$, respectively. A dramatic dependence on lens f−number is seen for all correlation functions, in fact significantly more dramatic than any effect caused by the correlation functions themselves. For example, comparing the final deformation step for $f/22$ with the $f/8$ the random error is 0.73 with an intensity correlation value of 0.72 as compared with 0.02–0.04 and 0.90 for the larger aperture. This dramatic effect is caused by the double effect of enlarging the speckle size and enlarging the speckle decorrelation. In comparing the performance of the different algorithms for the same aperture settings the same trends as shown in Figure 3 are found. The random errors for the smaller speckle sizes and larger correlation values are about a factor of two smaller for the combined correlation functions as compared with the pure intensity correlation. The function combining all three pure correlations performs the best and the pure Hessian correlation the worst. The effect decreases for the larger speckle sizes and for lower correlation values. Finally, it was noted that the reliability turned out to be close to unity for all analyzed images with all correlation functions in contrast to the reliability of the simulated patterns. The reason for this discrepancy is however unknown.

In conclusion, the performance of seven different correlation functions applied in DIC have been investigated using simulated and experimentally acquired laser speckle patterns. The correlation functions were constructed as combinations of the pure intensity correlation function, the gradient correlation function and the Hessian correlation function, respectively. It was found that the correlation function that was constructed as the product of all three pure correlation functions performed best for the small speckle sizes and large correlation values, respectively, but that the difference between the different functions disappeared as the speckle size increase and the correlation value drops. On average the random error of the combined correlation function was half that of the traditional intensity correlation function within the optimum region. It was also found that for the small speckle sizes, all combined correlation functions involving the Hessian correlation function appear to be more robust against speckle decorrelation down to correlation values of roughly 0.85. The reason for this has not been investigated in detail but a good guess is that the more well-defined peak provided by the Hessian correlation makes its position more defined and less susceptible to noise. This effect disappears for smaller correlation values and for larger speckles. In addition, the monumental dependence of the imaging f−number on the accuracy of DIC using laser speckles is demonstrated experimentally. This dependence appears for all seven correlation functions and practically dominates the performance of the calculations. While the difference between the most optimum correlation function and the worst is in the order of three, the difference between results using different image apertures may be five times as large. Modern digital detectors with pixel sizes in the order of μm may therefore open up for a renaissance of laser speckles in experimental mechanics, in particular in situations where a good random pattern is difficult to apply.

Conflicts of Interest: The authors declare no conflict of interest.

References

1.   Sutton, M.; Wolters, W.; Peters, W.; Ranson, W.; McNeill, S. Determination of displacements using an improved digital correlation method. *Image Vis. Comput.* **1983**, *1*, 133–139. [CrossRef]

2. Bruck, H.; McNeill, S.; Sutton, M.; Peters, W. Digital image correlation using Newton-Raphson method of partial differential correction. *Exp. Mech.* **1989**, *29*, 261–267. [CrossRef]

3. Willert, C.; Gharib, M. Digital particle image velocimetry. *Exp. Fluids* **1991**, *10*, 181–193. [CrossRef]

4. Sjödahl, M. Electronic speckle photography: Increased accuracy by nonintegral pixel shifting. *Appl. Opt.* **1994**, *33*, 6667–6673. [CrossRef] [PubMed]

5. Hild, F.; Roux, S. Digital image correlation: From displacement measurement to identification of elastic properties—A review. *Strain* **2006**, *42*, 69–80. [CrossRef]

6. Hild, F.; Roux, S. Comparison of local and global approaches to digital image correlation. *Exp. Mech.* **2012**, *52*, 1503–1519. [CrossRef]

7. Holenstein, C.; Lendi, C.; Wili, N.; Snedeker, J. Simulation and evaluation of 3D traction force microscopy. *Comput. Methods Biomech. Biomed. Eng.* **2019**. [CrossRef] [PubMed]

8. Alhaddad, M.; Dewhirst, M.; Soga, K.; Divriendt, M. A new photogrammetric system for high-precision monitoring of tunnel deformations. *Proc. Inst. Civ.-Eng.-Transp.* **2019**, *172*, 81–93. [CrossRef]

9. Sjöberg, T.; Kajberg, J.; Oldenburg, M. Calibration and validation of three fracture criteria for alloy 718 subjected to high strain rates and elevated temperatures. *Eur. J. Mech. A/Solids* **2018**, *71*, 34–50. [CrossRef]

10. Castillo, E.; Allen, T.; Henry, R.; Griffith, M.; Ingham, J. Digital image correlation (DIC) for measurement of strains and displacements in coarse, low volume-fraction FRP composites used in civil infrastructure. *Compos. Struct.* **2019**, *212*, 43–57. [CrossRef]

11. Mouheb, N.; Montillet, A.; Solliec, C.; Havlica, J.; Legentilhomme, P.; Comiti, J.; Tihon, J. Flow characterization in T-shaped and cross-shaped micromixers. *Microfluid. Nanofluid* **2011**, *10*, 1185–1197. [CrossRef]

12. Sjödahl, M. Accuracy in electronic speckle photography. *Appl. Opt.* **1997**, *36*, 2875–2885. [CrossRef] [PubMed]

13. Neggers, J.; Blaysat, B.; Hoefnagels, J.; Geers, M. On image gradients in digital image correlation. *Int. J. Numer. Meth. Eng.* **2016**, *105*, 243–260. [CrossRef]

14. O'Callaghan, R.; Haga, T. Robust Change-Detection by Normalised Gradient-Correlation. In Proceedings of the 2007 IEEE Conference on Computer Vision and Pattern Recognition, Minneapolis, MN, USA, 17–22 June 2007.

15. Goodman, J.W. Statistical properties of laser speckle patterns. In *Laser Speckle and Related Phenomena*; Dainty, J.C., Ed.; Springer: Berlin, Germany, 1975; pp. 9–75.

16. Yamaguchi, I. Speckle displacement and decorrelation in the diffraction and image fields for small object deformation. *Opt. Acta* **1981**, *28*, 1359–1376. [CrossRef]

17. Sjödahl, M.; Benckert, L. Electronic speckle photography: Analysis of an algorithm giving the displacement with subpixel accuracy. *Appl. Opt.* **1993**, *32*, 2278–2284. [CrossRef] [PubMed]

Article

Image Classification for Automated Image Cross-Correlation Applications in the Geosciences

Niccolò Dematteis, Daniele Giordan * and Paolo Allasia

Research Institute for Hydrogeological Protection, National Council of Research of Italy, 10135 Turin, Italy; niccolo.dematteis@irpi.cnr.it (N.D.); paolo.allasia@irpi.cnr.it (P.A.)
* Correspondence: daniele.giordan@irpi.cnr.it

Received: 1 April 2019; Accepted: 4 June 2019; Published: 8 June 2019

Featured Application: This study proposes a method that enables the automatic application of image cross-correlation when monitoring any displacements in the geosciences. As such, it solves one of the main current image processing issues: The requirement of manual image selection. The method reduces the need for extensive financial and human resources when conducting surveys, and it can be applied in a preventive warning context.

Abstract: In Earth Science, image cross-correlation (ICC) can be used to identify the evolution of active processes. However, this technology can be ineffective, because it is sometimes difficult to visualize certain phenomena, and surface roughness can cause shadows. In such instances, manual image selection is required to select images that are suitably illuminated, and in which visibility is adequate. This impedes the development of an autonomous system applied to ICC in monitoring applications. In this paper, the uncertainty introduced by the presence of shadows is quantitatively analysed, and a method suitable for ICC applications is proposed: The method automatically selects images, and is based on a supervised classification of images using the support vector machine. According to visual and illumination conditions, the images are divided into three classes: (i) No visibility, (ii) direct illumination and (iii) diffuse illumination. Images belonging to the diffuse illumination class are used in cross-correlation processing. Finally, an operative procedure is presented for applying the automated ICC processing chain in geoscience monitoring applications.

Keywords: image cross-correlation; monitoring; geosciences; automated systems; machine learning; image classification; image shadowing

---

1. Introduction

Image cross-correlation (ICC) is a well-known methodology used in the geoscience field to measure earth surface dynamics and deformation phenomena [1–4]. Ground-based ICC applications enable observations of relatively fast natural processes at a medium range and at a high spatiotemporal resolution. In addition, the low costs, minimal equipment required to conduct photographic surveys and the certain degree of automation used in data processing, make ground-based ICC applications valuable tools for use in monitoring phenomena, even in harsh environments [3]. The adoption of an automated procedure for processing monitoring data allows high-frequency-updated results to be obtained without the need for continuous human supervision, and can be applied in early warning system applications, thereby reducing human and economic costs [5]. Different automatic approaches used to collect and process monitoring data have been reported in literature, such as those for the inclinometer [6], total station [7,8], ground-based SAR [9] and integrated systems [10].

However, the major limitation of the technology used in the ICC approach for natural phenomena monitoring, is its dependence on visual conditions, since adequate illumination (i.e., sunlight) and a complete view of the scene (i.e., clear sky) are required. This prevents an automated ICC procedure being applied in near-real-time that could be used for warning purposes, as a human intervention to selecting suitable images is required. In addition, the presence of shadows formed by the interaction between direct illumination and surface roughness can cause errors in the cross-correlation (CC) process, and this problem has often been considered in literature [1,11–16]. The most popular solution used to avoid such effects is to consider pictures that have a similar visual appearance, and have been acquired in a similar light (usually in the middle of the day [12,17,18] when the length of shadows is reduced, due to the higher sun elevation angle). However, Ahn and Box [12], and Giordan et al. [15] proposed the use of images acquired in conditions of diffuse illumination (such as the evening hours) when any shadows are minimal or absent. Both of these approaches involve the manual selection of images, and images with a partial or absent view (caused by fog or the presence of obstacles), and with non-homogeneous illumination, are thus discarded. In this respect, Gabrieli et al. [14] developed a method for automatically discarding images taken during adverse meteorological conditions; this involved analysing the mean and standard deviation of colour values along predefined lines in the image. Hadhri et al. [19] also automatically identified images containing artefacts, or where vision was obscured, using a posteriori statistical analysis based on image entropy. Furthermore, Schwalbe and Maas [16] developed a method for automatically detecting shadowed areas, and then removing them from the CC computation.

In our work, we describe a method of classifying images according to visible and illumination conditions that enables the selection of images with diffuse illumination. To correctly classify the images, we adopted a support vector machine (SVM) approach. This supervised machine learning method was originally developed by Boser et al. [20] and Cortes and Vapnik [21]; it is a well-known methodology used in remote sensing and geoscience, and it has been adopted in a wide range of fields and applications (please, see reviews in [22–25]). In addition, we present the use of completely automated processing in monitoring active gravitational processes through ICC. The principal innovation relies on applying autonomous image selection to ICC according to the a posteriori probability that an image belongs to a certain class of illumination.

2. Methods

The objective of this study was to develop a procedure that conducts image cross-correlation (ICC) autonomously, and which can be implemented for monitoring geophysical processes. The ICC method is briefly described in Section 2.1. As already mentioned, one of the main sources of uncertainty in ICC results relates to the presence of the shadows; therefore, the impact of shadows on the ICC is analysed here to demonstrate how they negatively affect any results (Section 2.2). With respect to this inherent problem, a method that autonomously selects images acquired with diffuse illumination is developed, and to achieve this, a support vector machine (SVM) is trained to distinguish between three classes of images in accordance with the presenting illumination (Section 2.3). Finally, an operative and autonomous procedure is designed to conduct ICC, with the aim of being used in geoscience monitoring applications (Section 2.4).

2.1. Image Cross-Correlation Processing

Digital image correlation (DIC) developed following the advent of performant computer machines and digital photography in the early 1990s, and has been applied predominantly in fluid dynamics [26,27] and satellite imagery [1]. The rationale behind DIC is to determine the field of motion using spatial cross-correlation between corresponding subsets of two images (image cross-correlation, ICC), where ICC analyses the texture of an image rather than specific recognisable elements (as in feature tracking methods [28,29]).

ICC can be computed in the spatial domain (direct cross-correlation, DCC) [1,3,26] or in the frequency domain (phase cross-correlation, PCC) using the discrete Fourier transform (DFT). In the former case, a subset of the original (master) image slides into a larger interrogation area of the slave image. In this respect, a two-dimensional (2D) correlation coefficient is computed for each possible position, and a map of correlation values is obtained. The coordinates of the element with the maximum correlation coefficient correspond to the displacement of the slave image in relation to the master. Although DCC requires high computational costs, it suffers less from decorrelation associated with motion, because it computes correlations between the same tiles in different positions (the Lagrangian approach). Conversely, the PCC computes the correlation between master and slave tiles (that are in the same position) into the image (the Eulerian approach). Therefore, part of the original texture is removed with respect to motion from the interrogation area, and this causes possible decorrelation. The PCC is computed through a complex product involving the Fourier transforms of the master and slave images, and this is equivalent to conducting a spatial cross-correlation in accordance to the convolution theorem.

In our work, the ICC is operated in the phase domain, as it provides a faster computation [30,31] process than DCC, but a similar performance [32]. Specifically, we use the two-step computation method proposed by Guizar-Sicairos et al. [33]. In the first run, the CC is computed with integer precision, thereby avoiding the zero-padding operation, and the centre of the slave matrix is then translated according to the shifts obtained. In the second run, the DFT is applied only to a small subset of the data to obtain subpixel accuracy. This method has the advantage of strongly limiting the zero-padding operation, thereby reducing the computational costs, and limiting the loss of information intrinsic in the PCC due to the Eulerian approach used in measurements [16,31].

The processing chain adopted in this study is based upon the approach presented in Dematteis et al. [34], and can be divided into four steps. First, certain image enhancement operations are conducted [12–14]. The scene illuminant is computed using principal components analysis (PCA) [35] and then subtracted from the trichromatic (RGB) image. The removal of the illuminant allows the reduction of possible low-frequency signals caused by illumination changes that can affect the computation of the DFT [36]. Subsequently, the image is converted to greyscale, and a sharpen mask is applied to enhance the details in the image texture. Such an operation aims to reduce any possible defocusing or blurring effects that are frequent in images acquired with limited illumination; in this respect, Ahn and Box [12] stated that the application of a sharpen mask can improve the ICC results in hazy situations. Second, the images are coregistered with respect to a common image to correct possible misalignments. The CC is computed on an area that is assumed to be stable, and the images are planarly translated according to the shifts obtained. Third, a sliding window identifies corresponding tiles on both the master and the slave images; the CC is then computed on the tile couples. Adjacent windows are overlapped by 50% to improve the robustness of the results. Fourth, outlier removal is conducted. The usual approaches employed to correct possible outliers rely on heuristic and manual analysis [15,17], but Ahn and Box [12] identified outliers using cluster analysis during the CC procedure, and then applied a combined statistic-heuristic method in the post-interrogation phase. In this study, outlier correction is conducted using the universal outlier detection (UOD) approach [37] as this has the advantages of being automated, unsupervised, and non-subjective, which are fundamental requisites for automatic processing.

2.2. Analysis of Shadow Impact in ICC

Shadow lengths that change in relation to the sun's position during the course of the day can affect the CC results by producing apparent motion. However, their influence has been rarely quantified [13,16]. For example, Travelletti et al. [13] observed that changes in illumination can lead to an average CC offset $\mu > 1px$ for images acquired over a time lag of 1–3 h, and also noticed that the correlation coefficient diminishes at a low sun elevation angle.

In this study, the effects of shadows on the results of the CC are investigated by considering four groups of images that have different illumination conditions: (i) Diffuse illumination; (ii) direct illumination in images acquired at the same hour in the day (12:00), when the sun azimuth and elevation were almost the same; (iii) direct illumination in images taken at different times during the day (i.e., when the sun was in different positions); and (iv) synthetic images produced from an existing digital surface model using a set of shaded reliefs. In the latter, the position of the lighting point was changed, thereby simulating the motion of the sun during the course of a day.

For each dataset, ICC was conducted for each possible pair of images and 45 CC maps were produced. The mean and standard deviation of the shifts for each map were then analysed, as these are associated with measurement accuracy and precision, respectively.

The possibility that a shadow's presence could have differing effects on the CC results performed in phase or space domains was investigated, and the same analysis described above was conducted using both DCC and PCC. Results showed no relevant differences between the two techniques.

2.3. Image Classification

As previously mentioned, using images acquired in conditions that are diffusely illuminated enables a reduction in the apparent motion caused by changes in shadow shapes in relation to surface roughness. To this aim, it is fundamental that images are categorized in accordance with their visual appearance. In this study, three classes of visual and illumination states are considered: (i) Direct illumination (SunLight), (ii) diffuse illumination (DiffLight), and (iii) blocked or limited view (NoVis).

To identify the class of the images, SVMs are trained based on defining optimal decision hyper-planes that separate objects belonging to different classes [38]. In addition, the SVM assigns an a posteriori probability that each image belongs to a particular class. The objects are represented as vectors in an *n*-dimensional space, where the axes correspond to specific features that describe the objects. The elements of vectors represent feature values of each object that will be classified (for a didactic description of the SVM methods, please see Haykin [39]).

To describe the images, seven different features are identified as follows: (i) The number of RGB triplets (i.e., the values for R, G, and B bands) (*colorNum*), where the number of triplets is divided by the number of pixels in the image. The division is conducted to make the datum comparable for different image resolutions; (ii–iii) the number of pixels in the range 0–63 (*blackNum*) and 64–127 (*greyNum*) of the intensity histogram, which are determined using an 8-bit colour chart; (iv–v) the mean value of the hue (*hueMean*) and saturation (*satMean*) channels in the HSV colour space; (vi) the value of the maximum peak in the image intensity histogram (*maxPeak*), and (vii) the corresponding bin position (*posPeak*) (Figure 1). The image intensity (or greyscale) is defined as the mean value of the RGB channels.

We choose features that could be informative about the image appearance, and that were not correlated among them. Specifically, *colorNum* should be directly associated with the amount of light in the image, which enhances the colour contrast, while *blackNum* and *greyNum* indicate different levels of image darkness. *hueMean* is the average colour shade, and can be related to ambient skylight (i.e., diffuse illumination) [40], while *satMean* is high when the colours are vivid, and low when they tend toward grey. *posPeak* represents the dominant colour, while *maxPeak* is the magnitude of the dominant colour. Figure 1a shows an 8-bit (256 colour bins per RGB channel) NoVis class image. Only 0.5% of the possible RGB triplets are present, which is evident from the narrow spatial distribution of the RGB values shown in Figure 1d. The upper portion of the picture has an almost uniform and grey colour that implies low *satMean* (Figure 1b); this produces a prominent dark peak (high *maxPeak*, low *posPeak*) (Figure 1e).

To reduce the dimensionality of the decision plane (i.e., to reduce the number of features used to train the SVM), a Monte Carlo cross-validation is conducted [41,42], with the aim of selecting a group of features (i.e., the estimator) that minimize misclassification errors. The fundamental of that approach is to randomly divide the available known data into two groups, one to train the SVM (training set) and one to evaluate the misclassification error (test set), and the feature selection procedure is repeated many times. Features are sequentially added to the candidate estimator by selecting the feature or group of features that minimize the misclassification error (averaged on the multiple iterates). This procedure guarantees a higher generality of the SVM and enables overfitting to be avoided.

To maintain generality no a priori information on the probability of class occurrence is introduced during the tuning of the SVM.

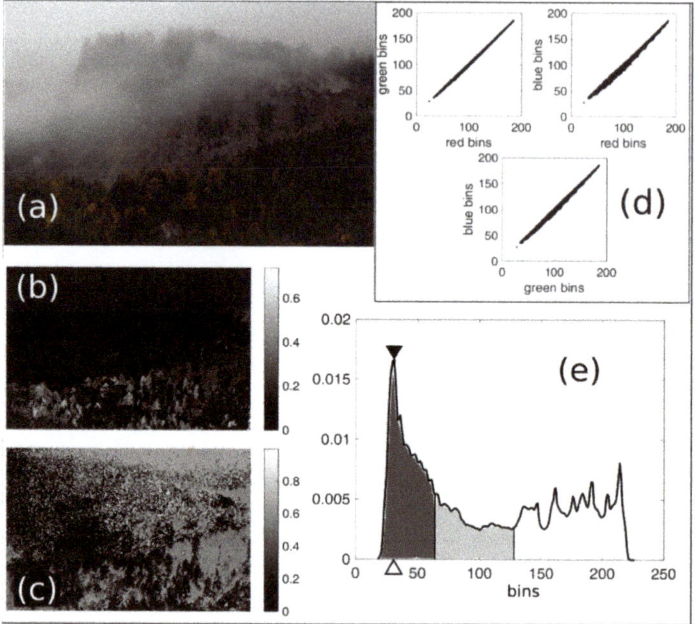

Figure 1. (**a**) Original 8-bit trichromatic (RGB) image of NoVis class; (**b**) saturation map; (**c**) hue map; (**d**) 8-bit RGB values, and (**e**) intensity histogram in which the ordinate value of the black triangle indicates the *maxPeaks* and the abscissa value of the white triangle represents *posPeaks* features. The dark and light grey areas are *blackNum* and *greyNum* features, respectively.

2.4. Automated ICC Procedure

The procedure presented here automatically selects images on which to perform ICC and obtains displacement maps of geophysical phenomena. This procedure was developed in 2018 and its effectiveness was tested in the monitoring of the Planpincieux glacier on the Italian side of the Mont Blanc massif. The processing chain is designed for automatic, continuous, and autonomous monitoring via ICC in the typical situation of daily acquisitions. The procedure is best suited for surveying natural phenomena at velocities greater than the common ICC uncertainty, i.e., $\sim 10^{-1}$ px [12,13]. It can thus be applied to observe relatively fast natural processes such as glacier flows and slope instabilities that are quite active. Depending on the phenomenon monitored, it is possible to vary the scheduled image acquisition and processing frequencies.

The procedure is conducted on a daily frequency and it is divided into four independent subroutines (Figure 2).

(i)   Image acquisition. First, the images are acquired at an hourly frequency by the monitoring system and are then transferred in real time to the data collection section of the operative server.

(ii)   Image classification and selection. The images acquired in the current day are classified through SVM according to the visual and illumination conditions. It is attributed a posteriori probability to belong to the classes. The results are saved in a list of classified images, and this list is subsequently browsed to search for the image with the highest probability score to belong to the DiffLight class. Such an image is labelled as suitable for ICC. If there are no new images, a warning is released and the process skips to the next subroutine.

(iii)   Image coregistration. This subroutine coregisters the new images acquired. A common reference image is used to coregister all the images and the coregistered images are saved. Such images are used for ICC and for possible manual vision inspection.

(iv) Image cross-correlation. The list of images suitable for ICC is browsed to check for possible new image couples. If identified, the new images are cross-correlated and displacement maps are computed.

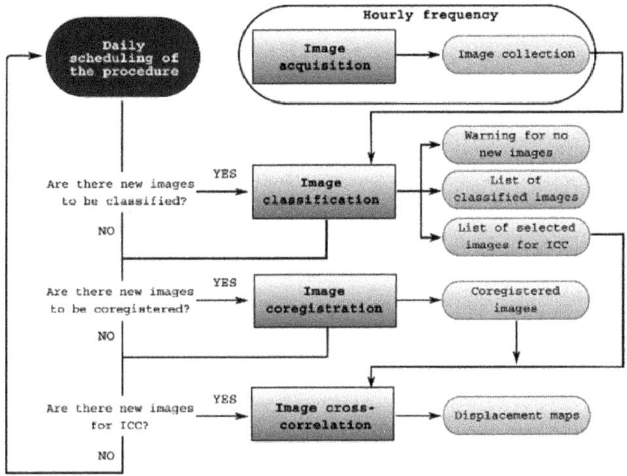

Figure 2. Workflow of an automated procedure used in image cross-correlation (ICC) applications.

To evaluate the performance of the automated image selection, we manually classified the images and selected the images suitable for ICC. On several occasions (i.e., approximately 10% of cases), manual classification of images was difficult, and there was a certain ambiguity between the classes. Therefore, three new bivalent classes were introduced relating to images that could be arbitrarily classified as either SunLight or DiffLight, SunLight or NoVis, and DiffLight or NoVis (respectively SD, SN, DN).

Moreover, the performances of automated selection through SVM were then compared with other possible simpler alternatives of automatic criteria. We analysed the cases of selecting images at a fixed hour of the day: At 12:00 (the SunLight class) and at 18:00 (the DiffLight class). To simulate such automated selection methods, photos were picked without applying quality control to verify whether the scene was adequately visible.

3. Dataset

To develop and test the analysis described in Section 2, various dedicated datasets were employed (Table 1). These comprise images of natural environment sites with heterogeneous geo-hydrological phenomena in north-western (NW) Italy (Figure 3). The generality of the method was thus strengthened by using a wide range of images.

Two different datasets comprise images from Monesi village (Liguria region, NW Italy), which is an Alpine village partially affected by a rotational slide activated during a flood in October 2016. MNS1 is the shaded relief extracted from a digital surface model (DSM) acquired using a drone and MNS2 is an RGB dataset acquired using a fixed camera installed on the opposite side of the valley.

The ACC dataset (Piemonte region, NW Italy) comprises a sequence of images acquired by a fixed camera installed to monitor the evolution of a large bedrock outcrop that has historically been affected by rock falls. During the monitoring period, there was no evidence of any slope instability activation in the area studied.

Table 1. Datasets adopted in the current study. The first column lists the type of analysis employed using the various datasets. Also presented is the site where images were acquired for each dataset, the type of environment and image, the number of elements, the image resolution and the symbol attributed to the dataset.

Analysis	Site	Environment	Image Type	# Images	Resolution [px]	Dataset Symbol
Shadow effect	Monesi	Alpine village	Shaded relief	10	656×875	MNS1
	Monesi	Rotational landslide	RGB	30	3456×5184	MNS2_a
Image classification	Acceglio	Bedrock outcrop	RGB	278	3456×5184	ACC
	La Saxe	Rockslide	RGB	159	3240×4320	SAXE1
	La Saxe	Rockslide	RGB	163	1536×2048	SAXE2
	La Saxe	Rockslide	RGB	268	3312×4416	SAXE3_a
	Monesi	Rotational landslide	RGB	117	3456×5184	MNS2_b
	Planpincieux	Glacier	RGB	225	3456×5184	PPCX1
	Planpincieux	Glacier	RGB	195	3456×5184	PPCX2_a
Automated ICC procedure	La Saxe	Rockslide	RGB	139	3312×4416	SAXE3_b
	Planpincieux	Glacier	RGB	1549	3456×5184	PPCX2_b

Three different datasets comprise images of the Mont de la Saxe rockslide (Valle d'Aosta region, NW of Italy); this is an active landslide with an estimated volume of 10^7 m^3 that endangers the underlying village and road, which road provides access to the Mont Blanc tunnel, an important communication route between Italy and France [5,43–45]. The three different datasets that were acquired from three different view perspectives for this rockslide are SAXE2 and SAXE3, which were acquired by cameras positioned outside the area of gravitational instability, and SAXE1, the camera for which was installed on the rockslide and controlled one of the most active sectors.

SAXE3 is divided into two subsamples: SAXE3_a, which includes pseudo-random images taken in 2014 and 2015, and SAXE3_b, which is a time-lapse relating to the period 15–30 April 2014, when activity was pronounced, and a large failure occurred. There are no images included in either SAXE3_a or SAXE3_b.

The last dataset consists of images of the Planpincieux glacier on the Italian side of the Mont Blanc massif. The glacier is characterised by relevant kinematics and dynamics; numerous detachments from the snout have endangered the underlying hamlet of Planpincieux. This glacier has therefore been the subject of various remote sensing surveys in the recent past [15,34,46]. Two different cameras monitor the evolution of the glacier: PPCX1 acquires images of the lower part of the glacier while PPCX2 acquires images of the right lower tongue, the most active part of the glacier. The PPCX2 dataset, like the SAXE3 dataset, is divided into two distinct subsamples: PPCX2_a, which comprises images acquired during the years 2014–2017; and PPCX2_b, which includes the complete dataset acquired between 18 May and 3 October 2018. During this period, the automated ICC procedure described in Section 2.4 was applied in real time and more than 1500 images were collected.

The following text describes and discusses the use of every dataset in the procedures described in Sections 2.2–2.4.

To evaluate the effect of shadows we conducted two different analyses: (1) A synthetic set of ten shaded reliefs of a DSM was produced by changing the lighting source position to simulate the sun's motion during the day. Shadows cast by obstacles were projected by computing the viewshed from the light source. Three of the ten shaded reliefs produced are shown in Figure 4a.1–a.3. (2) Ground-based RGB pictures of the same area acquired in different illumination conditions were considered. Three groups of ten images were selected that were acquired in (i) diffuse illumination (Figure 4(b.1)), (ii) direct illumination (Figure 4(b b.2)), and (iii) direct illumination at the same hour (at 12:00) (Figure 4(b b.3)), so that similar direct illumination was maintained in the photos.

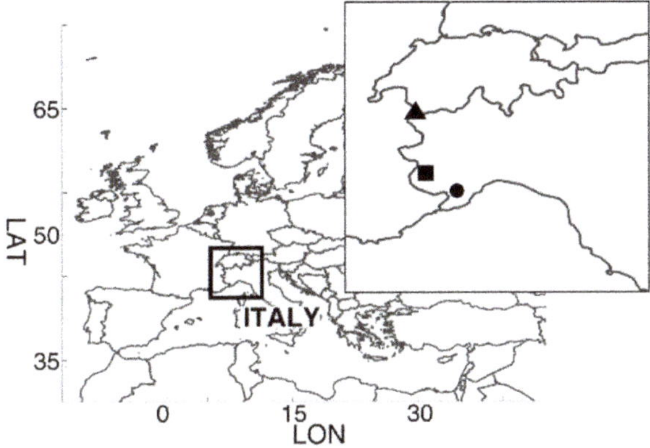

Figure 3. Locations of sites corresponding to the various datasets adopted. The black circle, square, and triangle correspond to Monesi di Mendatica, the Acceglio rock face, and the Mont de La Saxe landslide and Planpincieux glacier, respectively. This image was produced using the Matlab package borders [47].

An ensemble of seven datasets representing scenes from various natural environments was used to study the image classification methodology and to ensure that the method could be generalised. Specifically, datasets ACC, SAXE1, SAXE2, SAXE3_a, MNS2_b, PPCX1 and PPCX2_a were employed, and samples of images from these datasets acquired in different visual conditions are shown in Figure 4c1–i3. Each dataset comprises photos belonging to different classes (i.e., acquired in different illumination and visual conditions), as catalogued during the process described in Section 2.3. The numbers of images in each class employed from the different datasets are presented in Table 2.

The images included in all the datasets were acquired during different years and seasons. The results showed that the SVMs classified the images accurately, even when dissimilar environmental conditions and morphological changes were presented. However, although it is considered that SVM training by adopting photos taken during specific seasons could improve its performance, this could result in a loss of generality.

Table 2. Numbers of images belonging to different classes used in each dataset.

Dataset	SunLight	DiffLight	NoVis
A	94	97	87
B	77	45	37
C	53	60	50
D	95	96	77
E	59	31	27
F	75	75	75
G	65	65	65

The last application is automated ICC processing, and this was conducted using two-hourly time-lapses. The first dataset (PPCX2_b) comprised more than 1500 images, and it was collected to monitor the activity of the Planpincieux glacier from 18 May to 3 October. During this period, the automated ICC procedure described in Section 2.4 was applied in real time to monitor the glacier's evolution. The second dataset used was SAXE3_b, where 139 images of the Mont de La Saxe landslides were selected. Table 3 shows the number of images employed from PCCX2_b and SAXE3_b datasets and their distribution among the different classes (considering also the bivalent classes described in Section 2.4).

Figure 4. Examples of images taken from different datasets. Row (**a**) shows the 1st, 3rd and 7th shading reliefs with changes in the light source position; (**b.2**,**b.3**) were acquired in direct illumination at different hours of the day; and rows (**c–i**) show examples of each class from the datasets used in image classification analysis. The photographs in rows (**j**,**k**) were used to develop the automatic procedure.

Table 3. Number of images manually classified for datasets D1 and G1.

Dataset	SunLight	DiffLight	NoVis	SD	SN	DN
D1	24	78	11	16	0	10
G1	838	386	171	105	15	34

4. Results

A quantitative analysis of how the shadow effects affect the ICC performance is presented in Section 4.1, and this demonstrates the negative impact of that phenomenon. In Section 4.2, the results of the Monte Carlo cross-validation are presented, with the aim of determining the best features that can be used to train the SVMs. Finally, Section 4.3 presents the results of the automatic ICC processing chain with respect to two operative case studies.

4.1. Shadow Impacts of ICC

The MNS2_a dataset comprises three groups of ten images with different illumination states. As the acquired scene is fixed, it is valid to assume that any possible change in the results relates principally to the shadow effect. Figure 5 shows a boxplot of the mean (μ) and standard deviation (σ) of the single CC maps of each group; these were computed for all possible image couples, and differentiated for the vertical (y) and horizontal (x) directions. It is evident that the dispersion of the two variables is at a minimum for the images acquired with diffuse illumination. Moreover, the σ obtained with images acquired at the same hour and in direct illumination are approximately two and three times greater, on average, than that obtained for images acquired in diffuse illumination. It is also evident that the μ of the horizontal component has slightly greater variability, while there are no significant differences between the two components for the σ.

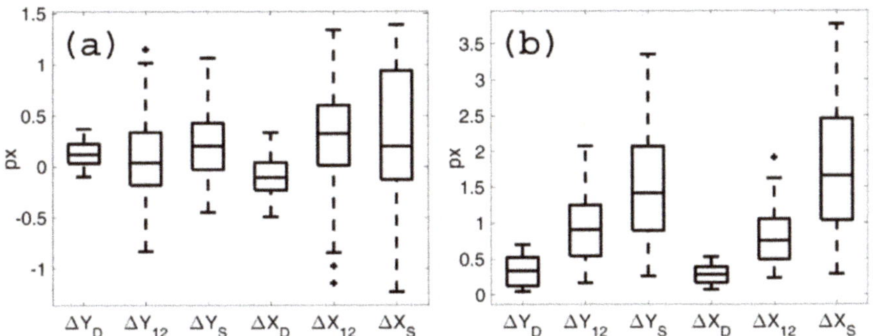

Figure 5. Boxplots of (**a**) mean and (**b**) standard deviation of CC. Subscripts D, 12 and S are as follows: D represents images acquired respectively with diffuse illumination, 12 denotes images with direct illumination taken at a fixed hour of the day (i.e., 12:00), and S concerns images taken in direct illumination with different sun azimuths and elevation angles.

Figure 6 presents the μ and σ of the ICC maps produced using all of the couples of the available images from the dataset MNS1, where the results are ordered according to the simulated position of the sun during the day. The results obtained with synthetic shaded relief show that both μ and σ increase with an increase in the difference between the azimuth and elevation of the sun's position. In addition, both values are higher for the horizontal direction, which indicates that the change in the sun's azimuth has a greater effect on CC results than the change in elevation.

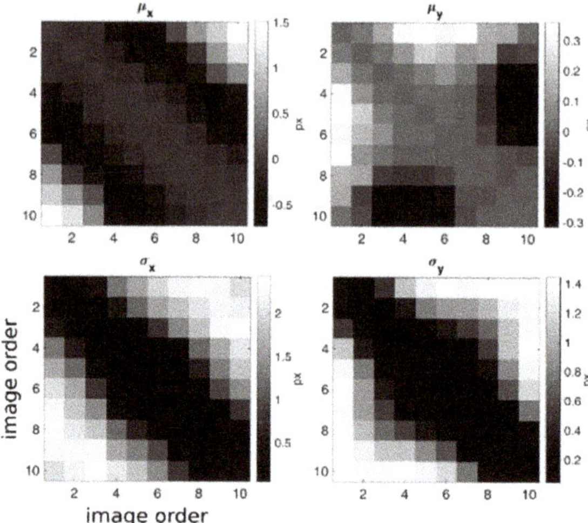

Figure 6. Mean and standard deviation of cross-correlation maps for all possible couples of shaded relief images. The two matrices on the left refer to the horizontal direction and those on the right refer to the vertical. The elements of the matrices correspond to the couples formed from the *i*th and *j*th images of a sequence of ten images that were ordered to simulate the motion of the sun during the day.

4.2. Cross-Validation and Image Classification

The results of the cross-validation (Table 4) show that the minimum classification curve is achieved by using three or four features to train the SVMs (Figure 7), where the minimum error obtained ranges between 0.03% and 7.9% (with a weighted mean of 5.2%). In contrast, greater errors are obtained when more than four features are employed; this can probably be ascribed to overfitting in the training phase, as the number of training elements is limited.

Table 4. Order of features identified during cross-validation analysis for each dataset; features in black are those selected for classification.

Dataset	I	II	III	IV	V	VI	VII
ACC	colorNum	hueMean	maxPeak	posPeak	satMean	greyNum	blackNum
SAXE1	colorNum	posPeak	hueMean	blackNum	maxPeak	greyNum	satMean
SAXE2	colorNum	maxPeak	posPeak	hueMean	greyNum	satMean	blackNum
SAXE3_a	colorNum	hueMean	greyNum	maxPeak	satMean	posPeak	blackNum
MNS2_b	colorNum	blackNum	maxPeak	greyNum	hueMean	satMean	posPeak
PPCX1	colorNum	hueMean	blackNum	posPeak	satMean	greyNum	maxPeak
PPCX2_a	colorNum	posPeak	hueMean	blackNum	maxPeak	greyNum	satMean

The *colorNum* is evidently the predominant and most relevant feature in the classification. For each dataset, it is found to be the most important feature; it alone provides a misclassification error in the range of 6% to 18%. *hueMean* is included in the group of relevant features in five of the datasets, whereas *greyNum* and *satMean* are never selected as relevant features. However, the features required for a proper training of the SVM differ between the datasets, and it is thus necessary to train a dedicated SVM for each specific scene.

For the MNS2_b dataset, a local minimum is achieved with only two features, while the absolute maximum is obtained with four features. Considering that this dataset comprises a limited number of images, training the SVM with two features is probably the best solution to avoid overfitting.

It is also worth noting that to select the minimum number of features, it may be a good criterion to use the features up to where the error curve decreases by less than 0.5% when one more feature is added. Using this approach, two or three features would be sufficient, depending on the dataset.

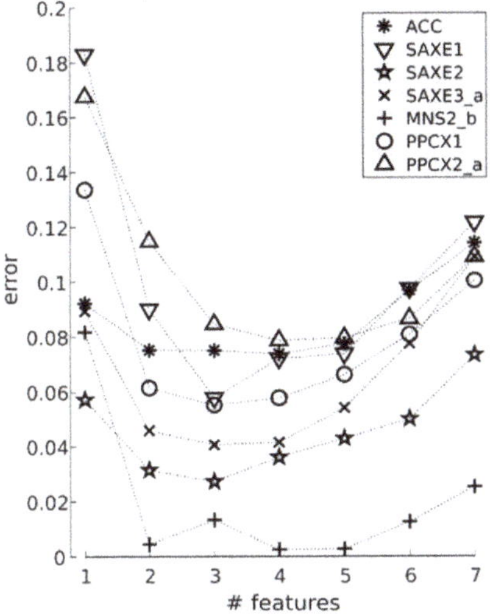

Figure 7. Error curves (expressed as fraction of misclassification) obtained from cross-validation of the support vector machine (SVM) training. The minimum was always reached by employing three or four features.

Figure 8 shows an example of image classification using the two of the most relevant features obtained from the cross-validation (Table 4), to illustrate the migration of feature values among the different visual conditions. From the graph, it can be seen how *colorNum* increases from the NoVis, DiffLight and SunLight classes (which is expected) while *hueMean* of the SunLight class is lower.

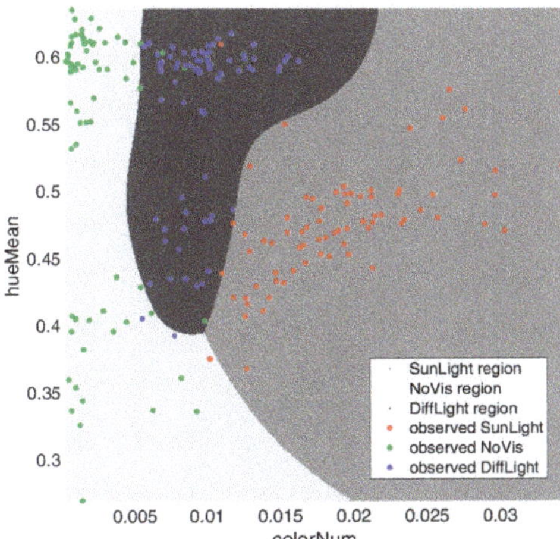

Figure 8. Image classification of PPCX_1 images using the two principal features *hueMean* and *colorNum*. The regions delimit the areas where new elements within are classified as belonging to the corresponding class.

4.3. Automated ICC Procedure

To evaluate the results of the automated ICC procedure, image classification and selection performances were analysed. Table 5 shows the results of automatic classification. In the case of images belonging to the bivalent class, an image was considered to be correctly classified if it was classified into one of the two classes that had been identified manually. Results therefore show that the classification error for dataset SAXE3_b is 4.3%, while the selection error is 0%. For dataset PPCX2_b, the misclassification is 7.7%, while the error in determining whether the correct image should be included in the ICC processing is 13%. For both datasets, the classification error is comparable with the value obtained during the cross-validation analysis. However, the selection error is related more to the occurrence of DiffLight false positives, which are higher in dataset PPCX2_b but considerably lower in SAXE3_b.

Table 5. Results of automatic classification and selection for datasets SAXE3_b and PPCX2_b. The number of images for each class are reported and correspond to exact, bivalent and erroneous classification. The percentage of correctly classified images is shown in the row below (in relation to the sum of exact and bivalent classifications). The two columns on the right show the percentage overall classification error and selection error, respectively.

Dataset	SunLight			NoVis			DiffLight			Classification Error	Selection Error
SAXE3_b	22	2	4	11	2	0	74	22	2	4.3%	0%
	86%			100%			98%				
PPCX2_b	781	45	52	166	23	26	335	79	42	7.7%	13%
	94%			88%			91%				

We also compared the ICC results obtained with automatic images selection through SVM, and the results of the automatic image selection using images selected at fixed hours, in conditions of direct and diffuse illumination. The correlation and mean absolute error (MAE) were computed with respect to results obtained using the manual selection of images (Table 6). The comparison was conducted pixel-by-pixel; therefore, more than $2 \cdot 10^5$ and $5 \cdot 10^5$ elements were considered for datasets SAXE3_b and PPCX2_b, respectively.

Table 6. Correlation ρ and MAE of two dimensions for dataset D1.

Dataset	Method	ρ_x	$MAE_x[px]$	ρ_y	$MAE_y[px]$
	SVM	0.69	0.18	0.84	0.24
SAXE3_b	SunLight	0.12	1.30	0.19	1.44
	DiffLight	0.38	0.44	0.69	0.57
	SVM	0.31	0.34	0.71	0.51
PPCX2_b	SunLight	0.02	1.78	0.14	2.21
	DiffLight	0.08	1.71	0.22	1.97

The results obtained with SVM selection show much higher correspondence with ICC results than the other two methods. Two principal factors provided lower performances in the other two selection methods: (1) As the state of view of the images was not controlled, it was possible to include images with obscured or partial views in the number of images that were processed; (2) the images with direct illumination suffered from the negative effects of shadows; therefore, any results for images with direct illumination are of a lower quality.

Furthermore, the correlation along the vertical direction was higher than that along the horizontal direction. As previously observed in the synthetic image analysis, the incidence of the sun's azimuth angle is greater because its value varies more than the sun's elevation, which influences the vertical direction. The MAEs of the vertical direction were thus greater because the values of displacement of these dimensions were greater.

Furthermore, a rather low correlation was noted between the horizontal direction of the images selected through SVM and those selected manually for the PPCX2_b dataset. An analysis of the absolute mean horizontal displacement value (Table 7) showed that this quantity was lower than the uncertainty. Therefore, the low correlation coefficient was likely due to the low signal-to-noise ratio.

Table 7. Mean absolute velocity and uncertainty δ of the two motion directions for two considered datasets; data are expressed in pixels.

Dataset	$\overline{|V_x|}$	$\overline{|V_y|}$	δ_x	δ_y
SAXE3_b	0.39	0.83	0.10	0.09
PPCX2_b	0.26	1.79	0.33	0.44

5. Conclusions

This study quantitatively analyses the disturbing effect caused by the shape and length of shadows on the results of image cross-correlation (ICC). First, the impact of shadows on the development of an automatized procedure for the selection of images was determined, and it was demonstrated that adopting images acquired in conditions of diffuse illumination enables the shadow-related uncertainty to be minimized. Therefore, a possible solution was developed to ensure that the images selected were taken in diffuse illumination.

A method for automatically classifying the images according to the illumination appearance was then proposed. Considering photos taken of various natural environments (e.g., landslides, glaciers, rock faces), SVMs were trained using different features that were representative of the photo's appearance. Of these, the number of RGB triplets was found to be the most significant feature used to discriminate between the classes. Training the SVMs with two or three features was sufficient for achieving an average misclassification error of 5.5%.

The processing chain used to automatically select images suitable for ICC application, which includes the abovementioned classification method, is also described herein. The application of the method was applied during a survey campaign in 2018 to monitor the activity of the Planpincieux glacier for civil protection purposes, and the results showed that a high correspondence was obtained with results obtained using images selected manually. The same experiment was then conducted using time-lapse photos acquired for the Mont de La Saxe landslide.

These results show the possibility of developing a monitoring system based on ICC that automatically selects and processes the pictures acquired. The use of image sequence analysis enables financial and human resources to be reduced in monitoring active geo-hydrological instabilities. The development of an automatic solution is a valuable tool for use in developing an autonomous monitoring system that supplies daily updates.

Author Contributions: Conceptualization, N.D. and D.G.; methodology and analysis, N.D.; data curation, P.A.; writing—original draft preparation, N.D.; writing—review and editing, D.G.

Funding: This research received no external funding.

References

1. Scambos, T.A.; Dutkiewicz, M.J.; Wilson, J.C.; Bindschadler, R.A. Application of image cross-correlation to the measurement of glacier velocity using satellite image data. *Remote Sens. Environ.* **1992**, *42*, 177–186. [CrossRef]
2. Duffy, G.P.; Hughes-Clarke, J.E. Application of spatial cross correlation to detection of migration of submarine sand dunes. *J. Geophys. Res. Earth Surf.* **2005**, *110*, F04S12. [CrossRef]
3. Delacourt, C.; Allemand, P.; Berthier, E.; Raucoules, D.; Casson, B.; Grandjean, P.; Pambrun, C.; Varel, E. Remote-sensing techniques for analysing landslide kinematics: A review. *Bull. Société Géologique Fr.* **2007**, *178*, 89–100. [CrossRef]

4. Leprince, S.; Berthier, E.; Ayoub, F.; Delacourt, C.; Avouac, J.-P. Monitoring earth surface dynamics with optical imagery. *EOS Trans. AGU* **2008**, *89*, 1–2. [CrossRef]

5. Wrzesniak, A.; Giordan, D. Development of an algorithm for automatic elaboration, representation and dissemination of landslide monitoring data. *Geomat. Nat. Hazards Risk* **2017**, *8*, 1898–1913. [CrossRef]

6. Allasia, P.; Lollino, G.; Godone, D.; Giordan, D. Deep displacements measured with a robotized inclinometer system. In Proceedings of the 10th International Symposium on Field Measurements in Geomechanics—FMGM2018, Rio de Janeiro, Brazil, 16–20 July 2018.

7. Manconi, A.; Giordan, D. Landslide early warning based on failure forecast models: The example of the Mt. de La Saxe rockslide, northern Italy. *Nat. Hazards Earth Syst. Sci.* **2015**, *15*, 1639–1644. [CrossRef]

8. Allasia, P.; Baldo, M.; Giordan, D.; Godone, D.; Wrzesniak, A.; Lollino, G. Near real time monitoring systems and periodic surveys using a multi sensors UAV: The case of Ponzano landslide. In Proceedings of the IAEG/AEG Annual Meeting Proceedings, San Francisco, CA, USA, 17–21 September 2018; Springer: New York, NY, USA, 2019; Volume 1, pp. 303–310.

9. Roedelsperger, S.; Becker, M.; Gerstenecker, C.; Laeufer, G. Near real-time monitoring of displacements with the ground based SAR IBIS-L. In Proceedings of the ESA Fringe Workshop, Frascati, Italy, 30 November–4 December 2009.

10. Giordan, D.; Wrzesniak, A.; Allasia, P. The importance of a dedicated monitoring solution and communication strategy for an effective management of complex active landslides in urbanized areas. *Sustainability* **2019**, *11*, 946. [CrossRef]

11. Berthier, E.; Vadon, H.; Baratoux, D.; Arnaud, Y.; Vincent, C.; Feigl, K.L.; Remy, F.; Legresy, B. Surface motion of mountain glaciers derived from satellite optical imagery. *Remote Sens. Environ.* **2005**, *95*, 14–28. [CrossRef]

12. Ahn, Y.; Box, J.E. Glacier velocities from time-lapse photos: Technique development and first results from the Extreme Ice Survey (EIS) in Greenland. *J. Glaciol.* **2010**, *56*, 723–734. [CrossRef]

13. Travelletti, J.; Delacourt, C.; Allemand, P.; Malet, J.-P.; Schmittbuhl, J.; Toussaint, R.; Bastard, M. Correlation of multi-temporal ground-based optical images for landslide monitoring: Application, potential and limitations. *ISPRS J. Photogramm. Remote Sens.* **2012**, *70*, 39–55. [CrossRef]

14. Gabrieli, F.; Corain, L.; Vettore, L. A low-cost landslide displacement activity assessment from time-lapse photogrammetry and rainfall data: Application to the Tessina landslide site. *Geomorphology* **2016**, *269*, 56–74. [CrossRef]

15. Giordan, D.; Allasia, P.; Dematteis, N.; Dell'Anese, F.; Vagliasindi, M.; Motta, E. A low-cost optical remote sensing application for glacier deformation monitoring in an alpine environment. *Sensors* **2016**, *16*, 1750. [CrossRef] [PubMed]

16. Schwalbe, E.; Maas, H.-G. The determination of high-resolution spatio-temporal glacier motion fields from time-lapse sequences. *Earth Surf. Dyn.* **2017**, *5*, 861–879. [CrossRef]

17. Debella-Gilo, M.; Kääb, A. Sub-pixel precision image matching for measuring surface displacements on mass movements using normalized cross-correlation. *Remote Sens. Environ.* **2011**, *115*, 130–142. [CrossRef]

18. Messerli, A.; Grinsted, A. Image georectification and feature tracking toolbox: ImGRAFT. *Geosci. Instrum. Methods Data Syst.* **2015**, *4*, 23–34. [CrossRef]

19. Hadhri, H.; Vernier, F.; Atto, A.M.; Trouvé, E. Time-lapse optical flow regularization for geophysical complex phenomena monitoring. *ISPRS J. Photogramm. Remote Sens.* **2019**, *150*, 135–156. [CrossRef]

20. Boser, B.E.; Guyon, I.M.; Vapnik, V.N. A training algorithm for optimal margin classifiers. In Proceedings of the Fifth Annual Workshop on Computational Learning Theory, Pittsburgh, PA, USA, 27–29 July 1992; ACM: New York, NY, USA, 1992; pp. 144–152.

21. Cortes, C.; Vapnik, V. Support-vector networks. *Mach. Learn.* **1995**, *20*, 273–297. [CrossRef]

22. Pal, M.; Mather, P.M. Support vector machines for classification in remote sensing. *Int. J. Remote Sens.* **2005**, *26*, 1007–1011. [CrossRef]

23. Mountrakis, G.; Im, J.; Ogole, C. Support vector machines in remote sensing: A review. *ISPRS J. Photogramm. Remote Sens.* **2011**, *66*, 247–259. [CrossRef]

24. Lary, D.J.; Alavi, A.H.; Gandomi, A.H.; Walker, A.L. Machine learning in geosciences and remote sensing. *Geosci. Front.* **2016**, *7*, 3–10. [CrossRef]

25. Maxwell, A.E.; Warner, T.A.; Fang, F. Implementation of machine-learning classification in remote sensing: An applied review. *Int. J. Remote Sens.* **2018**, *39*, 2784–2817. [CrossRef]

26. Willert, C.E.; Gharib, M. Digital particle image velocimetry. *Exp. Fluids* **1991**, *10*, 181–193. [CrossRef]

27. Westerweel, J. Fundamentals of digital particle image velocimetry. *Meas. Sci. Technol.* **1997**, *8*, 1379. [CrossRef]
28. Lowe, D.G. Object recognition from local scale-invariant features. In Proceedings of the Seventh IEEE International Conference on Computer Vision, Kerkyra, Greece, 20–27 September 1999; Volume 2, pp. 1150–1157.
29. Bay, H.; Ess, A.; Tuytelaars, T.; Van Gool, L. Speeded-up robust features (SURF). *Comput. Vis. Image Underst.* **2008**, *110*, 346–359. [CrossRef]
30. Pust, O. Piv: Direct cross-correlation compared with fft-based cross-correlation. In Proceedings of the 10th International Symposium on Applications of Laser Techniques to Fluid Mechanics, Lisbon, Portugal, 10–13 July 2000; Volume 27, p. 114.
31. Thielicke, W.; Stamhuis, E. PIVlab—Towards user-friendly, affordable and accurate digital particle image velocimetry in MATLAB. *J. Open Res. Softw.* **2014**, *2*, e30. [CrossRef]
32. Bickel, V.; Manconi, A.; Amann, F. Quantitative assessment of digital image correlation methods to detect and monitor surface displacements of large slope instabilities. *Remote Sens.* **2018**, *10*, 865. [CrossRef]
33. Guizar-Sicairos, M.; Thurman, S.T.; Fienup, J.R. Efficient subpixel image registration algorithms. *Opt. Lett.* **2008**, *33*, 156–158. [CrossRef] [PubMed]
34. Dematteis, N.; Giordan, D.; Zucca, F.; Luzi, G.; Allasia, P. 4D surface kinematics monitoring through terrestrial radar interferometry and image cross-correlation coupling. *ISPRS J. Photogramm. Remote Sens.* **2018**, *142*, 38–50. [CrossRef]
35. Cheng, D.; Prasad, D.K.; Brown, M.S. Illuminant estimation for color constancy: Why spatial-domain methods work and the role of the color distribution. *JOSA A* **2014**, *31*, 1049–1058. [CrossRef]
36. Manduchi, R.; Mian, G.A. Accuracy analysis for correlation-based image registration algorithms. In Proceedings of the 1993 IEEE International Symposium on Circuits and Systems, Chicago, IL, USA, 3–6 May 1993; pp. 834–837.
37. Westerweel, J.; Scarano, F. Universal outlier detection for PIV data. *Exp. Fluids* **2005**, *39*, 1096–1100. [CrossRef]
38. Lary, D.J. Artificial intelligence in geoscience and remote sensing. In *Geoscience and Remote Sensing New Achievements*; Imperatore, P., Riccio, D., Eds.; InTech: Vienna, Austria, 2010; ISBN 978-953-7619-97-8.
39. Haykin, S. *Neural Network. A Comprehensive Foundation*, 2nd ed.; Prentice Hall: Upper Saddle River, NJ, USA, 1999; ISBN 0-13-273350-1.
40. Sunkavalli, K.; Romeiro, F.; Matusik, W.; Zickler, T.; Pfister, H. What do color changes reveal about an outdoor scene? In Proceedings of the 2008 IEEE Conference on Computer Vision and Pattern Recognition, Anchorage, AK, USA, 23–28 June 2008; pp. 1–8.
41. Picard, R.R.; Cook, R.D. Cross-validation of regression models. *J. Am. Stat. Assoc.* **1984**, *79*, 575–583. [CrossRef]
42. Xu, Q.-S.; Liang, Y.-Z. Monte Carlo cross validation. *Chemom. Intell. Lab. Syst.* **2001**, *56*, 1–11. [CrossRef]
43. Crosta, G.B.; Di Prisco, C.; Frattini, P.; Frigerio, G.; Castellanza, R.; Agliardi, F. Chasing a complete understanding of the triggering mechanisms of a large rapidly evolving rockslide. *Landslides* **2014**, *11*, 747–764. [CrossRef]
44. Crosta, G.B.; Lollino, G.; Paolo, F.; Giordan, D.; Andrea, T.; Carlo, R.; Davide, B. Rockslide monitoring through multi-temporal LiDAR DEM and TLS data analysis. In *Engineering Geology for Society and Territory-Volume 2*; Springer: New York, NY, USA, 2015; pp. 613–617.
45. Manconi, A.; Giordan, D. Landslide failure forecast in near-real-time. *Geomat. Nat. Hazards Risk* **2016**, *7*, 639–648. [CrossRef]
46. Dematteis, N.; Luzi, G.; Giordan, D.; Zucca, F.; Allasia, P. Monitoring Alpine glacier surface deformations with GB-SAR. *Remote Sens. Lett.* **2017**, *8*, 947–956. [CrossRef]
47. Greene, C. Borders. In *MATLAB Central File Exchange*; MATLAB: Natick, MA, USA, 2019.

Article

Measurement of Interlaminar Tensile Strength and Elastic Properties of Composites Using Open-Hole Compression Testing and Digital Image Correlation

Guillaume Seon [1,*], Andrew Makeev [1], Joseph D. Schaefer [2] and Brian Justusson [2]

[1] Department of Aerospace and Mechanical Engineering, The University of Texas at Arlington, Arlington, TX 76019, USA
[2] Boeing Research & Technology, The Boeing Company, St. Louis, MO 63134, USA
* Correspondence: seon@uta.edu

Received: 15 May 2019; Accepted: 26 June 2019; Published: 29 June 2019

Abstract: Advanced polymeric composites are increasingly used in high-performance aircraft structures to reduce weight and improve efficiency. However, a major challenge delaying the implementation of the advanced composites is the lack of accurate methods for material characterization. Accurate measurement of three-dimensional mechanical properties of composites, stress–strain response, strength, fatigue, and toughness properties, is essential in the development of validated analysis techniques accelerating design and certification of composite structures. In particular, accurate measurement of the through-thickness constitutive properties and interlaminar tensile (ILT) strength is needed to capture delamination failure, which is one of the primary failure modes in composite aircraft structures. A major technical challenge to accurate measurement of ILT properties is their strong sensitivity to manufacturing defects that often leads to unacceptable scatter in standard test results. Unacceptable failure mode in standard test methods is another common obstacle to accurate ILT strength measurement. Characterization methods based on non-contact full-field measurement of deformation have emerged as attractive alternative techniques allowing more flexibility in test configuration to address some of the limitations inherent to strain gauge-based standard testing. In this work, a method based on full-field digital image correlation (DIC) measurement of surface deformation in unidirectional open-hole compression (OHC) specimens is proposed and investigated as a viable alternative to assessing ILT stress–strain, strength, and fatigue properties. Inverse identification using a finite element model updating (FEMU) method is used for simultaneous measurement of through-thickness elastic constants with recovery of the maximum ILT stress at failure for characterization of strength and fatigue *S–N* curves.

Keywords: characterization of composite materials; interlaminar tensile strength; digital image correlation; inverse method; finite element model updating

---

1. Introduction

Advanced polymeric composites are playing a major role in designing high-performance and lightweight aircraft structures. However, a major challenge delaying the implementation of the advanced composites is the lack of accurate methods for material characterization [1]. Accurate measurement of three-dimensional mechanical properties of composites, stress–strain response, strength, fatigue, and toughness properties, is essential in the development of validated analysis techniques accelerating design and certification of composite structures [2–4]. In particular, accurate measurement of interlaminar tensile (ILT) strength is needed to capture delamination failure, which is one of the primary failure modes in composite aircraft structures.

A major technical challenge to accurate measurement of ILT properties is their extreme sensitivity to manufacturing defects, including porosity, which could lead to unacceptable scatter in the test results. Unacceptable failure mode in standard test methods that use transverse tension specimens is another challenge to accurate measurement of the ILT properties. Currently, American Society for Testing and Materials (ASTM) D 6415 curved-beam (CB) method is a standard practice for measurement of the ILT strength [5]. However, large scatter typically found in ASTM D 6415 ILT strength raises doubts concerning the adequacy of the curved-beam test as a coupon independent test for measuring ILT material strength. Large scatter in the ILT strength test results may indeed reflect the manufacturing quality as much as materials properties. In particular, the ASTM D6415 CB strength is extremely sensitive to porosity/voids in the radius area. In [6], Jackson and Martin observed a drastic (up to a factor of four) CB strength reduction in low-quality CB specimens due to large (macroscopic) voids detected based on a cross-section cut using a diamond saw. In [7], the authors of this work measured the ASTM D 6415 CB strength using high-quality 6.6 mm (0.26 inch) thick CB coupons manufactured from Hexcel IM7/8552 unidirectional tape and cured per manufacturer's specifications. ASTM D 6415 ILT strength values ranging between 37.7 MPa (5.5 ksi) and 94.6 MPa (13.7 ksi) were reported with a coefficient of variation higher than 26%. Similarly, typical high-quality IM7/8552 ASTM D 6415 CB specimens were tested in fatigue in [8] and showed large variation in fatigue life and correlated poorly with the power-law least-squares fit commonly used for description of the S–N curves of carbon-epoxy materials. CT scans of the CB specimens tested in [7,8] revealed the presence of individual critical voids at ply interfaces that were correlated with the knockdown and large scatter found in the ASTM D 6415 CB strength and ILT CB fatigue life. It is worth noting that a test method that relies on extensive CT measurement may not be suitable for efficient characterization.

Another method relevant to the assessment of the ILT strength is ASTM D 7291 [9], which applies a tensile force normal to the plane of the composite laminate using adhesively bonded thick metal end-tabs. It was noted in the ASTM Standard D 7291 that through-thickness strength results using this method will, in general, not be comparable to ASTM D 6415 since ASTM D 7291 subjects a relatively large volume of material to an almost uniform stress field, while ASTM D 6415 subjects a small volume of material to a non-uniform stress field. Also, characterization of ILT strength using ASTM D 7291 is perhaps even more challenging than ASTM D 6415 as the failure could occur not in the composite material but at the bond lines between the composite and the metal end-tabs used to transfer the load to the composite.

Many of the challenges in standard testing of composite materials, including the difficulties mentioned previously in measurement of the ILT strength, stem from their inherent anisotropy and heterogeneous nature. Historically, standard practices for assessment of material stress–strain constitutive relations and strength properties have been based on resistance strain gauge measurements [10]. As a strain gauge measures the average strain over the gauge area, the requirement to achieve uniform strain distribution within the gauge area imposes constraints on the test configuration and test specimen design. Furthermore, standard test methods typically rely on the assumption of uniform stress distribution over the entire specimen cross-section for derivation of simple closed-form expressions used in stress and strength calculation. In the case of anisotropic composite materials, such simplifying assumptions may not be valid nor easy to satisfy. Similarly, material heterogeneities, including manufacturing defects common to composites, might act as stress raisers and lead to inaccurate and non-conservative strength calculation using standard closed-form approximations, which are only applicable to homogenous and pristine material.

In contrast, non-standard characterization methods using non-contact full-field measurement methods, such as the digital image correlation (DIC) techniques, do not require uniform and homogeneous strain fields, which enables additional flexibility compared to conventional strain gauge methods [11]. In particular, inverse methods based on full-field deformation measurement have emerged as attractive alternative methods for accurate characterization of composites using non-standard test setups, enabling measurement of matrix-dominated non-linear properties and extraction of multiple material parameters from a single experiment [12–15]. Among strategies developed to solve the inverse problem of determining the material constitutive parameters using the full field information, the finite element model updating (FEMU) method has been the most common approach [15–18], and the virtual fields method (VFM) the most notable alternative [19–23]. Improvement in digital camera resolution has also led to a recent interest in using the DIC technique for observation and characterization of interlaminar bond line failures. For example, a consumer digital camera was used in [24] for full-field measurement of deformation within the thin epoxy bond lines in single lap joints with thick aluminum adherends. Researchers have shown that high-resolution DIC measurement of crack tip separation in bonded joints can be used for direct characterization of the cohesive laws in thin adhesive layers [25–27].

Taking advantage of the added flexibility from high-resolution full-field DIC measurements, a methodology using unidirectional open-hole compression (OHC) specimens is proposed in this work and investigated as an alternative to assessing interlaminar tensile strength and fatigue properties, with the potential to alleviate the challenges previously mentioned using ASTM D6415 and ASTM D7291 standard tests. Under compression loading, transverse tensile stress concentrations aligned with the loading direction develop around the hole in OHC coupons away from test fixture grips, as opposed to stress concentrations typically located at specimen tabs in ASTM D7291. OHC test specimens used in this work were sliced from flat thick unidirectional panels, allowing development of through-thickness tensile stress concentration. Porosity is much easier to control in flat panels, as opposed to strong susceptibility to porosity defects in the radius area of the curved panels typically used for machining ASTM D6415 coupons. Upon further compressive loading in OHC coupons, consistent delamination failure under interlaminar transverse tension was observed, provided that holes were free of drilling damage. Full-field DIC strain data were used in combination with a finite element model updating (FEMU) method for inverse identification of the four elastic constants characterizing the orthotropic composite material under plane stress conditions, allowing simultaneous assessment of stress–strain and strength properties. The method was used for characterization of generic carbon/epoxy tape and fabric materials and validated by comparing with results from standard test methods and previously validated non-standard test methods. The method was also extended to fatigue loading for assessment of ILT *S–N* curves.

2. Open-Hole Compression Testing

OHC specimens used in this work are machined from thick unidirectional panels, as illustrated in Figure 1. Unidirectional panels made from carbon/epoxy IM7/8552 tape prepreg material, a generic carbon/epoxy tape prepreg material, and a generic carbon/epoxy fabric prepreg material were considered in this work. All panels were cured per manufacturer's specifications. IM7/8552 unidirectional panels were fabricated from a stack of 72 prepreg plies with a cured ply thickness of 178 μm (0.007 in) resulting in a total panel thickness of approximately 12.8 mm (0.5 in). The coupon dimensions for the IM7/8552 specimens are 80.8 × 12.8 × 1.02 mm (3.18 × 0.5 × 0.04 in), in the fiber, out-of-plane, and transverse directions, respectively. After machining the specimens, a 6.35 mm (0.25 in) diameter hole is drilled at the center on the test specimens, as illustrated in Figure 1. In order to minimize drilling damage, the holes were machined using high-quality diamond dust-coated drill bits and a Plexiglas backing plate. A smaller diameter drill was employed first, and the pre-hole was then enlarged to its final diameter using a 6.35 mm (0.25 in) diameter drill bit.

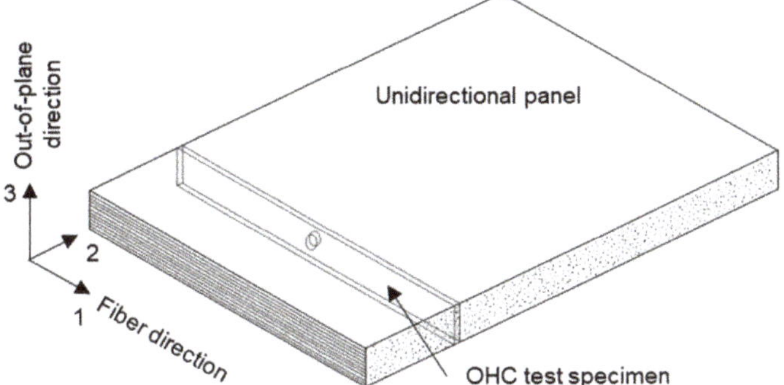

Figure 1. Open-hole compression (OHC) test specimen machined from a 12.8 mm (0.5 in) thick unidirectional panel.

The specimens are placed in a modified SACMA SRM/Boeing ASTM D695 test fixture, which uses a Plexiglas plate to provide lateral support and prevent buckling while allowing for digital image correlation (DIC)-based measurement of the surface deformation. The test setup is illustrated in Figure 2 showing a unidirectional OHC test specimen installed in the test fixture including the anti-buckling Plexiglas plates. One of the two cameras used for DIC monitoring of surface deformation is also visible in the foreground in Figure 2. A servohydraulic load frame was used for quasi-static loading of the test specimen at a 0.021 mm/s (0.05 in/min) crosshead displacement rate until specimen failure.

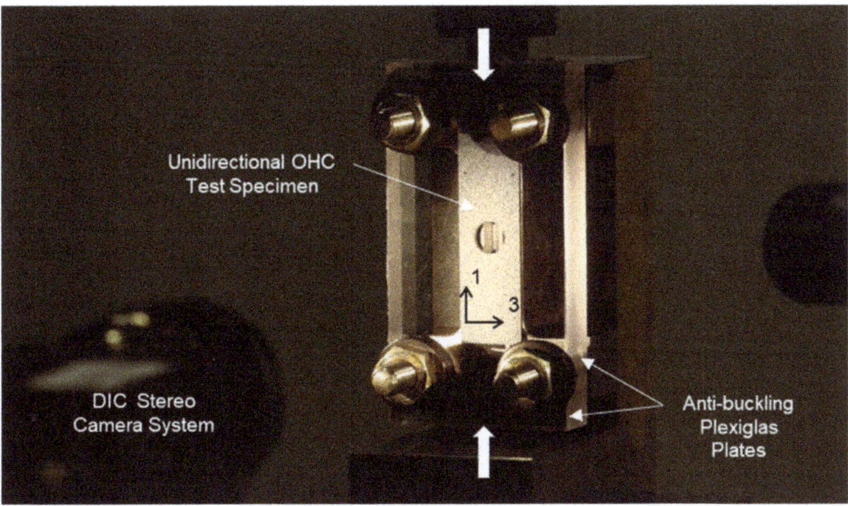

Figure 2. A test setup for measurement of interlaminar tensile (ILT) strength in a unidirectional OHC test specimen.

The DIC strain assessment is based on quantifying locations of a random texture on specimen surface [28]. Figure 3 shows such an example of random texture, also referred to as speckle pattern, created on the surface of an OHC test specimen using black and white spray paints.

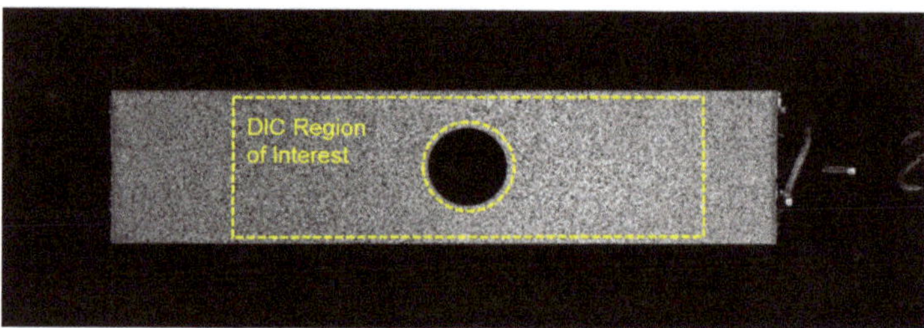

Figure 3. An example of speckle pattern created on the surface of an OHC test specimen and region of interest for digital image correlation (DIC) analysis.

A 16-megapixel stereo camera system is used to generate a sequence of images while the specimen is subject to loading. A sequence of about 20 DIC images taken throughout specimen loading was used for FEMU characterization. The angle between the two cameras was 20 degrees, and two LED-based illumination sources were used. The light sources were placed such that light reflection on the Plexiglas plate was as limited as possible. Figure 3 shows an example of DIC speckle image with no visible light reflection. Correlated solution image acquisition software VIC-Snap [29] was used for DIC data acquisition and synchronization with load data from the testing machine. The VIC-3D software [29] was used to determine the 3D positions during deformation by tracking the gray value pattern in small subsets throughout the acquired stereo image sequence. Further, 45 × 45 pixel subsets were used for DIC analysis of the OHC specimens over an approximately 38 mm × 10 mm (1.5 × 0.4 in) rectangular region of interest, as illustrated in Figure 3. The displacement data were obtained on 9-pixel centers, resulting in about 15,000 data points per load cases. The displacement fields obtained in this manner were then numerically differentiated using the strain computation algorithm in the VIC-3D software [28,29] to compute the Lagrangian surface strain tensor components. Quality of the subset patterns, as well as proper selection of the analysis parameters, were verified similarly to details provided in Reference [30].

It is worth noting that 3D DIC measurements were used in this work, despite using only in-plane strain components for inverse characterization in the FEMU approach. One of the most prominent reason for using 3D DIC measurement instead of 2D DIC was verification that the anti-buckling Plexiglas plates were working as intended.

Under compression loading, ILT stresses develop at the 12 and 6 o'clock locations around the hole and initiate delamination failure. After initiation, delamination progresses unstably towards the specimen extremities. The OHC coupon is oriented along the vertical loading direction, as shown in Figure 2. Figure 4a shows an example of interlaminar normal strain distribution on the surface of a unidirectional IM7/8552 OHC specimen measured prior to failure using the DIC technique and showing ILT strain concentrations around the hole. Figure 4b shows the delamination failure that initiates upon further loading at the 12 and 6 o'clock locations due to interlaminar tension.

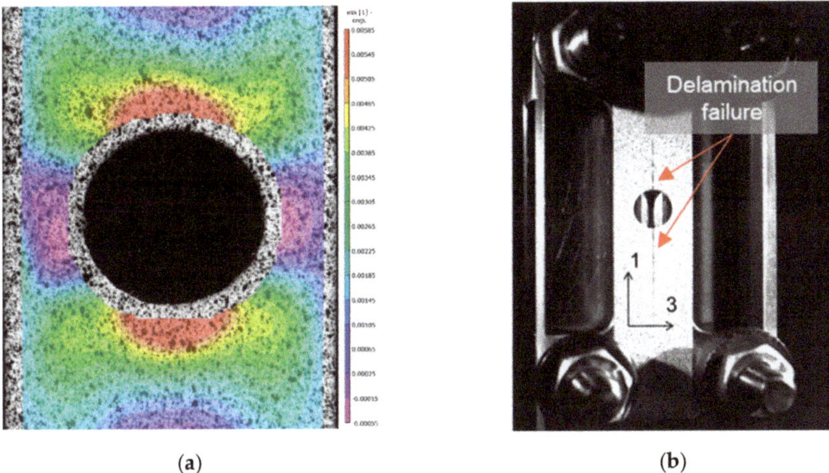

(a)   (b)

Figure 4. An example of (**a**) DIC-measured ILT strains prior to failure and (**b**) ILT delamination failure initiating at the 12 and 6 o'clock locations around the hole in a unidirectional IM7/8552 OHC specimen.

Figure 5 shows the full-field DIC strain data measured in a generic composite fabric OHC specimen considered in this work, with contour plots for the interlaminar normal ε_{33}, fiber-direction normal ε_{11}, and interlaminar shear strain γ_{13} fields, respectively. Due to small specimen thickness compared to the other dimensions and relative out-of-plane compliance of the anti-buckling device, deformation of the OHC coupons can be considered under a plane-stress approximation.

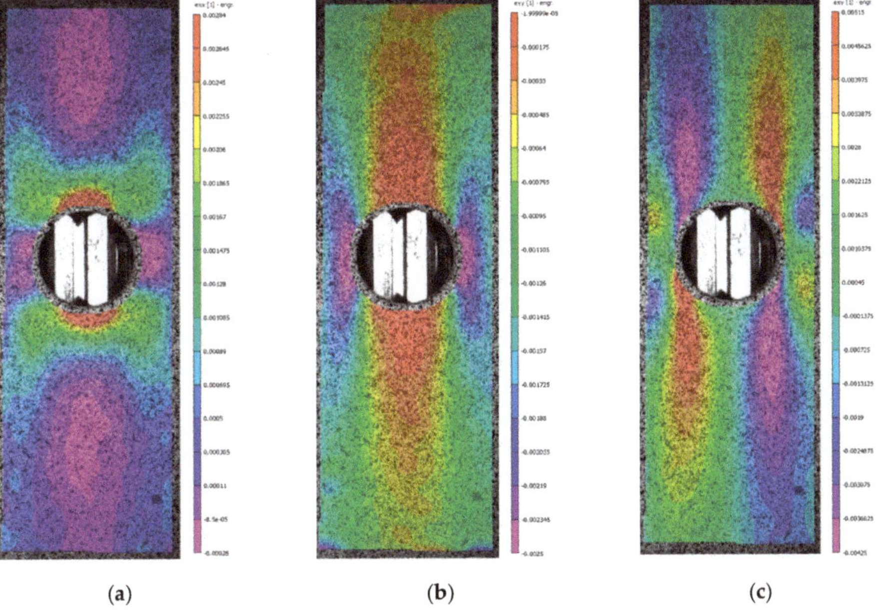

(a)   (b)   (c)

Figure 5. Example of full-field DIC strain measurement in a fabric OHC specimen considered in this work with contour plots of (**a**) transverse through-thickness ε_{33}, (**b**) longitudinal fiber-direction ε_{11}, and (**c**) interlaminar shear γ_{13} strains, respectively.

As shown in Figure 5, the longitudinal and shear strains at the 12 and 6 o'clock locations are several orders of magnitude lower than the maximum through-thickness tensile (ILT) strains, validating the assumption of ILT delamination failure under pure mode I conditions. It is worth noting that a concentration of longitudinal fiber-direction compressive stresses also develops at the three and nine o'clock locations, as well as interlaminar shear stresses. Alternate failure modes, including fiber-direction compressive failure and interlaminar shear failure, could occur depending on the combination of material constitutive anisotropy, material strength properties, and coupon geometry. However, both for the IM7/8552 tape, generic carbon/epoxy tape, and carbon/epoxy fabric materials considered in this work, delamination failure at the 12 and 6 o'clock locations always occurred before any other failure mode was observed (except in presence of drill damage, as discussed later in Section 5).

3. Methodology

3.1. Measurement of ILT Strength

A relationship between the peak load measured at failure using the testing machine load cell and the maximum ILT stress in the specimen is needed for assessment of the material ILT strength. In standard testing, the common design constraint of uniform stress distribution over the specimen cross-section allows using convenient closed-form relations for recovery of the maximum stress at failure.

It is worth noting that the maximum stress concentration factor K_t^{inf} for a circular hole in an infinite anisotropic plate subjected to normal far-field compressive stress can be derived analytically as [31]:

$$K_t^{inf} = \sqrt{\frac{E_{33}}{E_{11}}} \tag{1}$$

where E_{11} and E_{33} are the composite material Young's moduli in the fiber and thickness direction, respectively.

As shown in Equation (1), the maximum stress concentration factor K_t^{inf} for a circular hole in an infinite anisotropic plate depends on material properties.

In the OHC unidirectional specimens considered in this work, the width-to-hole-diameter ratio is equal to two (R = 2) and the effect of finite specimen's width, in addition to the effects of material anisotropy, must be included for evaluation of the stress concentration factor at the hole. To the best of the author's knowledge, there is no evident analytical closed-form solution for such a problem. However, when material constitutive stress–strain properties are known, the maximum ILT stress at failure can be easily determined from a simple 2D plane stress FE analysis. The stress concentration around the hole occurs far from the test fixture tabs, and it was verified that it was relatively insensitive to specimen boundary conditions for the test coupon geometry considered.

An illustration of the 2D FE mesh used in this work for stress analysis and FEMU inverse characterization is shown in Figure 6. The FE model is generated for Abaqus FEM analysis and includes about 5000 four-node bilinear 2D plane stress elements CPS4 from Abaqus element library. The total number of nodes is about 5300 and the total number of degrees of freedom is approximately 11,000. A linear elastic homogenized orthotropic material model is used for implicit static stress/displacement analysis using Abaqus standard solver. Mesh refinements are assigned to ensure proper convergence of the stress field around the hole. A mesh sensitivity study showing convergence of the ILT stress field around the hole with mesh refinement is provided in Appendix A.

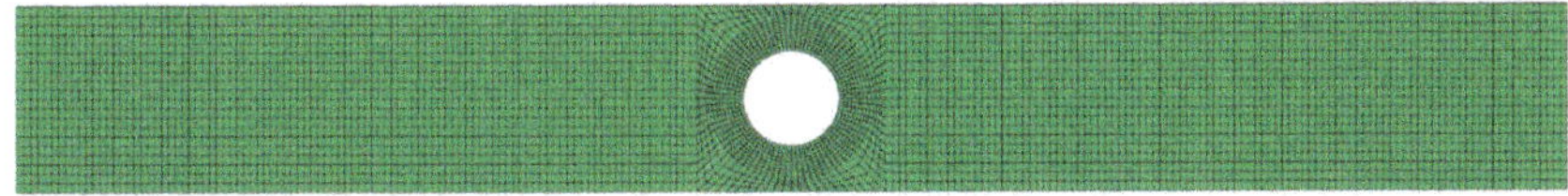

Figure 6. 2D plane stress FE mesh used for stress analysis and finite element model updating (FEMU) inverse characterization using Abaqus FE software.

Figure 7 shows FE contour plots of the normalized interlaminar normal stress component in the carbon/epoxy tape (Figure 7a) and carbon/epoxy fabric (Figure 7b) OHC coupons used in this work. The interlaminar stress is normalized by the far-field longitudinal normal compression stress, which is equal to the applied force divided by the specimen rectangular cross-section area. The two materials have different homogenized orthotropic properties, which result in different values for K_t^{inf} calculated using Equation (1), as shown in Figure 7.

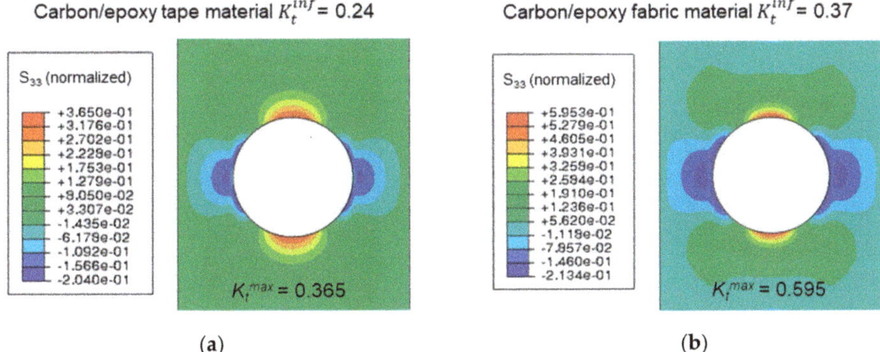

(a)   (b)

Figure 7. ILT stress concentration in a typical unidirectional OHC specimen using FE analysis for (**a**) the carbon/epoxy tape and (**b**) the carbon/epoxy fabric materials considered in this work.

OHC coupons ILT stress concentration factors K_t^{max} differ from K_t^{inf} for an infinite plate since they account for the effect of finite width, for which there is no obvious closed-form solution for anisotropic materials as previously mentioned. In these examples, a maximum ILT stress concentration factor K_t^{max} of 0.365 is obtained for the tape material, versus 0.595 for the fabric material. These values can be compared to $K_t^{inf} = 0.24$ and $K_t^{inf} = 0.37$ for an infinite plate with a circular hole for the tape material and fabric material, respectively.

3.2. Inverse Characterization of Elastic Constants

This work considers a self-sufficient methodology based on DIC strain measurements and finite element model updating (FEMU) allowing simultaneous measurement of the material interlaminar constitutive properties (E_{11}, E_{33}, ν_{13}, G_{13}) along with the ILT strength. The idea is to take advantage of the full heterogeneous strain field measured on the surface of the OHC specimens, as illustrated in Figure 5, and evaluate if such data can be used in inverse characterization. At the end of the FEMU procedure, optimized material parameters can be used in an additional FE analysis at failure load for assessment of the ILT strength properties. Such advancement would eliminate reliance on additional tests to determine material constitutive properties, which could be unknown prior to the ILT testing. This DIC data-driven OHC ILT test method does not require prior knowledge of any other additional material properties.

The DIC data-driven FEMU method is based on the nonlinear minimization of the least square error between DIC-measured strain field and FEM-predicted strains by fine-tuning the parameters of the material constitutive model. The concept of the FEMU algorithm using DIC data from unidirectional OHC specimens is shown in Figure 8.

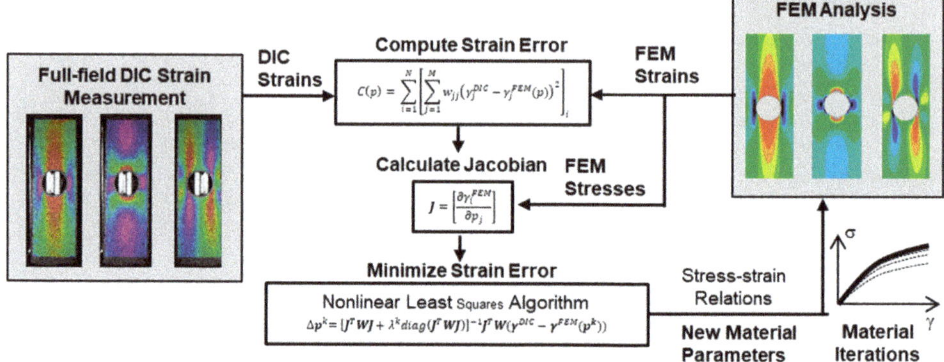

Figure 8. Finite element model updating (FEMU) algorithm for DIC data-driven modeling in unidirectional OHC specimens.

The FEMU methodology uses a weighted Levenberg–Marquardt optimization algorithm [32] where the cost function is defined as the weighted root mean square (RMS) error between FEM-computed and DIC-measured surface strains. The reader is referred to Reference [15] for more details on the optimization methodology, including iterative update of the material parameters using the Levenberg–Marquardt algorithm with smooth damping parameter variation [33]. Strain field optimization is carried out for about 20 DIC images taken throughout the loading history and up to ultimate failure. The two normal strain components ε_{11} and ε_{33} and the interlaminar shear strain component γ_{13} are considered in the optimization algorithm. The weights used in the cost function are defined at each node as a function of the local and maximum strain quantities:

$$
w_{ii}{}^{j} = \begin{cases} \left(\dfrac{\varepsilon_{ii}{}^{j}}{max_{j}\{|\varepsilon_{ii}|^{j}\}}\right)^{2} & j \text{ node index} \\ 0 \; if \; ii = 11 \; and \; |\varepsilon_{ii}|^{j} < 0.25 max_{j}\{|\varepsilon_{ii}|^{j}\} & ii = \{11, 33 \text{ or } 13\} \text{ strain component index} \end{cases} \tag{2}
$$

As shown in Equation (2), the weights are selected, such as more weight is given to data points with large strain values. For longitudinal fiber-direction strains, which are typically very small due to anisotropy of the composite material, a minimum threshold of 25% of the absolute maximum strain value is also used. This definition of weighted residuals is one way of reducing the sensitivity of the optimization algorithm to measurement noise.

Four optimization parameters (E_{11}, E_{33}, ν_{13}, G_{13}) are considered, which are the longitudinal and interlaminar Young's Moduli, the Poisson's ratio, and the shear modulus, respectively. A finite difference method is used to compute the sensitivity matrix used in the optimization algorithm. It is worth noting that using finite difference provides an accurate approximation of the sensitivity matrix, which can be also updated at each FEMU iterations, and results in optimum convergence. However, the finite difference approach can also be very costly in terms of computation times since multiple FE analyses are required at each iterations to compute the coefficients of the sensitivity matrix. In this work, 2D FEM is used and the runtime is limited to a few seconds, allowing for efficient use of the finite difference method for computation of the sensitivity matrix.

3.3. Verification and Robustness

The FEMU procedure is first validated using "virtual test" strain field data generated from FE simulation with a set of target material parameters (E_{11}^{target}, E_{33}^{target}, v_{13}^{target}, G_{13}^{target}). An initial approximation of the material parameters is needed to initiate the non-linear least-squares optimization algorithm. For example, Figure 9 shows convergence of the parameters to their target values and the reduction of the weighted strain error with the number of iterations starting from an initial approximation of the material parameters is shown in Figure 10. Results in Figure 9 have been normalized to one by the respective target parameter. The following initial approximation of the material parameters was used, which corresponds to ± 40% of relative error compared to the target parameters.

$$
\begin{aligned}
E_{11}^{[0]} &= 0.6\ E_{11}^{target} \\
E_{33}^{[0]} &= 1.4\ E_{33}^{target} \\
v_{13}^{[0]} &= 0.6\ v_{13}^{target} \\
G_{13}^{[0]} &= 1.4\ G_{13}^{target}
\end{aligned}
\tag{3}
$$

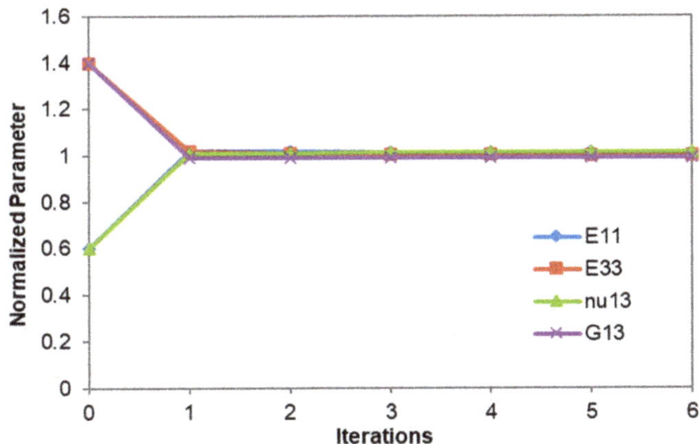

Figure 9. Convergence of normalized material parameters to their target value.

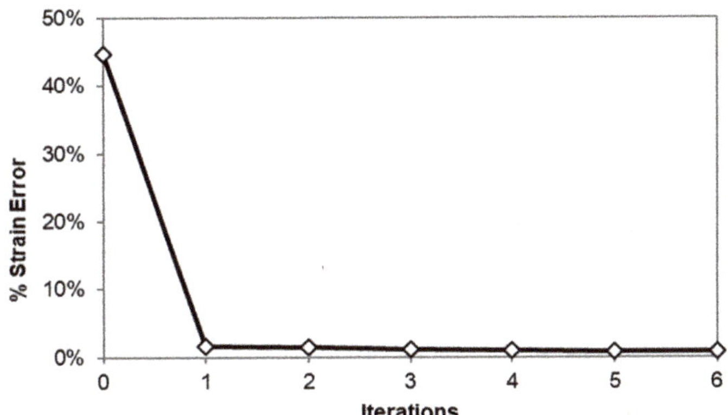

Figure 10. Weighted RMS strain error for the OHC FEMU methodology using FEM-generated virtual test strain data.

Similar convergence for different initial parameters up to 50% of initial relative error was verified.

The FEMU updating procedure is stopped when the relative change in the RMS average of the parameters between two iterations is less than 0.3%. As shown in Figures 9 and 10, convergence is quickly achieved and all parameters converge to their desired and expected target value.

Figure 11 shows an example of convergence of the material parameters for the FEMU procedure using experimental DIC strain data from OHC testing of a generic carbon-epoxy unidirectional specimen. The threshold for convergence is reached after nine iterations, however effective convergence is pretty much achieved after two or three iterations, as illustrated. In this example, convergence is illustrated for a 50% maximum relative error in the initial approximation of the interlaminar Young's Modulus $E_{33}^{[0]}$ compared to its final converged value after nine iterations $E_{33}^{[9]}$. Initial relative error for the other parameters is between 37% and 43%.

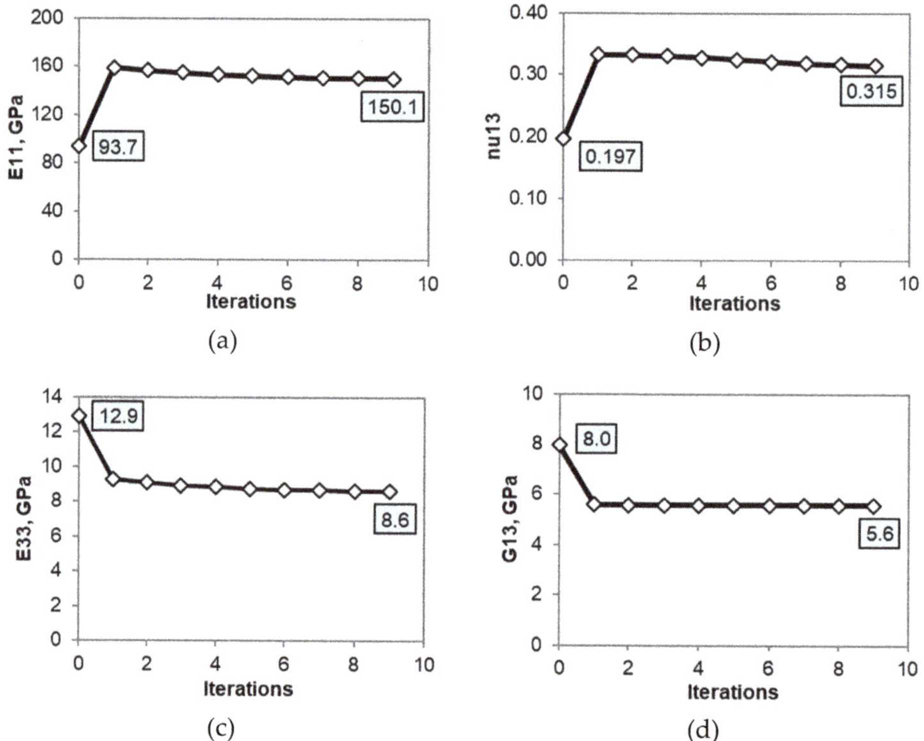

Figure 11. Convergence of the four material parameters (**a**) E_{11}, (**b**) v_{13}, (**c**) E_{33} and (**d**) G_{13} extracted with the OHC-ILT FEMU methodology using DIC-measured strain data from a generic unidirectional carbon/epoxy OHC test specimen.

Figures 12 and 13 compare the DIC-measured longitudinal normal and shear strain fields, respectively, with FEM results at the beginning and end of the updating process. It is worth noting that the DIC strain data have been mapped on the FEM node grid, as part of the FEMU methodology [15]. Interpolation of the DIC strains on the FEM nodes is realized at the beginning of the FEMU procedure using Abaqus Python scripting interface and the natural neighbor interpolation method implemented in Python function *griddata* of module *SciPy*.

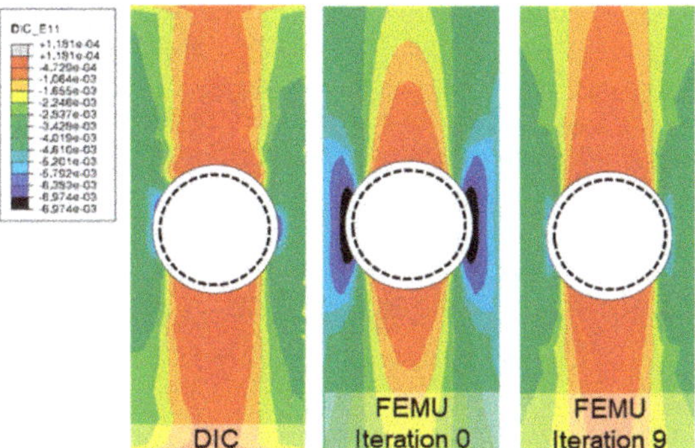

Figure 12. Comparison of DIC-measured longitudinal strains with FE results from the OHC-ILT FEMU procedure before (iteration 0) and after optimization (iteration 9) of material parameters.

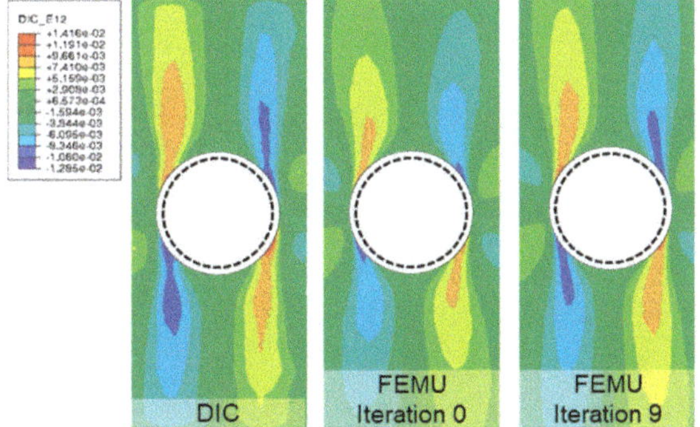

Figure 13. Comparison of DIC-measured interlaminar shear strains with FE results from the OHC-ILT FEMU procedure before (iteration 0) and after optimization (iteration 9) of material parameters.

Figure 14a shows the evolution of relative absolute strain error between FEM and DIC during the FEMU iterative procedure for the minimum longitudinal strain value in the contour plots shown in Figure 12. Figure 14b shows the relative absolute strain error for both the minimum and maximum shear strain value for the shear strain contour plots shown in Figure 13.

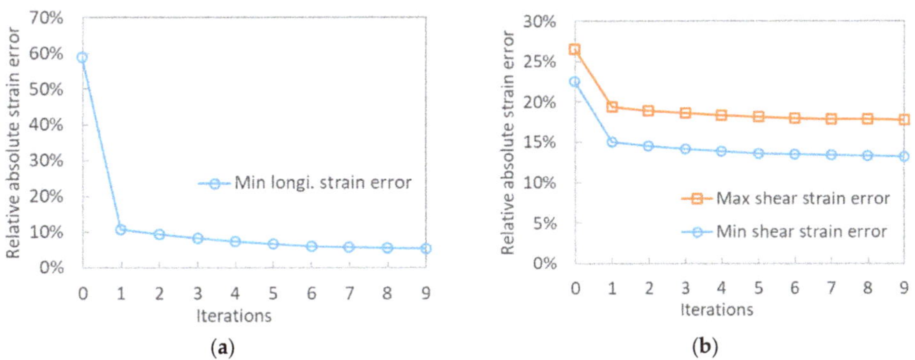

Figure 14. Relative absolute strain error between FEM and DIC data for (**a**) the minimum longitudinal strain and (**b**) the maximum and minimum shear strain value in the DIC region of interest.

As shown in Figure 14, an overall reduction of the point-local strain error is documented for both the longitudinal and shear strains. In particular, a significant reduction of the minimum longitudinal strain error is shown in Figure 14a. Figure 14b shows that the converged point-local relative error for both the maximum and minimum shear value is still relatively high at the end of the FEMU process (respectively 17.8% and 13.3%). However, this point-local error should be compared to the total weighted strain error, which takes into account the full-field strain data for all DIC images taken throughout the loading. The weights for calculation of the RMS strain error are defined in Equation (2). The evolution of the total weighted RMS during FEMU iterations is shown in Figure 15 for the strain data presented in Figures 12 and 13. As shown in Figure 15, the total RMS strain error is reduced from 32% to about 6%. Among the reasons for the difference in error reduction between the point-local strain error and the overall total weighted RMS strain error, small misalignments from the mapping of the DIC data on the FEM grid and local material heterogeneities are worth mentioning. This highlights the importance of A) careful DIC strain mapping and B) using full-field data for FEMU characterization, rather than point-local strain data.

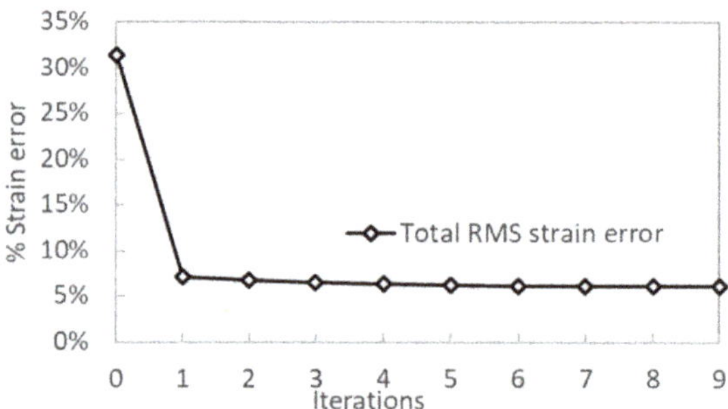

Figure 15. FEMU total weighted RMS strain error for the strain data of the generic unidirectional carbon/epoxy OHC test specimen shown in Figures 12 and 13.

After the FEMU procedure has converged, the material parameters (E_{11}, E_{33}, v_{13}, G_{13}) are used in a 2D plane stress static FE analysis run at failure load to obtain the maximum interlaminar tensile strength at the 12 and 6 o'clock locations, which corresponds to the ILT strength. For fatigue analysis, the maximum amplitude of the fatigue load is applied and the maximum ILT stress is used to generate the fatigue ILT *S–N* curves.

4. Results

Both static and fatigue testing was performed. Static testing was performed at quasi-static strain rates ($\varepsilon < 10^{-3}\mathrm{s}^{-1}$) and fatigue testing was used to develop *S–N* data for R = 0.1 load ratio.

4.1. Elastic Properties and ILT Strength

Static results are given in Tables 1 and 2 for the unidirectional generic carbon/epoxy and fabric materials tested, respectively. Five OHC coupons were tested for each material. Full-field DIC strain data recorded during specimen loading is used in the FEMU procedure to extract material parameters (E_{11}, E_{33}, v_{13}, G_{13}). After convergence, the ILTS is obtained from FE stress analysis at the failure load. Tables 1 and 2 also provide the average material parameters and ILT strength values and the associated normalized coefficient of variation (CV).

Table 1. Elastic constitutive properties and ILT strength for the carbon/epoxy unidirectional tape OHC-ILT specimens.

Specimen	E_{11}, GPa (Msi)	E_{33}, GPa (Msi)	ν_{13}	G_{13}, GPa (Msi)	ILTS S_{33}, MPa (ksi)
US-1	139 (20.2)	7.86 (1.14)	0.295	5.47 (0.794)	99.3 (14.4)
US-2	135 (19.6)	8.00 (1.16)	0.300	5.10 (0.740)	103 (14.9)
US-3	137 (19.9)	8.41 (1.22)	0.345	4.68 (0.679)	90.3 (13.1)
US-4	147 (21.3)	9.17 (1.33)	0.329	4.85 (0.704)	112 (16.2)
US-5	145 (21.0)	8.62 (1.25)	0.320	5.21 (0.756)	102 (14.8)
AVG	140 (20.4)	8.41 (1.22)	0.318	5.06 (0.734)	101 (14.7)
CV	3.6%	6.1%	6.5%	6.1%	7.6%

Table 2. Elastic constitutive properties and ILT strength for the carbon/epoxy fabric OHC-ILT specimens.

Specimen	E_{11}, GPa (Msi)	E_{33}, GPa (Msi)	ν_{13}	G_{13}, GPa (Msi)	ILTS S_{33}, MPa (ksi)
FS-1	62.4 (9.05)	5.93 (0.86)	0.344	4.22 (0.612)	26.2 (4.80) (early failure due to defects)
FS-2	70.3 (10.2)	9.03 (1.31)	0.396	3.82 (0.554)	67.1 (9.72)
FS-3	66.7 (9.68)	9.51 (1.38)	0.378	3.75 (0.544)	56.8 (8.24)
FS-4	72.4 (10.5)	7.79 (1.13)	0.346	3.55 (0.515)	42.3 (6.13)
FS-5	64.7 (9.39)	7.93 (1.15)	0.309	4.30 (0.623)	57.0 (8.27)
AVG	68.5 (9.94)	8.55 (1.24)	0.357	3.85 (0.559)	55.8 (8.09)
CV	5.1%	9.7%	10.7%	8.2%	18.2%

The failure mode for all OHC specimen tested in static was consistent with a 6 o'clock and 12 o'clock delamination failure. The average ILT strength obtained from the carbon/epoxy tape specimens was 101 MPa (14.7 ksi), which was consistent with the interlaminar tensile strength measured by the authors in similar legacy carbon/epoxy materials during other programs. The average ILT strength obtained for the carbon/epoxy fabric material was 55.8 MPa (8.09 ksi). The first fabric specimen tested was an outlier that failed earlier compared to the other four specimens tested. Early failure was attributed to the strong presence of defects.

The average elastic material constitutive properties obtained using the OHC-ILT method are compared with results from standard tests methods for E_{11}, E_{33}, and ν_{13} and from the small-plate twist method [15] for G_{13} in Tables 3 and 4. As shown in Tables 3 and 4, the average OHC-ILT results are very consistent with results obtained independently for the same materials, but from different test methods and test specimen geometries.

Table 3. Comparison of average constitutive properties obtained using the OHC-ILT test with other test methods for the unidirectional tape material.

Material Property	OHC-ILT FEMU Test Method	Other Test Method (Standards + Small Plate Twist)
E_{11}, GPa (Msi)	141 (20.4)	144 (20.9)
E_{33}, GPa (Msi)	8.41 (1.22)	8.76 (1.27)
ν_{13}	0.318	0.329
G_{13}, GPa (Msi)	5.06 (0.734)	5.22 (0.757)

Table 4. Comparison of average constitutive properties obtained using the OHC-ILT test with other test methods for the fabric material.

Material Property	OHC-ILT FEMU Test Method	Other Test Method (Standards + Small Plate Twist)
E_{11}, GPa (Msi)	68.5 (9.94)	74.5 (10.8)
E_{33}, GPa (Msi)	8.55 (1.24)	8.62 (1.25)
G_{13}, GPa (Msi)	3.85 (0.559)	4.17 (0.605)

4.2. ILT Fatigue S–N Curves

The fatigue *S–N* curves obtained using the OHC-ILT method for the carbon/epoxy unidirectional tape and fabric materials considered in this work are shown in Figures 16 and 17. It was noted that due to the existence of initial drill damage in some specimens, premature failure occurred in a limited set of unidirectional tape specimens before reaching the prescribed maximum threshold load. Figures 12 and 13 list the coefficients for a typical power-law fit of the *S–N* data, as well as the R^2 coefficient of correlation.

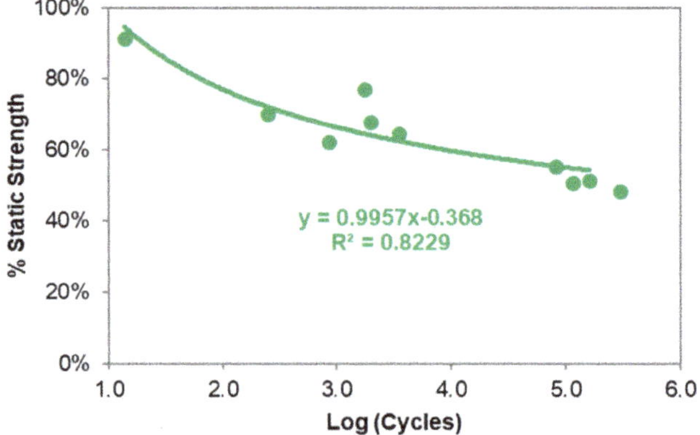

Figure 16. Fabric ILT fatigue. R = 0.1, 4 Hz.

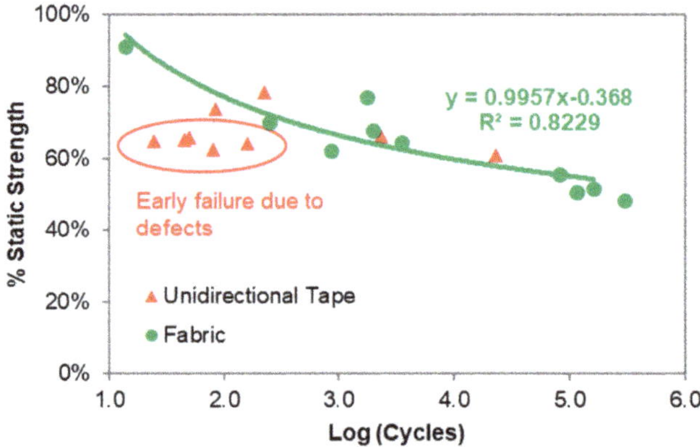

Figure 17. Unidirectional tape and fabric ILT fatigue; R = 0.1, 4 Hz.

5. Discussion

As mentioned previously, hole quality had an impact on failure mode and accuracy of the ILT strength and fatigue measurement, especially in the carbon/epoxy tape specimens tested. It was observed that if the hole is free of drilling damage, OHC specimens exhibit a consistent failure mode that is a crack at 6 and 12 o'clock locations. For example, Figure 18a shows a cross-sectional CT slice of a specimen with acceptable failure mode obtained from non-destructive CT inspection prior to testing. As illustrated, a clean hole free of significant drilling damage is documented. In such specimens, a symmetrical ILT strain distribution and a consistent ILT failure was observed as shown in Figure 18b,c.

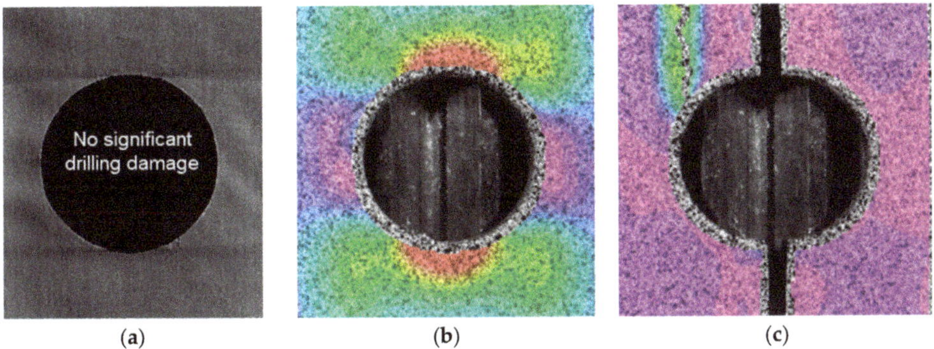

(a)   (b)   (c)

Figure 18. (**a**) ross-sectional CT data prior to testing, (**b**) DIC-measured ILT strain distribution and (**c**) failure mode in a unidirectional tape specimen with acceptable failure mode.

On the other hand, CT data in Figure 19a show a specimen with significant drilling damage detected prior to testing. Figure 19c shows that delamination failure in this specimen initiated at the location of the initial flaw, instead of initiating at the 12 and 6 o'clock location for pristine specimens. In some specimens, surface strain distribution was visibly affected by the drilling damage, as suggested by the asymmetry of ILT strain field shown in Figure 19b.

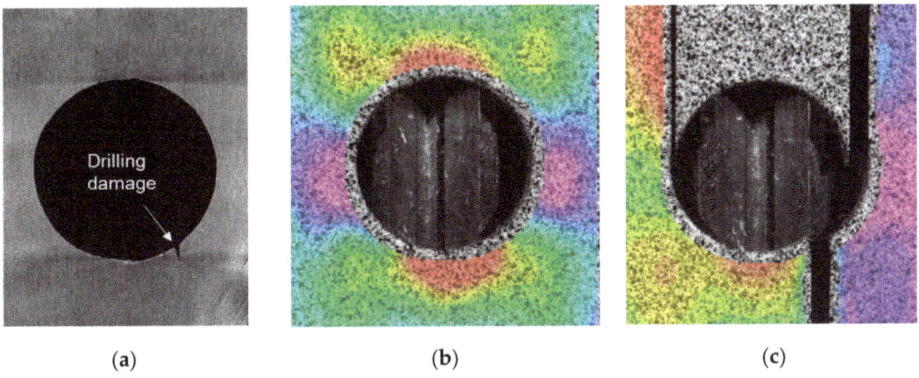

(a)   (b)   (c)

Figure 19. (**a**) Cross-sectional CT data prior to testing, (**b**) DIC-measured ILT strain distribution and (**c**) failure mode in a unidirectional tape specimen with non-acceptable failure mode due to the effects of drilling damage.

The ILT failure mode in the fabric specimens seemed to be less affected by the hole quality, which might be attributed to the micro-structure of the fabric material. ILT failure mode occurred in all fabric OHC coupons despite the drilling damage. Prior to testing, all OHC ILT samples were CT scanned to identify as-manufactured state. Specimens showing significant drilling damage, and inconstant failure modes were discarded. It is worth noting that drilling process can be much improved after this preliminary testing to allow for clean holes.

As indicated in the results listed in Tables 2 and 3, scatter in the extracted ILTS was higher for the fabric material compared to ILTS results for the unidirectional tape material, despite constituent failure mode. In particular, an 18.2% CV in average ILT strength was obtained for the fabric, compared to 7.6% CV tape material. It is worth noting that the sample size for the fabric specimens was four coupons versus five coupons for the tape material, which might contribute to the higher coefficient of variation. The larger scatter might also be attributed to the out-of-plane irregularity inherent to the woven prepreg fabric material. For example, Figure 20 shows a cross-section image of a fabric OHC specimen after delamination failure under interlaminar tension. As shown in Figure 16, delamination failure did not initiate exactly at the 6 o'clock and 12 o'clock location and the locus for failure initiation and delamination path seem to be correlated with the out-of-plane waviness of the fabric material. The FE model used in the FEMU procedure assumes a perfectly homogenized orthotropic material for the fabric material. Material heterogeneity and small variations of failure locus due to random out-of-plane irregularities are therefore likely to be the source of larger scatter in ILTS for the fabric material.

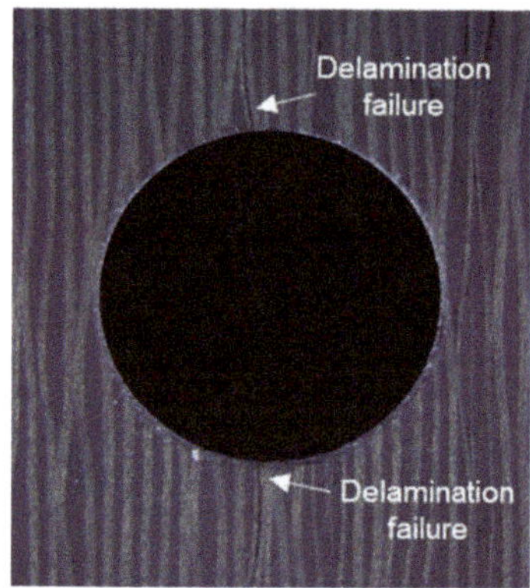

Figure 20. Cross-section image of an OHC ILT fabric specimen showing out-of-plane irregularities and correlation with the location of delamination failure.

6. Conclusions

A new methodology for measurement of interlaminar stress–strain, ILT strength, and ILT fatigue properties of composites based on a unidirectional open-hole compression specimen and DIC-based inverse characterization was proposed and investigated. The method was successfully used for measurement of elastic properties, ILT strength and fatigue, ILT *S–N* curves of a carbon/epoxy unidirectional tape material, and a carbon/epoxy fabric material. It was found that consistent ILT properties can be measured using the OHC-ILT method although significant effect of the hole drilling damage has been found in tape specimens. ILT failure mode occurred in all fabric OHC coupons despite the drilling damage.

In the authors' experience, controlling hole quality in unidirectional OHC specimens is easier than controlling porosity in ASTM D 6415 curved-beam specimens or early tab failure in ASTM D7291 testing for ILT properties that typically result in large scatter and underestimated ILT strength properties. Therefore, the OHC-ILT method is a viable alternative methodology for accurate measurement of ILT properties.

The OHC-ILT test does require FE analysis and prior knowledge of material for calculation of the maximum ILT stress at failure, since a closed-form solution is not available and the stress concentration factor depends on material properties. This work showed that the OHC-ILT test can be combined with DIC measurement of surface deformation and finite element model updating (FEMU) for simultaneous assessment of the elastic material constitutive properties and ILT strength properties. Such self-sufficient methodology was validated using independent measurements of the elastic constitutive properties obtained through other existing test methods. Simultaneous measurement of multiple properties could significantly accelerate material characterization, which currently suffers from the lack of efficient and cost-effective qualification.

Another advantage from simultaneous measurement of strength and elastic properties is that potential variability in constitutive properties between individual coupons is taken into account in an "average" sense and decoupled from other sources of variability in stress measurement, which might lead to more accurate assessment of strength properties. Furthermore, there is a practical interest in using an open-hole test specimen for material characterization, since open-hole laminate coupons are typically used for verification of failure criteria and failure models used in progressive damage and failure analysis. Constitutive properties of composites measured using standard tests might be coupon or test configuration dependent. Such dependency is oftentimes overlooked, which might delay the development and implementation of most advanced failure models for composites that rely on accurate material characterization.

Author Contributions: Conceptualization, A.M.; methodology, A.M. and G.S.; software, G.S.; writing—original draft preparation, G.S.; writing—review and editing, A.M., J.D.S., B.J.; supervision, A.M.; project administration, A.M., J.D.S., B.J.

Funding: This work was part of an effort funded by the Boeing Company as part of the "CALE Project 4: Assessing the Durability and Damage Tolerance of Advanced Composite Structural Features", which was performed under Contract # FA8650-17-C-2700 to the Air Force Research Laboratory (AFRL).

Acknowledgments: The authors are grateful to Dick Holzwarth (AFRL) for his guidance and advice during the course of the work. The authors also thank Brian Shonkwiler and Bastiaan Van Der Vossen at the University of Texas at Arlington for their assistance with specimen testing.

Conflicts of Interest: The authors declare no conflict of interest. The views and conclusions contained in this article should not be interpreted as representing the official policies, either expressed or implied, of the U.S. Government.

Appendix A

Accurate FEM stress analysis is required for calculation of the ILT strength and inverse characterization using the FEMU approach. In particular, the FE mesh used in this study must be refined such as stress concentrations around the hole in the OHC specimens are correctly captured. Appendix A provides results of a mesh sensitivity study demonstrating convergence of the interlaminar normal stress field as the mesh is refined.

An example of unidirectional OHC coupon 2D plane-stress FE mesh and contour plot of the interlaminar normal stress component S_{33} normalized by the far field longitudinal normal compression stress is shown in Figure A1. In this example, orthotropic linear elastic properties for the generic carbon/epoxy tape material considered in this work are used. The mesh is refined as illustrated in Figure A2, where mesh density is defined as a function of the number of elements along the hole radius. Figure A2 shows details of the resulting FE mesh for a circumferential mesh density of 12, 16, 80 and 160 elements. The full mesh shown in Figure A1 corresponds to a circumferential mesh density of 80.

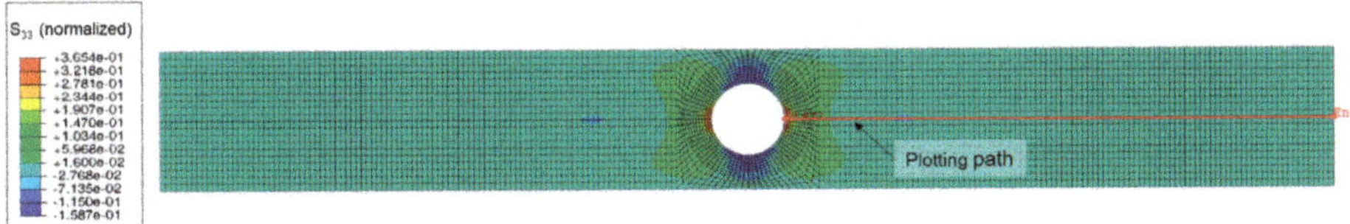

Figure A1. An example of unidirectional OHC coupon 2D FE mesh, contour plot of the normalized interlaminar normal stress component for the generic carbon/epoxy tape material and plotting path for verification of convergence of the stress field with mesh refinements.

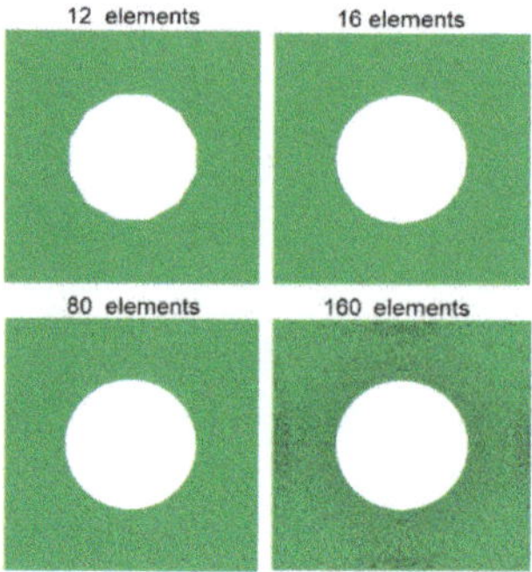

Figure A2. Details of the mesh refinements around the hole in the unidirectional OHC coupon 2D FE mesh for a circumferential mesh density of 12, 16, 80 and 160 elements.

The normalized interlaminar normal stress is plotted along the path illustrated on Figure A1 for verification of stress convergence. Figure A3a shows the stress plots along the selected path for 12, 16, 40, 80 and 240 circumferential element densities. Figure A3b shows convergence of the interlaminar tensile stress concentration factor K_t^{max}, which is equal to the maximum value of the normalized interlaminar tensile stress at the hole boundary, as a function of circumferential mesh density.

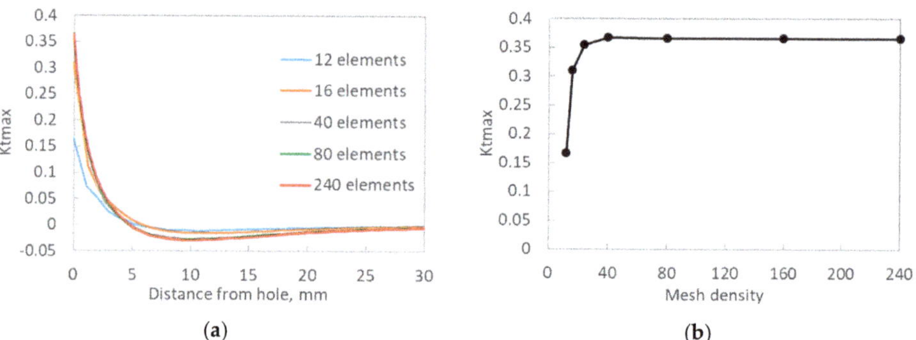

Figure A3. Convergence of (**a**) the normalized interlaminar normal stress distribution along the plotting path shown in Figure A1 and (**b**) the interlaminar tensile stress concentration factor K_t^{max} with mesh refinements.

As shown in Figure A3, convergence of the stress field around the hole is rapidly achieved. In particular, the relative absolute error between K_t^{max} obtained using a circumferential mesh density of 40 and K_t^{max} for a mesh density of 240 is only about 0.4%. For a circumferential mesh density of 80, this error drops below 0.1%.

For all the OHC coupons FE meshes used in this work, a circumferential mesh density of 80 elements was used. A stress field convergence similar to the results shown in Figure A3 was observed using homogenized orthotropic material properties representative of the fabric material considered in this work.

References

1. Clay, S.; Ranatunga, V.; Wilkinson, M.; Knoth, P. Challenges in Experimental Validation of Composite Damage Progression Models. In Proceedings of the American Society for Composites 30th Technical Conference, East Lansing, MI, USA, 28–30 September 2015.
2. Schaefer, J.D.; Werner, B.T.; Daniel, I.M. Progressive Failure Analysis of Multi-Directional Composite Laminates Based on the Strain-Rate-Dependent Northwestern Failure Theory. In *Mechanics of Composite and Multi-functional Materials*; Thakre, P., Singh, R., Slipher, G., Eds.; Springer: Cham, Switzerland, 2018; Volume 6, pp. 197–214.
3. Makeev, A.; Seon, G.; Nikishkov, Y.; Nguyen, D.; Mathews, P.; Robeson, M. Analysis Methods for Improving Confidence in Material Qualification for Laminated Composites. *J. Am. Helicopter Soc.* **2019**, *64*, 1–13. [CrossRef]
4. Rousseau, C.; Engelstad, S.; Clay, S. Data Requirements for Progressive Damage Analysis of Composite Structures. In Proceedings of the 58th AIAA/ASCE/AHS/ASC Structures, Structural Dynamics, and Materials Conference, Grapevine, TX, USA, 9–13 January 2017.
5. ASTM D6415/D6415M. *Standard Test Method for Measuring the Curved Beam Strength of a Fiber-Reinforced Polymer-Matrix Composite*; ASTM International: West Conshohocken, PA, USA, 2013.
6. Jackson, A.; Martin, R. An Interlaminar Tensile Strength Specimen. In *Composite Materials: Testing and Design*; Camponeshi, E.T., Ed.; STP1206-EB; ASTM International: West Conshohocken, PA, USA, 1993; Volume 11, pp. 333–354.
7. Makeev, A.; Seon, G.; Nikishkov, Y.; Lee, E. Methods for Assessment of Interlaminar Tensile Strength of Composite Materials. *J. Compos. Mater.* **2015**, *49*, 783–794. [CrossRef]
8. Seon, G.; Makeev, A.; Nikishkov, Y.; Lee, E. Effects of Defects on Interlaminar Tensile Fatigue Behavior of Carbon/epoxy Composites. *Compos. Sci. Technol.* **2013**, *89*, 194–201. [CrossRef]
9. ASTM D7291/D7291M. *Standard Test Method for Through-Thickness Flatwise Tensile Strength and Elastic Modulus of a Fiber-Reinforced Polymer Matrix Composite Material*; ASTM International: West Conshohocken, PA, USA, 2015.
10. MIL-HDBK-17-1F. *Composite Materials Handbook. Polymer Matrix Composites Guidelines for Characterization of Structural Materials*; U.S. Department of Defense: Virginia, WV, USA, 2002; Volume 1.
11. Avril, S.; Bonnet, M.; Bretelle, A.S.; Grediac, M.; Hild, F.; Ienny, P.; Latourte, F.; Lemosse, D.; Pagano, S.; Pagnacco, E.; et al. Overview of Identification Methods of Mechanical Parameters Based on Full-field Measurements. *Exp. Mech.* **2008**, *48*, 381. [CrossRef]
12. Grediac, M. The Use of Full-field Measurement Methods in Composite Material Characterization: Interest and Limitations. *Compos. Part A* **2004**, *35*, 751–761. [CrossRef]
13. Bruno, L. Mechanical Characterization of Composite Materials by Optical Techniques: A review. *Opt. Lasers Eng.* **2018**, *104*, 192–203. [CrossRef]
14. Makeev, A.; He, Y.; Carpentier, P.; Shonkwiler, B. A Method for Measurement of Multiple Constitutive Properties for Composite Materials. *Compos. Part A* **2012**, *43*, 2199–2210. [CrossRef]
15. Seon, G.; Makeev, A.; Cline, J.; Shonkwiler, B. Assessing 3D Shear Stress–strain Properties of Composites using Digital Image Correlation and Finite Element Analysis Based Optimization. *Compos. Sci. Technol.* **2015**, *117*, 371–378. [CrossRef]
16. Ienny, P.; Caro-Bretelle, A.S.; Pagnacco, E. Identification from Measurements of Mechanical Fields by Finite Element Model Updating Strategies: A Review. *Eur. J. Comput. Mech.* **2009**, *18*, 353–376. [CrossRef]
17. Lecompte, D.; Smits, A.; Sol, H.; Vantomme, J.; Van Hemelrijck, D. Mixed Numerical-experimental Technique for Orthotropic Parameter Identification using Biaxial Tensile Tests on Cruciform Specimens. *Int. J. Solids Struct.* **2007**, *44*, 1643–1656. [CrossRef]
18. Passieux, J.C.; Bugarin, F.; David, C.; Périé, J.N.; Robert, L. Multiscale Displacement Field Measurement using Digital Image Correlation: Application to the Identification of Elastic Properties. *Exp. Mech.* **2015**, *55*, 121–137. [CrossRef]
19. Grédiac, M.; Pierron, F.; Surrel, Y. Novel Procedure for Complete in-plane Composite Characterization using a Single T-shaped Specimen. *Exp. Mech.* **1999**, *39*, 142–149. [CrossRef]
20. Moulart, R.; Avril, S.; Pierron, F. Identification of the Through-thickness Orthotropic Stiffness of Composite Tubes from Full-field Measurements. *Appl. Mech. Mater.* **2005**, *3*, 161–166. [CrossRef]

Appl. Sci. **2019**, *9*, 2647

21. Chalal, H.; Avril, S.; Pierron, F.; Meraghni, F. Experimental Identification of a Nonlinear Model for Composites using the Grid Technique Coupled to the Virtual Fields Method. *Compos. Part A* **2006**, *37*, 315–325. [CrossRef]
22. Grédiac, M.; Pierron, F.; Avril, S.; Toussaint, E. The Virtual Fields Method for Extracting Constitutive Parameters from Full-field Measurements: A Review. *Strain* **2006**, *42*, 233–253. [CrossRef]
23. Rahmani, B.; Villemure, I.; Levesque, M. Regularized Virtual Fields Method for Mechanical Properties Identification of Composite Materials. *Comput. Methods Appl. Mech. Eng.* **2014**, *278*, 543–566. [CrossRef]
24. Kosmann, J.; Völkerink, O.; Schollerer, M.J.; Holzhüter, D.; Hühne, C. Digital Image Correlation Strain Measurement of Thick Adherend Shear Test Specimen Joined with an Epoxy Film Adhesive. *Int. J. Adhes. Adhes.* **2019**, *90*, 32–37. [CrossRef]
25. Sarrado, C.; Turon, A.; Costa, J.; Renart, J. An Experimental Analysis of the Fracture Behavior of Composite Bonded Joints in Terms of Cohesive laws. *Compos. Part A* **2016**, *90*, 234–242. [CrossRef]
26. Rajan, S.; Sutton, M.; Fuerte, R.; Kidane, A. Traction-separation Relationship for Polymer-Modified Bitumen under Mode I Loading: Double Cantilever Beam Experiment with Stereo Digital Image Correlation. *Eng. Fract. Mech.* **2018**, *187*, 404–421. [CrossRef]
27. Perrella, M.; Berardi, V.P.; Cricrì, G. A Novel Methodology for Shear Cohesive Law Identification of Bonded Reinforcements. *Compos. Part B* **2018**, *144*, 126–133. [CrossRef]
28. Sutton, M.A.; Orteu, J.J.; Schreier, H. *Image Correlation for Shape, Motion and Deformation Measurements: Basic Concepts, Theory and Applications*; Springer: New York, NY, USA, 2009.
29. VIC3D-7. Available online: http://www.correlatedsolutions.com (accessed on 28 June 2019).
30. Makeev, A.; He, Y.; Schreier, H. Short-beam Shear Method for Assessment of Stress–Strain Curves for Fibre-reinforced Polymer Matrix Composite Materials. *Strain* **2013**, *49*, 440–450. [CrossRef]
31. Sitzer, M.R.; Stavsky, Y. Stress Concentrations around Holes in Anisotropic Laminated Plates. *Z. Für Angew. Math. Phys. ZAMP* **1982**, *33*, 684–692. [CrossRef]
32. Marquardt, D.W. An Algorithm for Least-squares Estimation of Nonlinear Parameters. *J. Soc. Ind. Appl. Math.* **1963**, *11*, 431–441. [CrossRef]
33. Nielsen, H.B. *Damping Parameter in Marquardt's Method. Informatics and Mathematical Modelling*; Technical University of Denmark, DTU: Copenhagen, Tammy, 1999.

Article

Digital Image Correlation Applications in Composite Automated Manufacturing, Inspection, and Testing

Farjad Shadmehri * and Suong Van Hoa *

Department of Mechanical, Industrial & Aerospace Engineering, Concordia Center for Composites (CONCOM), Concordia University, Research Center for High Performance Polymer and Composite Systems (CREPEC), Montreal, QC H3G 1M8, Canada
* Correspondence: farjad.shadmehri@concordia.ca (F.S.); suong.hoa@concordia.ca (S.V.H.);
  Tel.: +1-514-848-2424 (ext. 7037) (F.S.); +1-514-848-2424 (ext. 3139) (S.V.H.)

Received: 15 March 2019; Accepted: 26 June 2019; Published: 5 July 2019

Abstract: Since its advent in the 1970s, digital image correlation (DIC) applications have been rapidly growing in different engineering fields including composite material testing and analysis. DIC combined with a stereo camera system offers full-field measurements of three-dimensional shapes, deformations (i.e., in-plane and out-of-plane deformations), and surface strains, which are of most interest in many structural testing applications. DIC systems have been used in many conventional structural testing applications in composite structures. However, DIC applications in automated composite manufacturing and inspection are scarce. There are challenges in inspection of a composite ply during automated manufacturing of composites and in measuring transient strain during in-situ manufacturing of thermoplastic composites. This article presents methodologies using DIC techniques to address these challenges. First, a few case studies where DIC was used in composite structural testing are presented, followed by development of new applications for DIC in composite manufacturing and inspection.

Keywords: Digital image correlation (DIC); composite structures; structural testing; experimental mechanics; composite materials; automated composite manufacturing; composite inspection; automated fiber placement (AFP)

---

1. Introduction

First conceived in the early 1970s, digital image correlation (DIC) is a technique that captures images of an object of interest and delivers full-field measurements on the object of interest using image analysis. Formerly, DIC was used for measuring deformations and strains on a planar object subjected mainly to in-plane loadings. This was referred to as two-dimensional (2D) DIC, which to date remains an important technique for 2D deformations and strain measurements in material testing [1–10]. However, nonplanar objects subjected to out-of-plane loading and deformations are mostly unavoidable in practice. In the 1990s, 2D DIC was extended to three-dimensional (3D) measurements through stereovision systems [11–13]. Referred to as 3D DIC, the system consists of two or more cameras to capture digital images of the object of interest from two or more perspectives.

Using a stereoscopic sensor setup, the position of each point in the area of interest is focused on a specific pixel in the camera plane. If the orientation of the sensors with respect to each other (extrinsic parameters) and the magnifications of the lenses and all imaging parameters (intrinsic parameters) are known, the 3D position of any point in the area of interest can be calculated by applying image correlation algorithms. This process determines the shift and/or rotation and distortion of little facet elements in the reference image [14].

DIC offers full-field and noncontact measurements and has seen significant growth in recent years in different fields including aerospace, microscale measurements, bio materials, etc. The focus of

---

this article is on the application of DIC in manufacturing and inspection of composite materials and structures. First, a few case studies where 3D DIC was used in testing composite structures at Concordia Center for Composites (CONCOM) labs are presented in the "Conventional DIC Applications" section. Subsequently, the development of two new DIC applications in manufacturing and inspecting of composites made by an automated fiber placement (AFP) process is discussed in the "Development of New Applications for DIC" section. One of the challenges for in situ manufacturing of thermoplastic composites using AFP is the formation of residual strain and distortion during manufacturing. However, because of the high temperature and pressure involved in in situ manufacturing of thermoplastic composites, strain gages cannot be used to monitor and measure strain and deformation during manufacturing. In Section 3.1, it is discussed how DIC can be employed to overcome this challenge. Another main challenge in automated manufacturing of composites using AFP is the inspection of the ply during manufacturing. Conventional manual inspection techniques are tedious, time consuming, and operator-dependent. In Section 3.2, methodology and preliminary results are presented for the first time that show how DIC can be used for inspection purposes during AFP.

2. Conventional Digital Image Correlation (DIC) Applications

2.1. DIC- and Gage-Measured Strain Comparisons

Two experiments were designed with the main goal of evaluating the accuracy of DIC strain measurements in composite structures. In both experiments, strain gages were used as a reference to evaluate the strain measurements obtained by DIC.

In the first experiment, buckling under an axial compression test on a composite cylinder was performed, and a stereo DIC system (from LaVision Company, Ypsilanti, MI, USA called "StrainMaster") was used to measure strains. The composite cylinder was made of graphite/epoxy prepreg using hand layup and autoclave processes. Preparation of the composite cylinder included applying some random speckles on the cylinder's surface using a spray paint technique (speckle size 3–5 pixels in the area of interest, density about 50%). The test setup is shown in Figure 1. Before the start of the test, calibration of the cameras was performed by moving a standard calibration panel in front of the cameras.

Figure 1. Axial buckling test setup for evaluating DIC [15].

The composite cylinder was loaded through the aluminum end plates at a rate of 1 mm/min, and buckling occurred after 2 min from the start. The axial strain measured by a strain gage installed at the middle of the cylinder and the DIC system before and after buckling occurred is plotted in Figure 2a,b, respectively. As can be seen from Figure 2, before buckling, the measured axial strains by strain gage and the DIC system were in good agreement. However, after buckling occurred (around T = 120 s), the results from two techniques (i.e., strain age vs. DIC) did not agree quantitatively, although they had a similar trend. This could be attributed to the fact that after buckling, due to large local deformation, debonding between the strain gage and composite surface may have occurred. Accuracy for a specific test setup depends on many factors including speckle pattern size and quality, resolution of the camera used, lighting, stereo angle, lens selection, etc. For the setup used in this experiment (5 Megapixel camera system from LaVision, Ypsilanti, MI, USA), the accuracy for measuring the local strain could be expected to be around 200 microstrains [15].

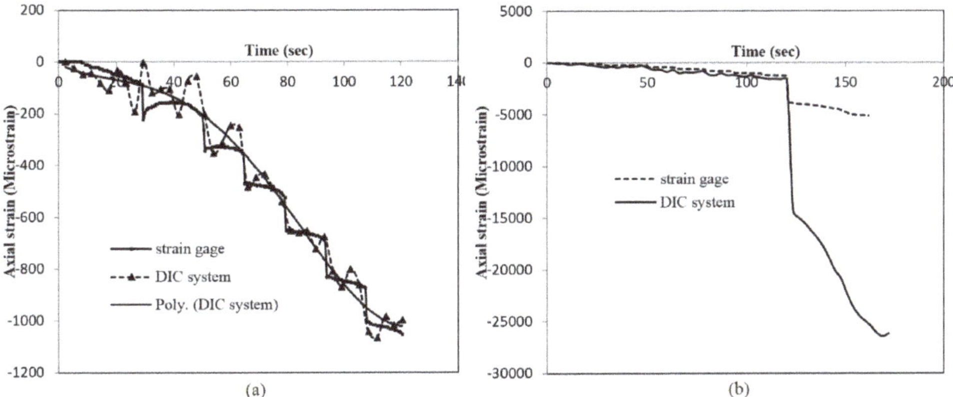

Figure 2. Comparison between strain gage and DIC (**a**) before and (**b**) after buckling [15].

In the second experiment, with the goal of comparing DIC strain measurements with strain gage measurements, an L-shape composite angle was tested under tension using a universal testing machine. The experiment involved large deformations of the composite angle under tension. A unidirectional strain gauge was installed on the outer surface of the composite angle, and a random pattern was created on the composite angle surface using permanent marker (Figure 3).

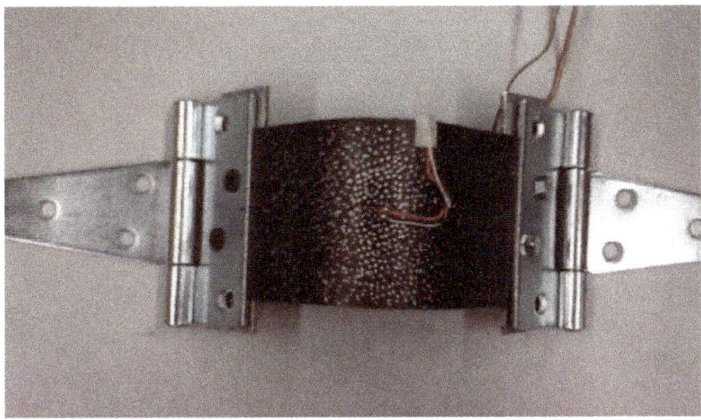

Figure 3. L-shape composite angle with a random pattern and installed strain gage [16].

A stereo DIC system (from Correlated Solutions Company, Irmo, SC, USA called "VIC-3D") was used to measure strains. The strains measured by both strain gage and DIC are compared in Figure 4. As can be seen, DIC results show an overall good agreement with strain gage results. The deviation between DIC and the strain gage measurements was between 50 to 200 s of the test, which may be due to the large out-of-plane deformation of the composite angle as it was flattened. During the large out-of-plane deformation, the motion of the points was perpendicular to the image plane, which may cause error in DIC results. However, after 200 s of the test, as the L-shape angle flattened, an excellent agreement between DIC results and strain gage results was observed [16].

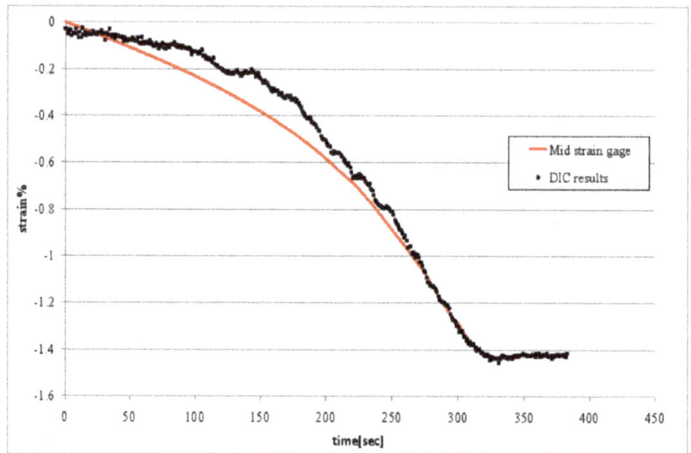

Figure 4. DIC strain measurement versus strain gage results for L-shape composite angle [16].

2.2. Bending and Buckling Analyses of Composite Conical Shells for Helicopter Tail-Boom Applications

2.2.1. Synopsis

In this case study, bucking and bending behaviors of composite conical shells were studied experimentally and theoretically. In the theoretical approach, a first-order shear deformation shell theory was proposed to study buckling and bending behaviors of composite conical shells, and a semi-analytical solution was developed to study buckling under bending of composite conical shells. In the experimental approach, a pure bending test setup was designed and developed to study buckling under bending of composite shells. The setup was equipped with a 3D DIC system to measure three-dimensional deformation and surface strain of the test article. A thermoplastic composite cone, a full-scale section of a helicopter tail-boom, was tested using this setup in pure bending, and the experimental buckling moment was measured, which agreed well with the theoretical one [15].

2.2.2. Experimental Work

Sample manufacturing and preparation: The test article, a composite conical shell, which was a full-scale section of a helicopter tail-boom, was made out of advanced thermoplastic composite material (carbon fiber and poly-ether-ether-ketone (PEEK) with the commercial name of AS4/APC-2 from Cytec Engineered Materials, Anaheim, CA, USA). It was manufactured at National Research Council Canada's facility in Montreal using an automated fiber placement (AFP) machine on a steel mandrel with an internal heating system (Figure 5).

Figure 5. Six-axis automated fiber placement (AFP) machine with a steel mandrel (courtesy of National Research Council Canada) [15].

Preparation of the composite cone for bending tests included applying the end tabs, creating a random pattern (Figure 6a) on the surface of the cone (for strain and deformation measurements using DIC), installing strain gages, and potting the composite cone inside the installation rings (Figure 6c) using low melting temperature point alloy (LMPA) (Figure 6b). The surface of the thermoplastic composite cone was fairly black after manufacturing, and, to have maximum contrast, a white random pattern was applied on the surface of the composite cone using permanent markers by hand. The size of speckles ranged between 3 to 7 pixels with approximately 50% area of interest covered by white speckles. A 3D DIC system from Correlated Solutions, Irmo, SC, USA equipped with two 5 Megapixel cameras (monochrome charge-coupled device (CCD) with 2/3-inch sensor from Point Grey) attached to Schneider 17 mm lenses were used in this experiment, and results were obtained using VIC-3D software.

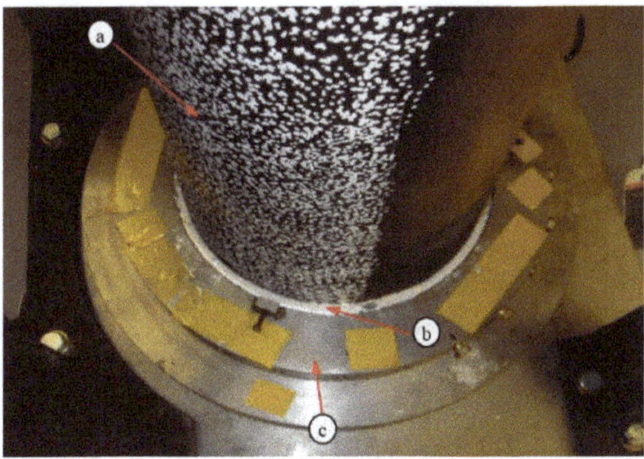

Figure 6. Test article preparation: **(a)** random pattern, **(b)** low melting temperature point alloy (LMPA), and **(c)** installation ring [15].

Test setup: Despite the bending load being the dominant load on conical shells in many applications (e.g., helicopter tail-boom), experimental setups to study bending and buckling behaviors of cylindrical and conical shells under pure bending loads are scarce. In order to study bending and buckling behaviors of conical shells, a pure bending test setup was developed at Concordia Centre for Composites (CONCOM) laboratory (Figure 7). The test setup was equipped with a 3D DIC system that consisted of four cameras to cover the top and front sections of the test article for full-field deformation and strain measurements.

Figure 7. Bending test setup with a composite cone installed for the test [15].

2.2.3. DIC Measurements

The contour plot of the axial strain obtained by the camera pair, looking down at the top surface of the composite cone, just prior to and after buckling is shown in Figure 8a,b, respectively. As can be seen, prior to bucking, while the whole top surface of the cone was under compression, the strain concentration occurred near the small end of the composite cone with the maximum value of 7100 microstrains. This nonuniform strain distribution can be expected because the small end had less stiffness in comparison with the large end of the thermoplastic composite cone. Moreover, from Figure 8a, which shows the strain distribution prior to buckling, one can forecast the possible location of the failure based on the strain accumulation near the small end.

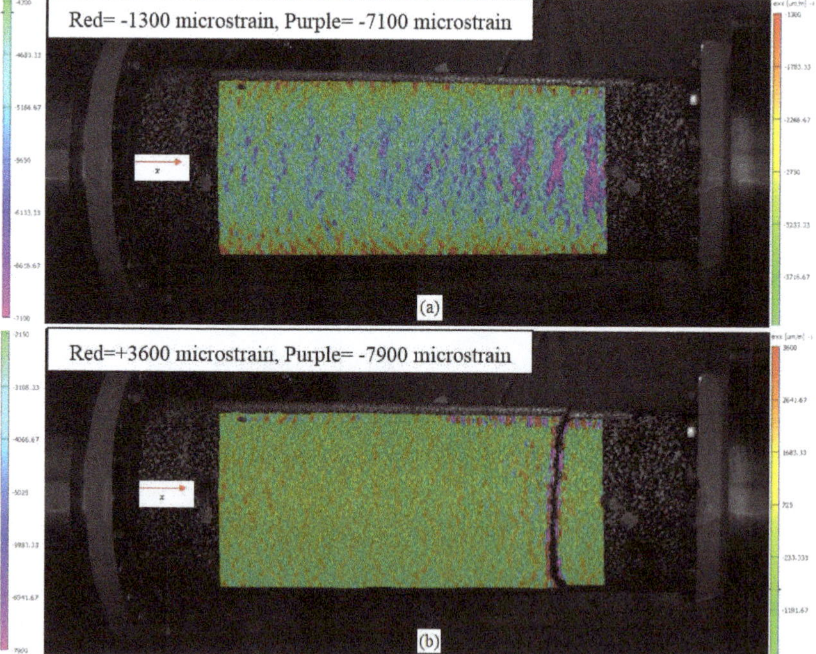

Figure 8. Axial strain (ε_{xx}) obtained by DIC before (**a**) and after (**b**) failure (top view) [15].

The contour plot of the axial strain obtained by the camera pair, looking to the front side of the composite cone, just prior to the buckling is shown in Figure 9. The purple color shows negative strain (compression) at the top, and the red color represents positive strain (tension) at the bottom of the composite cone. Once more, an axial strain concentration can be seen near the small end (left side of Figure 9) where the failure occurred [15].

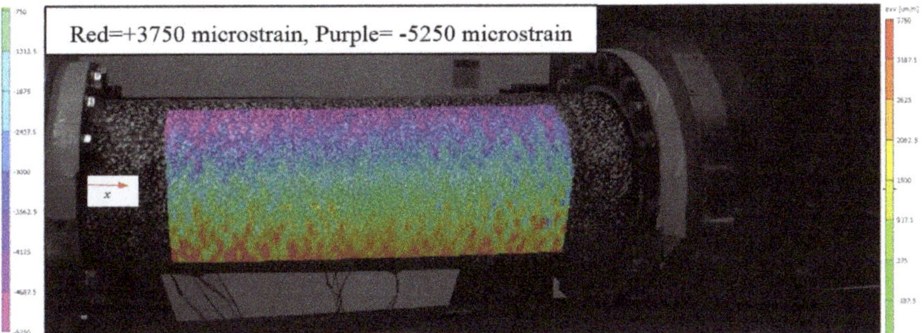

Figure 9. Before axial strain failure (ε_{xx}) obtained by DIC (side view) [15].

2.3. Bending Behavior of Thick-Walled Composite Tubes

2.3.1. Synopsis

In this case study, the bending behavior of thick-walled composite tubes was investigated with focus on the bending stiffness property of tubes. Both theoretical and experimental approaches were considered. The theoretical formulation was developed based on a 3D elasticity theory, and the bending stiffness calculation was validated by pure bending tests performed on thick-walled composite tubes. The pure bending test was shown to be a better alternative test, compared to the conventional three-point and four-point bending tests, to compare the theoretical result with the experiment for validation purposes. This was mainly because the effect of stress concentration at the loading point was less compared to three-point and four-point bending tests, and also the effect of shear was not present in the pure bending test. A 3D DIC system was used during the test of composite tubes to capture deformations and strains of the tubes until failure occurred [16].

2.3.2. Experimental Work

Two thick-walled thermoplastic composite tubes made of Carbon/PEEK material were manufactured using an automated fiber placement (AFP) process and were tested using a homemade pure bending test setup.

Sample manufacturing and preparation: The test articles were made of 0.25-inch wide unidirectional tows supplied by TenCate Advanced Composites Company, Nijverdal, Netherlands (commercial name "Cetex TC1200 PEEK AS4"). The tows were made of a semicrystalline poly-ether-ether-ketone thermoplastic resin with unidirectional carbon fibers. The tows were fed into a robotic-type AFP machine available at CONCOM for manufacturing two thick-walled composite tubes (Figure 10). The first tube was made with a laminate stacking sequence of $[(25)_{45}, (-25)_{45}]$ (i.e., the inner layer and out layer are oriented at 25 and −25 degrees with respect to the tube axis), while the second tube had a laminate stacking sequence of $[25, -25]_{45}$, which means it was made of $[25, -25]$ layers alternatively arranged. Both tubes had the same dimensions (outer diameter = 61.1 mm, inner diameter = 38.1 mm, thickness = 11.5 mm, and length = 1016 mm).

Figure 10. Manufacturing of the thermoplastic composite tube using AFP [16].

For preparation of the tubes for measuring deformations and strains using DIC, the surface of the tubes should have a random pattern with good contrast to be recognized by the cameras. The surface of the thermoplastic tubes were quite black, and, therefore, a white random pattern was applied on the specimens' surface by hand using permanent markers (Figure 11). According to the recommendation of the DIC system's manufacturer (Correlated Solutions, Irmo, SC, USA), the size of the speckles was kept between 3 to 5 pixels with about 50% density. Two 5 Megapixel CCD cameras (from Point Grey with 2/3-inch sensor) attached with Schneider 17 mm lenses were used in this experiment.

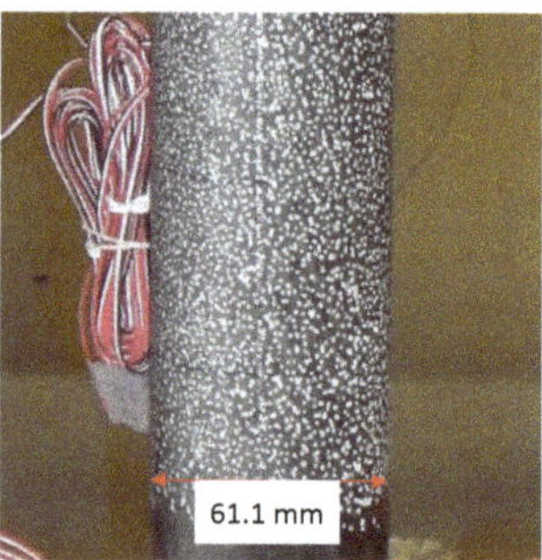

Figure 11. Random pattern on the surface of the thick composite tube [16].

Test setup: The bending test was carried out using a homemade pure bending test setup as shown in Figure 12. The DIC system used for the test consisted of two pairs of cameras; one pair was placed in front of the tube, and the other pair was mounted above the test setup. After calibrating the cameras and connecting all the gages to the data acquisition system, the test was started. End forces applied by two hydraulic cylinders were converted to the bending moment using two moment arms at two ends of the setup. The bending moments were transferred to the tube ends through a set of inner and outer rings filled with low melting temperature point alloy (LMPA) for smooth transition of the load to the tube. In order to prevent tube failure inside the rings, tabs were added at the tube ends.

Figure 12. Bending test setup including the DIC system and thick composite tube (after failure) [16].

2.3.3. DIC Measurements

Deformation results: Axial and hoop deformations captured by the first camera pair located in front of the tube is shown in Figure 13 just before failure. Axial deformation results (Figure 13a) showed that in the upper part of the tube, due to compression, both sides were moved towards each other, while in the lower part, due to tension, they were moved away from each other. This indicates the smooth transition of the load to the test article. The tube deformation in the hoop direction (Figure 13b) revealed that maximum deflection happened at the mid length of the tube, closer to the neutral axis than the lower part. This is an interesting finding and can be explained due to ovalization of the tube cross section.

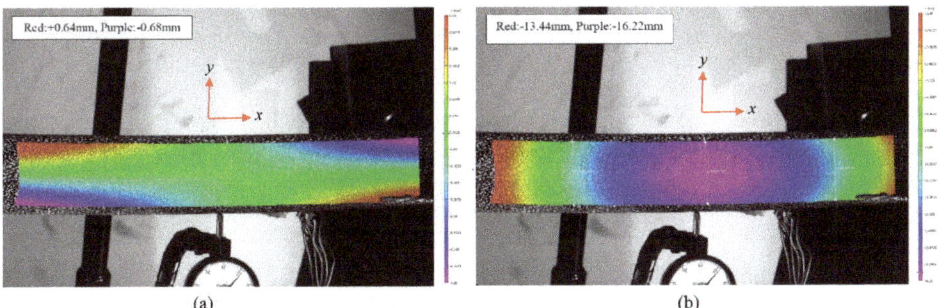

Figure 13. Deformation of the tube from the front view just before failure: (**a**) axial deformation u (mm), (**b**) hoop deformation v (mm) [16].

Axial and out-of-plane deformations captured by the second pair of cameras looking to the top of the tube is shown in Figure 14 just before and after failure. Figure 14a shows that both sides of the tube deformed toward each other, and the axial deformation was zero at the middle of the tube. Maximum out-of-plane deflection before failure occurred at the mid length of the tube as expected (Figure 14c). After failure, the location of the maximum deformation moved below the failure zone (Figure 14d).

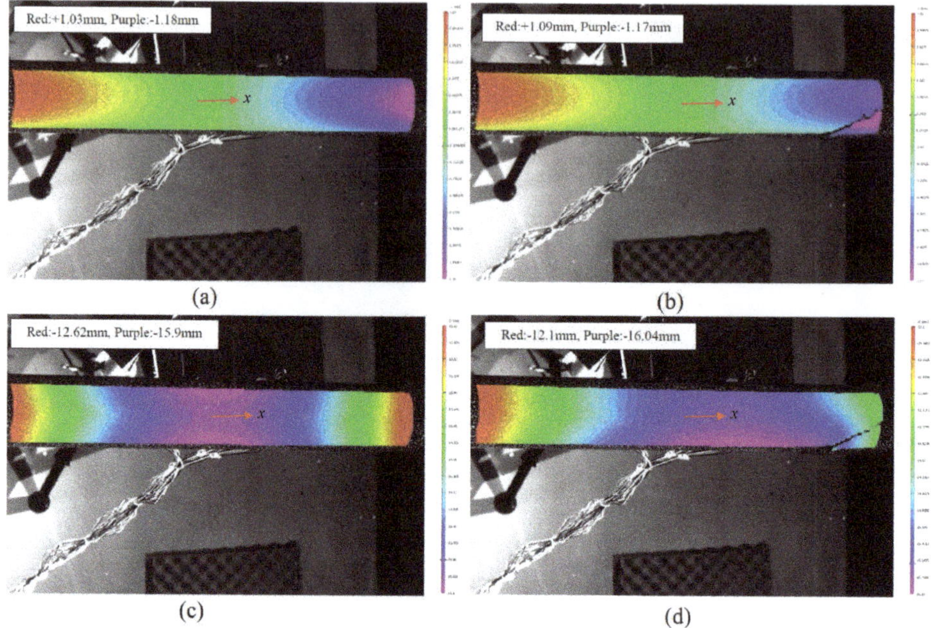

Figure 14. Axial deformation u (mm) just before (**a**) and after (**b**) failure. Out-of-plane deformation w (mm) just before (**c**) and after (**d**) failure [16].

Strain results: Figure 15 shows the contour plot of the axial strain just before and after failure measured by the first camera pair located in front of the tube. While it shows a negative axial strain in the upper part and a positive axial strain in the lower part, the absolute strain values at the top and the bottom side of the tube were nearly the same as expected in a typical pure bending case (Figure 15a). After failure (Figure 15b), the maximum axial strain was accumulated at the failure location on the upper right side of the tube.

Axial strain measured by the second camera pair located above the tube, just before and after failure, is shown in Figure 15. Just before failure (Figure 15c), the maximum compressive strain was measured at the top position in the tube, and one could expect the failure location by observing the accumulation of the axial strain at the right side of the tube (purple color pattern). After failure (Figure 15d), the maximum compressive strain accumulated around the crack that propagated parallel to the fiber directions of the outer layer [16]. The failure mechanism can be explained by a kink band mechanism [17,18]. Kink bands develop under compression stress due to plastic microbuckling induced by nonlinear matrix deformation. It is a complex phenomenon affected by many factors like fiber failure, matrix failure, fiber–matrix interface strength, etc. For the tube under bending (Figure 15c), delamination started under the outer layers (i.e., [25, −25] layer) at the top right side of the tube accompanied with matrix cracking causing the delaminated layers to slide parallel to the fiber direction (i.e., 25 degrees) towards the neutral axis of the tube. This sliding motion removes the constraint on the fibers in the layer below (i.e., −25 degrees) causing microbuckling [16].

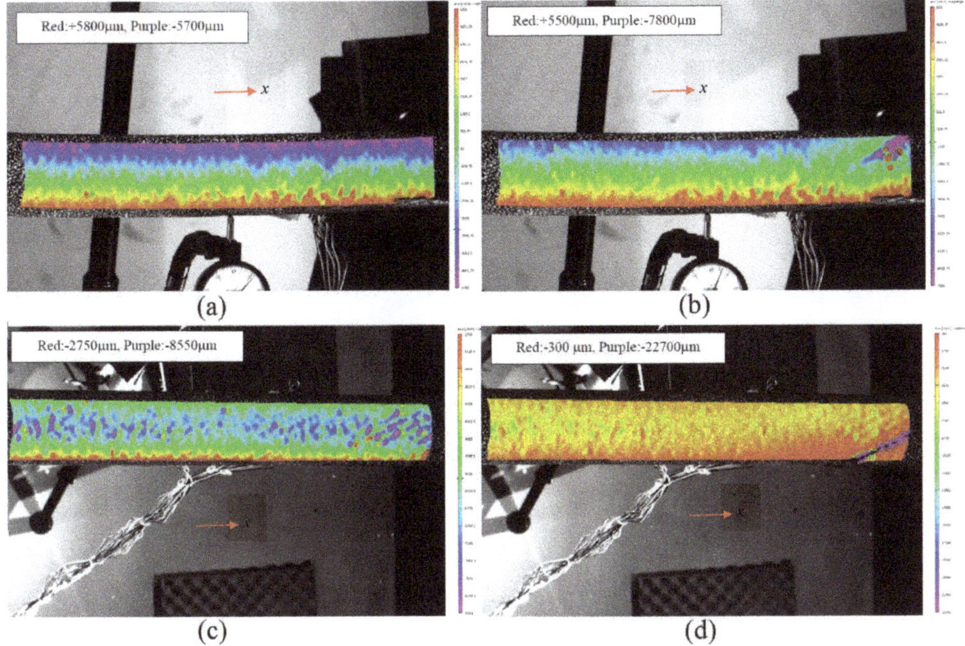

Figure 15. Axial strain (ε_{xx}) side view just before (**a**) and after (**b**) failure. Axial strain (ε_{xx}) top view just before (**c**) and after (**d**) failure [16].

3. Development of New Applications for DIC

3.1. Strain and Deformation Measurements for In-situ Manufacturing of Thermoplastic Composites Using Automated Fiber Placement (AFP)

3.1.1. Synopsis

Advanced thermoplastic composites are of special interest because of their superior properties, such as unlimited shelf life, high-fracture toughness, high-temperature resistance, high-fatigue performance, etc., in comparison to thermoset composites. They have a unique possibility to be manufactured in-situ using the automated fiber placement (AFP) process, which eliminates the need for a secondary process leading to significant savings in cost and energy. In AFP processing of thermoplastic composites, heat and pressure are applied simultaneously to a composite tow to achieve in-situ consolidation of the part. The composite tow consisting of fiber and a matrix is passed by a high-temperature heat source that melts the matrix and is placed by a pressure roller on the tool. High processing temperature and pressure combined with fast layup speed involved in AFP cause transient strain development, which in turn leads to residual stresses in the thermoplastic composites.

Conventional strain gages cannot be used to measure transient strain during AFP of thermoplastic composites because they are processed at a high temperature and pressure. In this case study, 3D DIC was proposed to measure full-field transient strain and deformation of the thermoplastic composite tow during manufacturing [19].

3.1.2. Experimental Work

To employ DIC it is necessary to have a random pattern with correctly sized speckles and good density on the composite tow surface. The main challenge in using DIC for strain measurement in this application was to make a fine speckle pattern that can survive high temperatures and pressures, as the composite tow would pass the heat source and would go under the compaction roller during AFP. Furthermore, due to the small width of the tow (one quarter of an inch), creating a correctly sized random pattern was a challenging task. In order to overcome these challenges, a high-temperature, abrasion-resistant paint with the commercial name of CP4040-S1 of Aremco Products, Valley Cottage, NY, USA was used. The random pattern was generated using a stiff bristle brush and a mesh metal sheet. The paint particles were thrown to the composite tow from a distance of 8 inches using the bristle brush, while a fine metal mesh was used to make sure only small paint particles could reach the tow surface. The size of speckles ranged from 0.1 mm (5 pixels) to 0.4 mm. The pattern density was about 50% (i.e., overall area covered by paint was about 50%). The random pattern was created on a composite tow made of carbon fiber and a poly-ether-ether-ketone (PEEK) matrix (commercial name TenCate Cetex TC1200). After applying the pattern on the tow, the paint was cured in an oven between two aluminum sheets for 45 min (Figure 16).

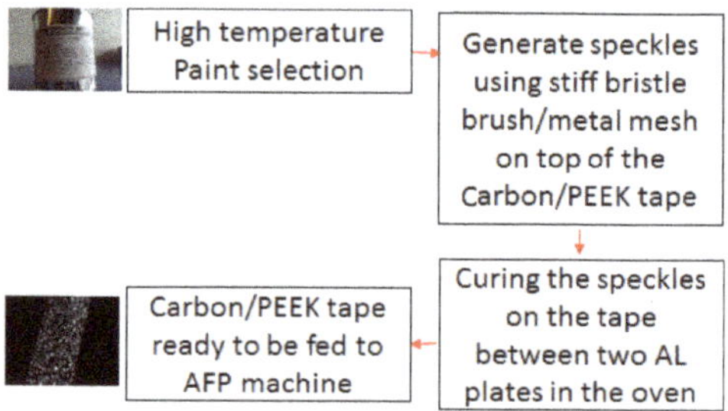

Figure 16. Preparation of carbon/poly-ether-ether-ketone (PEEK) tape with random pattern.

The test setup including the AFP head and the DIC system is shown in Figure 17. The cameras were distanced 8 inches apart and were positioned at about a 40-degree stereo angle. The light source was placed between the cameras to create proper lighting conditions without creating reflections in the pictures. As AFP laid the composite tows, a series of photos were taken at a rate of 10 pictures per second, providing the opportunity to measure transient strain from the initial layup moment until it was cooled down to about room temperature.

Figure 17. Setup for strain measurements during AFP manufacturing: (1) AFP head mounted on a robotic arm, (2) hot gas torch, (3) compaction roller, (4) DIC camera system, and (5) halogen lamp [19].

3.1.3. DIC Measurements

Strain measurement: To measure strain build-up after AFP layup, it was assumed that, at the moment in which the tow passed under the roller and came out, there was no strain along the width and length of the tow. For DIC calculations, VIC-3D software from Correlated Solutions, Irmo, SC, USA was used, and a subset size of 35 pixels (i.e., spatial resolution 35 × 35 of pixels) with a step size (i.e., the spacing of the points that are analyzed during correlation) of 3 pixels was selected. A default value of 0.05 for the pixel confidence margin was set for the matching process using the covariance matrix of the correlation equation in VIC-3D software. Two 5 Megapixel cameras (monochrome CCD with 2/3-inch sensor from Point Grey) attached to Schneider 28 mm lenses were used in this experiment. Figure 18a shows this moment; three virtual strain gages were located right after the roller within a distance of 0.5 inches from one another along the length of the tow. Just after passing the roller, strains started building up, as it is shown in Figure 18b, due to heat dissipation to the environment and the substrate. As one can expect, the highest lateral strain build up was measured at Gage 1 (about 10,000 microstrains) since the temperature drop (ΔT) would be higher than other two gages (i.e., Gage 2 and 3) after 60 s of layup. Fast strain build up can be seen in all three gages during the first 10 s after layup, which emphasizes the significance of the first few seconds just after the AFP head laid down the composite tow. It should be mentioned that in Figure 18b, photos that were taken between 5–7 s after layup were removed from the strain analysis because during this interval the AFP head moved up after finishing the layup, causing tool vibration and consequently creating errors in the photos [19].

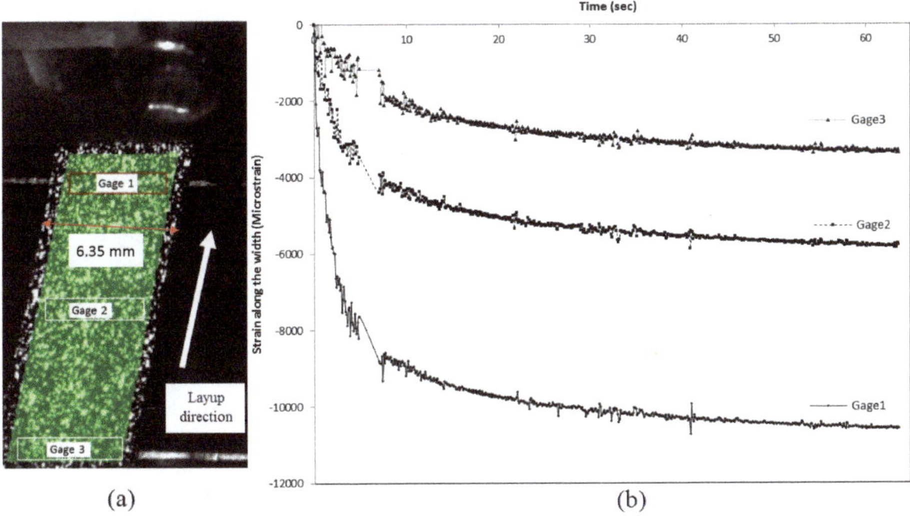

Figure 18. (**a**) Virtual gage locations just after layup. (**b**) Lateral strain variation after layup [19].

Deformation measurement: To measure lateral deformation after AFP layup, the initial condition ($T = 0$ s) was assumed to be at the moment the tow came out under the roller (i.e., all deformations were zero at this time). The distribution of lateral deformation across the width and along the length of the tow is shown in Figure 19a after 60 s of layup. As it can be seen, due to cooling of the composite tow, contraction happened across the width of the tow (i.e., positive deformation (red color) on the left side and negative deformation (purple color) on the right side of the tow). Along the length of the tow, the lateral deformation gradient was higher at the top side of the tow because it was closer to the AFP head, considering the layup direction, and consequently had higher temperature drop out (ΔT). The variation of lateral deformation during the first 60 s after layup across the width of the tow in Section 1 (see Figure 19a) is plotted in Figure 19b. As can be seen, lateral deformation increased rapidly during the first 10 s after layup and stabilized after about 60 s [19].

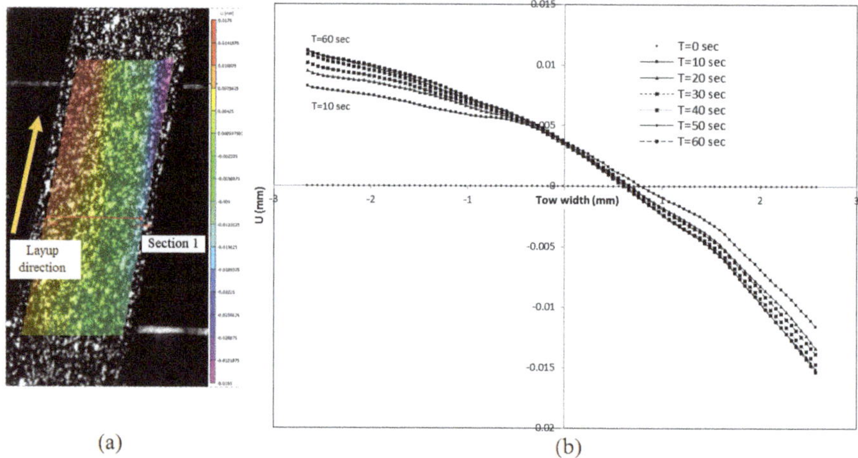

(a)   (b)

Figure 19. (**a**) Distribution of the lateral deformation along the width and length of the composite tow one minute after layup. (**b**) Variation of lateral deformation along the width of the composite tow one minute after layup [19].

3.2. Inspection of the Automated Fiber Placement (AFP) Process

3.2.1. Synopsis

Automated fiber placement (AFP) is a composite manufacturing technique in which narrow composite tows consisting of fiber and matrix are pushed against the tool surface. While the head of the AFP machine is laying the composite tows, heat and pressure are applied simultaneously to consolidate the laminate [20]. AFP is a relatively new manufacturing technique, and development of inspection techniques to assure the quality of the composite part during the layup process is the topic of ongoing research and development activities in the aerospace industry. Possible defects needed to be inspected are gaps and overlaps between tows, deviation in fiber orientation and tow location, etc. [21]. Gaps and overlaps are common defects during AFP, and several studies [22–24] have shown that they reduce structural performance of composite laminates. Figure 20 shows a typical gap between two composite tows. The gaps between tows make resin-rich areas and cause failure initiation points, while overlaps create thickness build-up and cause out-of-plane waviness in adjacent plies and stress concentration. In this study, the possibility of using DIC for inspecting gaps and overlaps was considered, and proof of concept was demonstrated.

Figure 20. Gap size defect.

3.2.2. Proof of Concept

In order to use DIC to measure the three-dimensional shape of composite tows and consequently extract any gaps and/or overlaps between them, there should be enough random textures with contrast on the tows and the substrate layer. However, composite tows made of carbon fiber and epoxy are usually black, and they are laid on a black substrate as well (previous layer), which makes it difficult, if not impossible, for DIC to capture the features of interest (i.e., gaps and overlaps). Applying a random pattern using a marker or spray paint is not a viable option during inspection, as it takes time to apply the pattern and it might introduce unwanted materials (i.e., marker ink) between composite layers. In order to overcome this difficulty, a random pattern was projected by a digital projector to a set of tows laid on a composite substrate. A regular digital projector (i.e., a computer data projector) with full High-definition (1920 × 1080 pixels) capability was placed at the correct distance from the sample under inspection and projected a pattern to the area of interest. Since only the shape measurement was of interest in this application, and not strain measurement, there was no need for a pattern to be actually applied on the surface, and only projection served this purpose. After a regular calibration procedure, stereo DIC was used to evaluate the possibility of detecting the tows' locations and the gap between them.

Figure 21a shows a flat panel consisting of a substrate layer and two tows, in which a random pattern was projected on them. As it can be seen from Figure 21c, DIC was able to detect both tows' locations and the gap between the ends of the tows. However, the shape measurement results were relatively noisy for inspection purposes; the main reason for this was due to the quality of the projected pattern on the composite layer. Since a digital projector was used, not only were the white dots projected, but the black background of the pattern was also projected, which made the pattern look noisy on the composite layer. It was suggested that an optical projector should be used instead of a digital projector to avoid this issue.

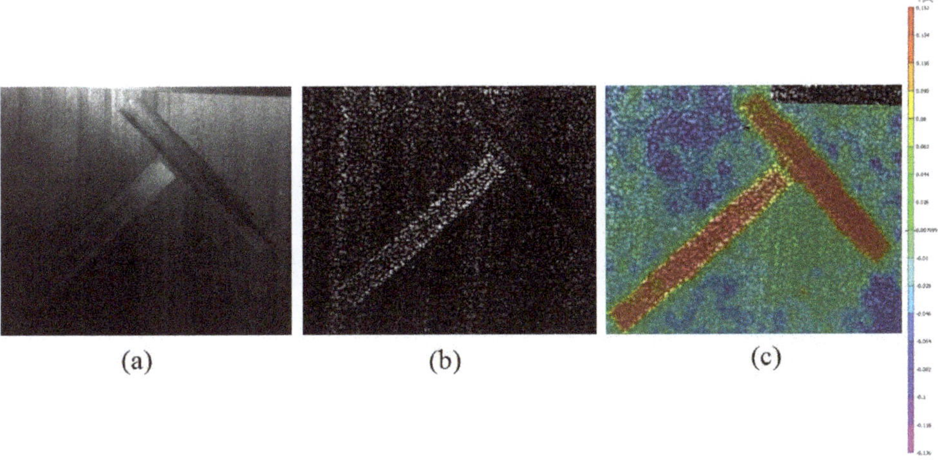

Figure 21. (**a**) Two composite tows on a composite substrate. (**b**) Random pattern projection. (**c**) DIC shape measurement results, z-axis represents thickness direction.

In order to evaluate the potential of DIC to measure gaps and overlaps, assumed to be of good quality and have a random pattern, a spray technique was used to generate white dots randomly on a gap and overlap features between two tows, as shown in Figure 22. For evaluation purposes, two tows were placed in such a way to create a continuously decreasing gap, and the other two tows were placed to simulate continuously decreasing overlap. It was found that, if a good random pattern quality is used, gaps and overlaps as small as about 0.4 mm could be detected.

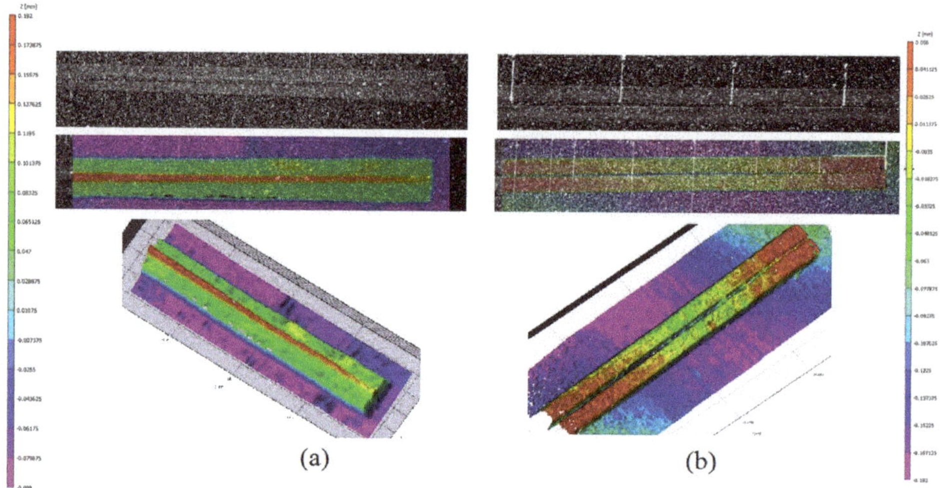

Figure 22. (**a**) Overlap between two tow detections. (**b**) Gap detection, z-axis represents thickness direction.

Furthermore, other types of defects beside gaps and overlap that might occur during AFP can be detected and inspected by DIC. Two examples of such defects are twisted tow and damaged tow. During AFP and when the AFP head places composite tows on the substrate, the placed tows might get twisted, which is referred to as a twisted tow defect, or the fibers might be pulled out by the compaction roller creating a damaged tow defect. Figure 23 shows the capability of DIC to detect such defects.

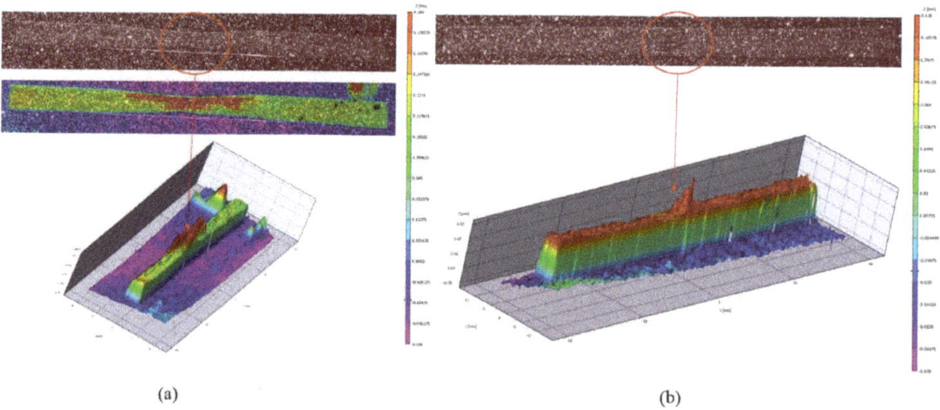

Figure 23. AFP defect detection by DIC: (**a**) twisted tow and (**b**) damaged tow.

4. Discussion

Application of 3D DIC in composite structural testing is presented using a few case studies performed at CONCOM over the years. It was found that the accuracy of the technique in strain measurements of composites was limited to about 200 microstrains; therefore, it may not be an appropriate measurement technique in structural testing in which few strains are expected.

Furthermore, two new applications for DIC in manufacturing and inspecting of thermoplastic composites developed at CONCOM are presented, and proof of concepts are demonstrated. It was demonstrated that 3D DIC can be used for composite ply inspection during manufacturing using AFP. However, an optical random pattern projector needs to be developed for this purpose.

Author Contributions: F.S.: conceptualization, methodology, investigation, analysis, writing—original draft preparation, and editing. S.V.H.: conceptualization, methodology, writing—review and editing, supervision, project administration, and funding acquisition.

Funding: The financial contributions from the Natural Sciences and Engineering Research Council of Canada (NSERC), industrial chair on Automated Composites, Center for Research in Polymers and Composites (CREPEC), Bell Helicopter Textron Canada Ltd., Bombardier Aerospace, and Concordia University are appreciated.

References

1. Anuta, P.E. Spatial registration of multispectral and multitemporal digital imagery using fast fourier transform techniques. *IEEE Trans. Geosci. Electron.* **1970**, *8*, 353–368. [CrossRef]
2. Keating, T.J.; Wolf, P.R.; Scarpace, F.L. An improved method of digital image correlation. *Photogramm. Eng. Remote Sens.* **1975**, *41*, 993–1002.
3. Peters, W.H.; Ranson, W.F. Digital imaging techniques in experimental stress analysis. *Opt. Eng.* **1982**, *21*, 427. [CrossRef]
4. Sutton, M.A.; Wolters, W.J.; Peters, W.H.; Ranson, W.F.; McNeill, S.R. Determination of displacements using an improved digital correlation method. *Image Vision Comput.* **1983**, *1*, 133. [CrossRef]
5. Sutton, M.A.; Cheng, M.; Peters, W.H.; Chao, Y.J.; McNeill, S.R. Application of an optimized digital image correlation method to planar deformation analysis. *Image Vision Comput.* **1986**, *4*, 143–150. [CrossRef]
6. Schreier, H.W.; Braasch, J.; Sutton, M.A. Systematic errors in digital image correlation caused by intensity interpolation. *Opt. Eng.* **2000**, *39*, 2915. [CrossRef]
7. Lu, H.; Cary, P.D. Deformation measurements by digital image correlation: Implementation of a second-order displacement gradient. *Exp. Mech.* **2000**, *40*, 393–400, 2000. [CrossRef]
8. Haddadi, H.; Belhabib, S. Use of rigid-body motion for the investigation and estimation of the measurement errors related to digital image correlation technique. *Opt. Lasers Eng.* **2008**, *46*, 185–196. [CrossRef]
9. Tiwari, V.; Sutton, M.; McNeill, S.R. Assessment of high speed imaging systems for 2D and 3D deformation measurements: Methodology development and validation. *Exp. Mech.* **2007**, *47*, 561–579. [CrossRef]
10. Sutton, M.A.; Yan, J.H.; Tiwari, V.; Schreier, H.W.; Orteu, J.J. The effect of out-of-plane motion on 2D and 3D digital image correlation measurements. *Opt. Lasers Eng.* **2008**, *46*, 746–757. [CrossRef]
11. Luo, P.F.; Chao, Y.J.; Sutton, M.A.; Peter III, W.H. Accurate measurement of three-dimensional deformations in deformable and rigid bodies using computer vision. *Exp. Mech.* **1993**, *33*, 23–132. [CrossRef]
12. Luo, P.F.; Chao, Y.J.; Sutton, M.A. Application of stereo vision to three-dimensional deformation analyses in fracture experiments. *Opt. Eng.* **1994**, *33*, 981. [CrossRef]
13. Helm, J.D.; McNeill, S.R.; Sutton, M.A. Improved three-dimensional image correlation for surface displacement measurement. *Opt. Eng.* **1996**, *35*, 1911. [CrossRef]
14. Herbst, C.; Splitthof, K. *Basics of 3D Digital Image Correlation*; T-Q-400-Basics-3DCORR-002a-EN; Dantec Dynamics GmbH: Ulm, Germany.
15. Shadmehri, F. Buckling of Laminated Composite Conical Shells; Theory and Experiment. Ph.D. Thesis, Concordia University, Montreal, QC, Canada, September 2012.
16. El-Geuchy, M.I. Bending Behavior of Thick-Walled Composite Tubes. Ph.D. Thesis, Concordia University, Montreal, QC, Canada, June 2013.
17. Patel, J.; Peralta, P. Mechanisms for Kink Band Evolution in Polymer Matrix Composites: A Digital Image Correlation and Finite Element Study. In *Proceeding of the ASME 2016 International Mechanical Engineering Congress and Exposition, Phoenix, AZ, USA, November 11–17 2016*; ASME: New York, NY, USA, 2016; Volume 9: Mechanics of Solids, Structures and Fluids; NDE, Diagnosis, and Prognosis, p. V009T12A055. [CrossRef]
18. Nizolek, T.J.; Begley, M.R.; McCabe, R.J.; Avallone, J.T.; Mara, N.A.; Beyerlein, I.J.; Pollock, T.M. Strain fields induced by kink band propagation in cunb nanolaminate composites. *Acta Mater.* **2017**, *133*, 303–315. [CrossRef]
19. Ghayoor, H.; Shadmehri, F.; Hoa, S.V. Development of experimental technique for measuring strain and deformation in manufacturing of thermoplastic composites using automated fiber placement (AFP). In Proceedings of the International SAMPE Technical Conference, Seattle, WA, USA, 2–5 June 2014.

Appl. Sci. **2019**, *9*, 2719

20. Hoa, S.V. *Principles of the Manufacturing of Composite Materials*, 2nd ed.; Destech Publications, Inc.: Lancaster, PA, USA, 2009.
21. Shadmehri, F.; Ioachim, O.; Pahud, O.; Brunel, J.-E.; Landry, A.; Hoa, S.V.; Hojjati, M. Laser-Vision Inspection System for Automated Fiber Placement (AFP) Process. In Proceedings of the 20th International Conference on Composite Materials (ICCM20), Copenhagen, Denmark, 19–24 July 2015.
22. Blom, A.W. Structural performance of fiber-placed, variable-stiffness composite conical and cylindrical shells. Ph.D. Thesis, Delft University of Technology, Delft, The Netherlands, November 2010.
23. Croft, K.; Lessard, L.; Pasini, D.; Hojjati, M.; Chen, J.; Yousefpour, A. Experimental study of the effect of automated fiber placement induced defects on performance of composite laminates. *Composites Part A* **2011**, *42*, 484–491. [CrossRef]
24. Cai, X. Determination of Process Parameters for the Manufacturing of Thermoplastic Composite Cones using Automated Fiber Placement. Master's Thesis, Concordia University, Montreal, QC, Canada, June 2012.

 *applied sciences*

Article

A Method for Calibrating a Digital Image Correlation System for Full-Field Strain Measurements during Large Deformations

Robert Blenkinsopp [1], Jon Roberts [1,*], Andy Harland [1], Paul Sherratt [1], Paul Smith [2] and Tim Lucas [2]

[1] Sports Technology Institute, Loughborough University, Loughborough Park, Loughborough LE11 3TU, UK
[2] Adidas AG, Adi-Dassler-Strasse 1, 91074 Herzogenaurach, Germany
* Correspondence: J.R.Roberts@lboro.ac.uk

Received: 16 May 2019; Accepted: 11 July 2019; Published: 16 July 2019

Abstract: Numerous variables can introduce errors into the measurement chain of a digital image correlation (DIC) system. These can be grouped into two categories: measurement quality and the correlation principle. Although previous studies have attempted to investigate each error source in isolation, there are still no comprehensive, standardized procedures for calibrating DIC systems for full-field strain measurement. The aim of this study, therefore, was to develop an applied experimental method that would enable a DIC practitioner to perform a traceable full-field measurement calibration to evaluate the accuracy of a particular system setup in a real-world environment related to their specific application. A sequence of Speckle Pattern Boards (SPB) that included artificial deformations of the speckle pattern were created, allowing for the calibration of in-plane deformations. Multiple deformation stages (from 10% to 50%) were created and measured using the GOM ARAMIS system; the results were analysed and statistical techniques were used to determine the accuracy. The measured strain was found to be slightly over-estimated (nominally by 0.02%), with a typical measurement error range of 0.34% strain at a 95% confidence interval. Location within the measurement volume did not have a significant effect on error distributions. It was concluded that the methods developed could be used to calibrate a DIC system in-situ for full-field measurements of large deformations. The approach could also be used to benchmark different DIC systems against each other or allow operators to better understand the influence of particular measurement variables on the measurement accuracy.

Keywords: DIC; traceable calibration; accuracy; error

1. Introduction

Digital image correlation (DIC) is one of a number of optical full-field technologies used to measure the shape and deformation characteristics of a wide range of materials. The theory and principles of DIC can be read in various publications [1,2]. Due to the multi-faceted nature of DIC systems, there are numerous variables that can introduce errors into the measurement chain, which can be grouped into two categories: measurement quality (imaging hardware, lighting, etc.) and the correlation principle (algorithm, processing variables, speckle pattern, etc.) [3].

All DIC systems require a *system calibration* to be performed before a measurement can commence and numerous articles have been published outlining the best practice [4–10]. A system calibration enables image points on the camera's CCD to be transformed to the corresponding 3D coordinates of that point and determines imaging parameters such as lens distortion, camera positions, and orientations [11]. The quality of the calibration process is usually reported by means of a calibration score, which is typically based on the difference between the reconstructed point and the extracted point [6]. This score, however, does not indicate the uncertainty in a subsequent strain measurement as it does not take into account correlation errors.

Numerous studies have attempted to investigate each error source in isolation; a study by Haddadi used a combination of experimental and numerical techniques to decouple sources of error, including the environment, lighting, speckle pattern, subset size, grid pitch, translations, and rotations of the sample [3]. Testing was based on the rigid-body motion of an undeformed sample, where measured strain was equated to measurement error as, theoretically, it should have been zero. Strain errors of up to 5×10^{-3} were reported for each source. The method used by Haddadi has the benefit of being simple; however, it does not take into account any errors associated with the distortion of subsets during large deformations.

The focus of this study, however, is to develop an applied experimental method that will enable a DIC practitioner to perform a traceable full-field measurement calibration to evaluate a particular equipment setup in a real-world environment related to their specific application. Even though a number of 2D and 3D DIC systems are commercially available and there has been considerable growth in the use of DIC, the procedures for calibrating DIC systems for full-field strain measurement have lagged behind [12]. Recent initiatives have led to the publication of A Good Practices Guide for Digital Image Correlation [10], which considers the influence of variance errors and bias errors on measurement uncertainty. The guide recommends computing spatial and temporal standard deviations of the quantity of interest from images of a static, undeformed test piece to quantify variance errors. It also suggests a number of options for investigating bias errors by analysing the rigid body motion of the test piece, but it acknowledges that these approaches are not sufficient to fully evaluate all bias errors. This is because the quantification of bias errors requires the true value of a quantity of interest to be known [10].

A *measurement calibration* is used to understand the performance of a measurement system through the comparison of measurements made against a reliable, calibrated source [12]. Calibrations can be used to explore the influence of variables in the measurement process on the accuracy of a measurement. Understanding these influences improves confidence in the reliability of both singular and comparative measurements, as well as supporting the refinement of experimental design. The challenges faced in the creation of experimental calibration methodologies for optical techniques include controlling the uniformity and intensity of strain fields [13], establishing a traceable calibration measurement, and calibrating measurements that are both full-field and dynamic [14].

In recent years, several attempts have been made to address this need through the employment of a traceable 'reference material' or 'material measure' suitable for a range of different optical measurement systems. Calibration is achieved through the comparison of optically measured deformations and theoretical predictions. A number of material measures have been proposed for evaluating both static, in-plane strains [15] and out-of-plane displacements during dynamic loading [16,17]. In all cases, however, a metallic material measure was used, which significantly limits the maximum strains and displacements that can be evaluated. Given the range of applications for which DIC is used, it would be beneficial to be able to evaluate strains associated with much larger deformations.

An alternative approach to creating a physical material measure is to apply artificial deformations to a reference image. One method is to use an image from an experiment as the reference image as this provides a genuine representation of a real speckle pattern. Synthetic patterns have also been created using software packages as this technique allows greater control of speckle characteristics [13,18]. Either the real or synthetic image is then deformed artificially in a known manner, although care needs to be taken to minimize additional errors introduced through the transformation procedure, such as the interpolation technique [19]. This approach has been successfully used to assess errors due to the correlation principle [20], but a physical embodiment of the deformed images would be required to evaluate a complete system. Fazzini et al. [21] attempted to do this by presenting synthetic images on an LCD screen that were then captured by a stereo camera system; however, any errors introduced in the presentation of the images are unknown.

The aim of this paper is to create an experimental method to enable a traceable measurement calibration to be established for a complete DIC system in a real-world environment. The proposed method will enable the calibration of full-field measurements of large deformations. Although surface deformations can occur in three dimensions, due to the difficulties in creating traceable non-planar strain states, this method will focus on calibrating for in-plane deformations only.

2. Materials and Methods

2.1. DIC System

The 3D-DIC system used in this work was the ARAMIS system from GOM (Braunschweig, Germany) employing two Photron (Tokyo, Japan) SA1.1 monochrome high-speed video cameras in a stereo arrangement. The use of a 3D system is recommended, even for planar test pieces undergoing planar deformation, to avoid the introduction of errors due to misalignment of the test piece [10]. The Photron cameras have a 1024 × 1024 pixel resolution, with a pixel dimension of 20 × 20 μm. High-speed cameras were not a necessity for this study, but were used due to availability. Titanar lenses with a fixed focal length of 50 mm were attached to each camera.

2.2. Material Measure Design

The basis of a calibration is to compare a known input with a measured output which, in a DIC application, involves measuring a set of defined deformation states. In order to establish confidence in the values delivered by the calibration methodology, the defined deformations against which measurements are compared must have a known accuracy. This is usually known through a traceability chain; a series of measurement standards that allow the accuracy of a measurement to be traced back, via a hierarchy of calibrations, to a national or international standard [14].

A material measure is "a device intended to reproduce or supply, in a permanent manner, values of a given quantity" [22]; in this case, deformation values. The material measure facilitates a meaningful measurement calibration by being traceable back to a suitable standard level, i.e., the inaccuracy of the material measure is known. The final uncertainty measurement that is obtained through a calibration is the sum of the inaccuracy of the material measure, plus the uncertainty contributions that are a result of the measurement process.

Strain, which is the quantity that will be used to reflect deformation, is derived from a relative change in length and DIC essentially measures length changes to compute strain; length, therefore, is the obvious measurement chain for traceability [12], especially for large deformations.

DIC systems make measurements by tracking a number of surface points created from subsets within the digital images of a surface undergoing a deformation. A non-periodic stochastic pattern, referred to as a 'speckle pattern', is usually applied to the surface of interest to create suitable intensity fields for data point creation and to aid in the unique correspondence of subsets between images. This pattern adheres to and deforms with the surface during deformation and measurements are thus derived from relative movements of the surface pattern.

The synthetic speckle pattern approach was deemed the most appropriate as it enabled full control of speckle characteristics and the pattern could be deformed in a known manner to create multiple stages of deformation. The patterns were printed onto separate rigid boards to produce high-resolution material measures that could be imaged individually by the DIC sensors; the stages of the deformation could be precisely controlled and, more importantly, measured and traced.

2.3. Pattern Creation

To determine the dimensions of the speckle pattern, its features, and the print resolution, the area of the object plane sampled by each pixel on the camera sensor needed to be established. A target field of view of approximately 360 × 360 mm at the centre of the measurement volume was specified, which could be achieved using the optical parameters provided in Table 1.

Table 1. Field of View Summary.

Lens Focal Length (mm)	Sensor Size (mm)	Lens-Object Distance (mm)	Lens-Image Distance (mm)	Angle of View (Degrees)	Field of View (mm)
50	20.48	930	52.84	21.9	360.5

Based on the pixel resolution (1024 × 1024) of the camera's charge-coupled device (CCD) and the target field of view, it was determined that each pixel at the image plane (CCD) would sample approximately 0.35 × 0.35 mm (0.12 mm^2) at the object plane; a plane parallel to the image plane at the centre of the measurement volume. Pattern features were defined to be a minimum of four times this area ($\approx$0.49 mm^2), so as to be oversampled to achieve an accurate measurement of deformations [23]. The print resolution used to print a single feature of 0.12 mm^2 was established to be approximately 72 dots per inch (dpi) (2.84 dpmm), hence a 2 × 2 pixel block, at this resolution, would achieve the pattern feature with an area of 0.49 mm^2.

Speckle patterns were created using a custom Matlab (Mathworks, USA) code that allowed the definition of pattern size, number and size of pattern features, and the greyscale value range of the pixels for each feature created. A pattern was created at a size of 200 × 200 pixels, as shown in Figure 1, which, at 72 dpi, equated to a 'test specimen' of approximately 70.55 × 70.55 mm at the reference stage prior to deformation.

(a)   (b)

Figure 1. Speckle pattern generated using MATLAB. (**a**) Complete pattern and (**b**) magnified view of area highlighted in (**a**).

2.4. Pattern Deformation

To aid in accurate deformation of the pattern, the resolution of the pattern image was increased by a factor of ten using Adobe Photoshop image editing software (Adobe, San Jose, CA, USA). The result was that every original individual pixel comprised a 10 × 10 block of identical smaller pixels, each pixel with dimensions of 0.012 × 0.012 mm ($\approx$0.049 mm^2), requiring a printing resolution of 720 dpi (28.3 dpmm).

Deformation stages of 0%, 10%, 20%, 30%, 40%, and 50% strain were created by deforming the created pattern in the vertical direction using the image editing software. To induce a simulated deformation, the size of the image was increased and then resampled after it had been resized. To create the 10% strain state, for example, the image size was increased in the vertical direction by 10% from 70.56 mm to 77.62 mm and the number of pixels was increased by 200 pixels from 2000 pixels to 2200 pixels.

Resampling was conducted using the 'nearest neighbour' interpolation algorithm within the software, meaning the cumulative effect of deforming each 10×10 pixel block by 10% increased the size of each original speckle feature by exactly one new pixel in the direction of deformation. Therefore, the exact deformation across the entire pattern could be maintained, whilst remaining within the acceptable limits of printing technology.

2.5. Embodiment

The patterns for each deformation stage were incorporated into a speckle pattern board (SPB) design, shown in Figure 2. The SPB included calibration lines around the perimeter of the image, added after 'deformation' in the imaging software, to facilitate the calibration of each board and circular markers for alignment. A complete set of SPBs are included as Supplementary Files. The SPB designs were printed on 100% cotton Hahnemühle fine art paper (308 g/sqm) with a matt finish using an Epson (Suwa, Japan) 11,880 inkjet printer containing UltraChrome Pro inks at 1440 dpi, twice the new pattern resolution, to attain a high print accuracy.

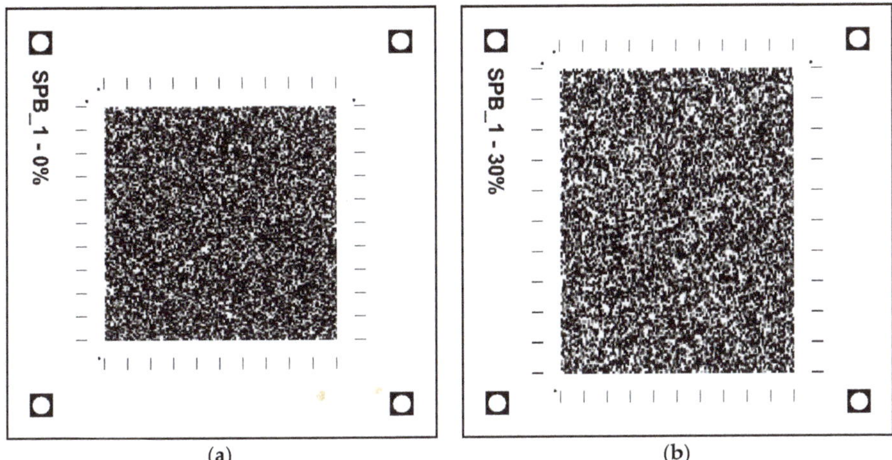

Figure 2. Speckle pattern board (SPB) design for (**a**) 0% and (**b**) 30% strain.

2.6. Material Measure Calibration

The material measure SPB panels for every deformation stage were calibrated through the measurement of the speckle pattern dimensions using a SmartScope Flash 200 multi-sensor optical measuring machine (OMM) (Rochester, NY, USA). The OMM had been calibrated with traceability by the National Institute of Standards and Technology (N.I.S.T), with the length measurement error being a function of the length measured (L), defined as

$$Machine\ Error\ (mm) = \left(0.002 + \frac{6L}{10^6}\right) \tag{1}$$

Ten separate measurements of the total pattern length were made along the axis of deformation for each deformation stage panel and an average pattern length was calculated, shown in Table 2. OMM measurements were corrected for the machine error, calculated using Equation (1) for each stage.

The results showed that each pattern appeared to be printed slightly longer than designed, by approximately 0.1–0.2 mm. It was necessary, therefore, to determine if the error was cumulative across the whole pattern, so that deformation values could be adjusted accordingly for the actual printed lengths.

The standard deviations for the pattern length measurements (Table 2) demonstrated that the metrology system was able to measure the pattern length to a high level of repeatability. However, inevitable bleeding of the printed ink on the paper substrate meant that there was error associated with identifying the edge of the pattern using the OMM. Determining the magnitude of this error would enable it to be accounted for in the calibration measurement.

Table 2. Pattern Length Measurements.

Deformation Stage (% Strain)	Average Measured Length (mm) (±1SD)	Machine Error (mm)	Corrected Average Length (mm)	Designed Length (mm)	Average Error (mm)
0	70.706 (±0.002)	0.002	70.704	70.556	0.148 (±0.002)
10	77.740 (±0.002)	0.002	77.738	77.611	0.127 (±0.002)
20	84.821 (±0.002)	0.003	84.818	84.667	0.151 (±0.002)
30	91.861 (±0.002)	0.003	91.858	91.722	0.136 (±0.002)
40	98.930 (±0.002)	0.003	98.927	98.778	0.149 (±0.002)
50	106.035 (±0.002)	0.003	106.032	105.833	0.199 (±0.002)

Distances between the incremental calibration lines printed around the perimeter of the patterns were measured on each board along the deformation axis. Measurements were made for the ten divisions created by the calibration lines and compared with the theoretical separation calculated as one tenth of the total pattern length. Any edge detection error incorporated as part of the whole pattern length measurement was divided by ten and was thus deemed negligible in the measurements. The errors in these measurements were found to be normally distributed (based on an Anderson–Darling statistical test, $p < 0.05$), with a mean error of 0 mm and a standard deviation of 0.013 mm. Consequently, with a 95% confidence, the maximum error in a length measurement as a result of the edge detection error would be ±0.026 mm which, at the print resolution of 1440 dpi, was the equivalent of approximately one pixel at either end of the pattern. These results also indicated that the error in the length of the printed pattern was evenly distributed across the pattern.

Detailed calibration results for each deformation stage are presented in Table 3. The results show relatively consistent error across all strain states, with mean strain values within 0.1% of the desired strain. This is small relative to the total strain and for applications involving the measurement of these larger strains, the material measures developed can be considered suitable for measurement calibration.

Table 3. Material measure calibration results.

Def. Stage (% Strain)	Measured Length (± Edge Detection) (mm)	Length Minus Edge Detection Error (mm)	Length Plus Edge Detection Error (mm)	Min Material Measure Deformation (% Strain)	Max Material Measure Deformation (% Strain)
0	70.704 (±0.026)	70.678	70.730	0	0
10	77.738 (±0.026)	77.712	77.764	9.87	10.02
20	84.818 (±0.026)	84.792	84.844	19.88	20.04
30	91.858 (±0.026)	91.832	91.884	29.83	30.00
40	98.927 (±0.026)	98.901	98.953	39.83	40.01
50	106.032 (±0.026)	106.006	106.058	49.87	50.06

2.7. System Calibration Test Procedure

The Photron cameras were set up to achieve the field of view outlined in Table 1 and aligned to share a common centre point at a distance of 930 mm from the lens. Two ARRILUX 400 'Pocket Par' (Munich, Germany) spotlights were positioned either side of the cameras to achieve appropriate lighting of the measurement volume. Lighting position and intensity were adjusted to achieve sufficient, uniform image contrast, comparable across both cameras when configured with a lens aperture of f16 and a camera shutter speed of 1/10,000 s. System calibration was completed in line with the manufacturer's guidelines. The SPB panels were mounted on a tripod at the centre of the camera views, as shown in Figure 3a, and sequential images of each deformation stage were captured by changing the SPB. SPBs were only placed in front of lights for short periods of time so as to minimize any effects of humidity and heating of the pattern boards. To determine if the accuracy was affected by the position within the measurement volume, the test was repeated with the SPBs placed at the locations illustrated in Figure 3b.

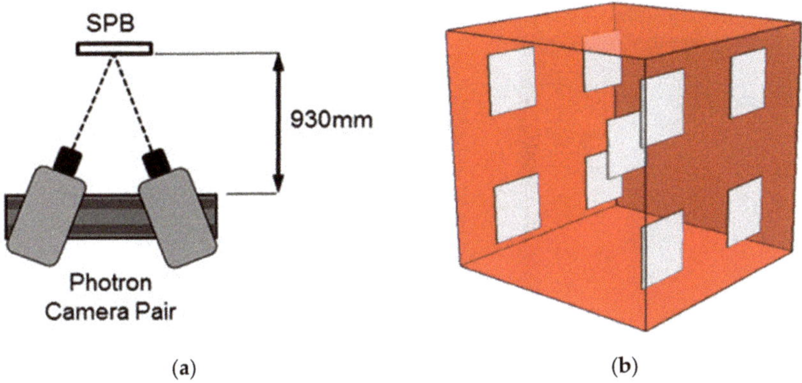

Figure 3. (**a**) Camera set-up and (**b**) measurement locations across the calibrated volume.

The pattern was positioned approximately parallel to the camera image plane; as a 3D system was being used, more precise alignment was not deemed necessary. Recorded images were imported into the ARAMIS software and processed with a square subset size of 20×20 pixels and a computational step size of 13 pixels.

2.8. Data Analysis

DIC is a full-field measurement generating a large number of data points and measurement calibration is thus not simply a task of comparing two values. An evaluation of the measurement accuracy was achieved by considering distributions of measurements and not individual values. Measurement accuracy was defined by the distribution of measurement error: the difference between deformation measurements and the SPB calibration deformation (from Table 3). An Anderson–Darling (A–D) statistical test was employed to test whether the measurement error data for each stage was normally distributed, with the null hypothesis rejected at a significance level of 5%.

A comparison of the error distributions from different measurements allowed the effect of position within the measurement volume on measurement accuracy to be established. A comparison of distributions was conducted using a two-sample Kolmogorov–Smirnov (K–S) test; a non-parametric test that compares the cumulative distributions of two data sets, testing the probability that the two data sets are sampled from the same distribution. A *p*-value is calculated from the maximum difference between the cumulative distributions (K–S statistic) and the sample size. The *p*-value gives the probability that if the two data sets were randomly sampled from the same population, the distributions would be as far apart as observed. The smaller the *p*-value, the more likely data sets are from populations with different distributions; conversely, the larger the *p*-value, the more likely data sets are from populations with the same distributions.

The null hypothesis that the data sets were drawn from populations with the same distribution was rejected for *p*-values equal to or below a 5% level of significance. If the null hypothesis was not rejected, it could then be concluded that the measured distributions have the same underlying distribution, and therefore, position in the volume did not affect the measurement accuracy.

3. Results

An example of a measured strain distribution is illustrated in Figure 4 for the 30% deformation stage. The measurement accuracy results for all deformation stages are presented in Table 4, which indicates that the measured mean was very close to the calibrated mean for each deformation stage, with similar standard deviation values. The *p*-values from the A–D tests (Table 4) show that the null hypothesis was accepted for all deformation stages, meaning the errors measured for each stage could be assumed to be normally distributed.

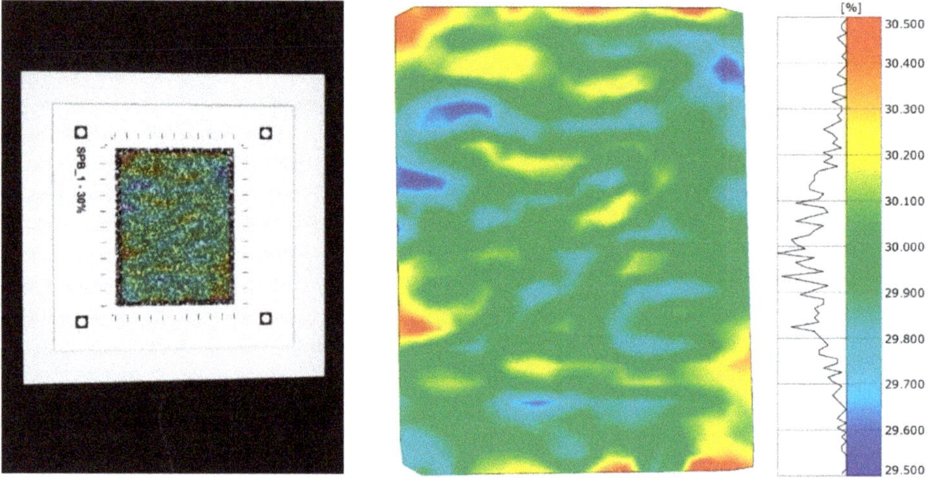

Figure 4. An example strain measurement distribution for the 30% deformation stage with the Speckle Pattern Board (SPB) located in the centre of the measurement volume.

Table 4. Measurement accuracy results calculated for each deformation stage with *p*-value results for the Anderson–Darling statistical test for normality.

Deformation Stage (% Strain)	Measured Mean (±1 SD) (% Strain)	Calibrated Deformation (% Strain)	Mean Measurement Error (±1 SD) (% Strain)	*p*-Value
10	9.96 (±0.11)	9.95	0.01 (±0.11)	0.575
20	20.01 (±0.13)	19.96	0.05 (±0.13)	0.069
30	29.96 (±0.13)	29.92	0.04 (±0.13)	0.096
40	39.93 (±0.15)	39.92	0.01 (±0.15)	0.166
50	49.95 (±0.15)	49.97	−0.02 (±0.15)	0.098

In a normal distribution, 95% of the deformation error measurements at each stage should fall within two standard deviations of the mean. Using this theory, the system measurement accuracy was calculated and the results are shown in Table 5. The results for the measurement accuracy are relatively consistent across all deformation stages, with the mean error being close to zero and within a ±0.5% strain at a 95% confidence interval.

Table 5. Measurement accuracy results for each deformation stage at a 95% confidence level for measured deformations and incorporating edge detection error for calibrated deformations.

Deformation Stage (% Strain)	Measured Mean (±2 SD) (% Strain)	Calibrated Mean (% Strain)	Mean Accuracy (% Strain)
0	0	0	0
10	9.96 (±0.22)	9.95 (±0.08)	0.01 (±0.30)
20	20.01 (±0.26)	19.96 (±0.08)	0.05 (±0.34)
30	29.96 (±0.26)	29.92 (±0.08)	0.04 (±0.34)
40	39.93 (±0.30)	39.92 (±0.08)	0.01 (±0.38)
50	49.95 (±0.30)	49.97 (±0.14)	−0.02 (±0.44)
		Mean Measurement Accuracy:	0.02 (±0.34)

Measurement error distributions from the 50% deformation stage were used to compare measurements from nine positions within the calibrated measurement volume. The statistical analysis (Table 6) shows that for positions at the extremities of the volume, when compared with the deformation occurring at the center, the null hypothesis was not rejected and, consequently, it could be assumed that all measurements were drawn from the same distribution. These results indicate that all measurements made were comparable and, therefore, measurement accuracy was unaffected by position within the volume.

Table 6. *p*-values from two sample Kolmogorov–Smirnov (K–S) tests comparing error distributions at the 50% deformation stage between measurements made at different locations within the measurement volume relative to the centre position measurement.

Position	Mean Error (% Strain)	SD (% Strain)	*p*-Value
Centre	−0.02	0.15	-
Front Top Left	−0.02	0.15	0.541
Front Top Right	0.00	0.14	0.387
Front Bottom Left	−0.03	0.13	0.502
Front Bottom Right	0.00	0.11	0.161
Back Top Left	−0.03	0.12	0.167
Back Top Right	−0.04	0.13	0.146
Back Bottom Left	−0.03	0.14	0.584
Back Bottom Right	0.00	0.132	0.522

4. Discussion

The aim of this paper was to develop an experimental method to enable a traceable measurement calibration to be established for a complete DIC system in a real-world environment. A synthetic speckle pattern, artificially deformed and printed onto speckle pattern boards, was deemed the most suitable approach as speckle characteristics could be controlled and large deformations could be studied.

Optical measurements of the printed boards indicated that the patterns were slightly longer than designed by approximately 0.1–0.2 mm. Errors in identifying the edges of patterns and line features using the OMM were ruled out as the main cause as they were found to be an order of magnitude smaller at ±0.026 mm. The error was found to be evenly distributed across the pattern and relatively consistent over multiple boards and is therefore likely to be related to the printer. By performing a traceable calibration on the actual printed length, this printing error was accounted for in the final DIC system evaluation.

Calibration of the GOM DIC system revealed minimal systematic error. The majority of the measurement error (±0.4% strain) was a result of the distribution of random errors across the full-field measurement. Unfortunately, it is not possible to compare these calibration values to other published work as full-field calibrations of large deformations have not been published to date. However, the measured errors are comparable in magnitude to those determined by Haddadi [3] using the rigid-body motion of an undeformed sample to determine the measurement error. GOM, the manufacturer of the DIC system, quote an accuracy of strain measurement up to 0.01% strain [24], but there is no disclosure of the set-up and parameters that should be used to achieve this value or whether this is a best case scenario.

The acceptability of the calibration achieved is very much dependent upon the application of the DIC system and will differ between applications, but is certainly encouraging for those interested in larger deformations. Improvements in the calibration can be achieved, as approximately one quarter of the total measurement error is a result of the edge detection error introduced as part of the material measure calibration. Reducing this error would improve the calibration results for each deformation stage. This could be achieved through utilizing a higher print resolution or a change in printing substrate to reduce ink bleeding.

A limitation of the proposed methodology is that it only considers planar deformations, whereas in real-world applications, deformations are much more complex. Simplification is a necessary first step for controlling the uniformity and intensity of strain fields and calibrating the material measure. Whilst an ideal material measure would include both in- and out-of-plane deformations, this presents significant challenges. Advances in 3D printing may subsequently enable more complex material measures to be created in the future, but, in the meantime, the proposed approach advances further the recommendations in the latest guide to good practice [10] and will be sufficient for most DIC practitioners wishing to establish a baseline measure of uncertainty.

The advantage of the developed calibration methodology is that it is a simple approach, and the SPBs are cheap and easy to produce and can be adjusted to match the requirements of a particular user or application. The material measure design could be further developed to include more complex planar deformations, for example, the inclusion of strain gradients or multi-axial planar strains and for larger overall deformations. Access to metrology facilities is required to calibrate the material measure, but this is necessary if the calibration is to be traceable.

The approach could be exploited in a number of different ways to understand the performance of 3D-DIC measurement techniques. The most obvious is in the benchmarking and comparison of DIC system performance. Having a traceable material measure would allow particular measurement variables to be isolated and its effect on the measurement accuracy at different deformation stages to be ascertained; for example, lens types or computational parameters. This would support the development of methodologies and set-ups, as well as promote a meaningful analysis of DIC results.

5. Conclusions

A method has been presented that allows the experimental calibration of a 3D-DIC system using a novel, traceable, material measure consisting of a synthetic speckle pattern, artificially deformed and printed onto speckle pattern boards. A traceable calibration was performed on the printed boards to enable any deviations in the printed pattern length from the designed length to be accounted for. Using these material measures, a commercially available DIC system was calibrated for planar deformations at five deformation stages up to s 50% strain and at nine locations within the measurement volume. Measured strain was found to be slightly overestimated, on average, by a nominal value of approximately 0.02% strain, with a typical measurement error range of ±0.34% strain at a 95% confidence interval. Location within the measurement volume was not found to have a significant effect on error distributions. The methodology has the potential to enable DIC practitioners to be able to assess the accuracy of their system in their particular working environment. It could also be applied in future research to enable system benchmarking and comparisons, as well as to further evaluate the performance of DIC measurement systems, particularly for those interested in larger deformations.

Supplementary Materials: Supplementary materials are available online at http://www.mdpi.com/2076-3417/9/14/2828/s1.

Author Contributions: Conceptualization, R.B., J.R., and A.H.; formal analysis, R.B., J.R., and A.H.; funding acquisition, J.R. and A.H.; investigation, R.B.; methodology, R.B., J.R., and A.H.; project administration, J.R., A.H., P.S., and T.L.; supervision, J.R., A.H., P.S., and T.L.; writing—original draft, R.B., J.R., and P.S.

Funding: This research was funded by adidas AG.

Conflicts of Interest: The authors declare no conflicts of interest.

References

1. Sutton, M.; Wolters, W.; Peters, W.; Ranson, W.; McNeill, S. Determination of displacements using an improved digital correlation method. *Image Vis. Comput.* **1983**, *1*, 133–139. [CrossRef]
2. Schreier, H.; Orteu, J.J.; Sutton, M.A. *Image Correlation for Shape, Motion and Deformation Measurements*; Boston, M.A., Ed.; Springer: Berlin/Heidelberg, Germany, 2009. [CrossRef]

3. Haddadi, H.; Belhabib, S. Use of rigid-body motion for the investigation and estimation of the measurement errors related to digital image correlation technique. *Opt. Lasers Eng.* **2008**, *46*, 185–196. [CrossRef]
4. Reu, P. Calibration: Care and Feeding of a Stereo-rig. *Exp. Tech.* **2014**, *38*, 1–2. [CrossRef]
5. Reu, P. Calibration: Sanity Checks. *Exp. Tech.* **2014**, *38*, 1–2. [CrossRef]
6. Reu, P. Calibration: Stereo Calibration. *Exp. Tech.* **2014**, *38*, 1–2. [CrossRef]
7. Reu, P. Calibration: A good calibration image. *Exp. Tech.* **2013**, *37*, 1–3. [CrossRef]
8. Reu, P. Calibration: 2D Calibration. *Exp. Tech.* **2013**, *37*, 1–2. [CrossRef]
9. Reu, P. Calibration: Pre-Calibration Routines. *Exp. Tech.* **2013**, *37*, 1–2. [CrossRef]
10. International Digital Image Correlation Society. *A Good Practices Guide for Digital Image Correlation*; Jones, E.M.C., Iadicola, M.A., Eds.; International Digital Image Correlation Society: Portland, OR, USA, 2018.
11. Becker, T.; Splitthof, K.; Siebert, T.; Kletting, P. Error estimations of 3D digital image correlation measurements. In Proceedings of the SPIE 6341, Speckle06: Speckles, From Grains to Flowers, 63410F, Nimes, France, 15 September 2006; Slangen, P., Cerruti, C., Eds.; [CrossRef]
12. Patterson, E.A.; Hack, E.; Brailly, P.; Burguete, R.L.; Saleem, Q.; Siebert, T.; Tomlinson, R.A.; Whelan, M.P. Calibration and evaluation of optical systems for full-field strain measurement. *Opt. Lasers Eng.* **2007**, *45*, 550–564. [CrossRef]
13. Amiot, F.; Bornert, M.; Doumalin, P.; Dupré, J.-C.; Fazzini, M.; Orteu, J.-J.; Poilâne, C.; Robert, L.; Rotinat, R.; Toussaint, E.; et al. Assessment of Digital Image Correlation Measurement Accuracy in the Ultimate Error Regime: Main Results of a Collaborative Benchmark. *Strain* **2013**, *49*, 483–496. [CrossRef]
14. Hack, E.; Burguete, R.; Siebert, T.; Davighi, A.; Mottershead, J.; Lampeas, G.; Ihle, A.; Patterson, E.A.; Pipino, A. Validation of full-field techniques: discussion of experiences. In Proceedings of the ICEM 14–14th International Conference on Experimental Mechanics, Poitiers, France, 4–9 July 2010; Volume 6, pp. 46004–46007. [CrossRef]
15. Sebastian, C.; Patterson, E.A. Calibration of a Digital Image Correlation System. *Exp. Tech.* **2015**, *39*, 21–29. [CrossRef]
16. Davighi, A.; Burguete, R.L.; Feligiotti, M.; Hack, E.; James, S.; A Patterson, E.; Siebert, T.; Whelan, M.P. The Development of a Reference Material for Calibration of Full-Field Optical Measurement Systems for Dynamic Deformation Measurements. *Appl. Mech. Mater.* **2011**, *70*, 33–38. [CrossRef]
17. Hack, E.; Lin, X.; A Patterson, E.; Sebastian, C.M. A reference material for establishing uncertainties in full-field displacement measurements. *Meas. Sci. Technol.* **2015**, *26*, 075004. [CrossRef]
18. Balcaen, R.; Wittevrongel, L.; Reu, P.L.; Lava, P.; Debruyne, D. Stereo-DIC Calibration and Speckle Image Generator Based on FE Formulations. *Exp. Mech.* **2017**, *57*, 703–718. [CrossRef]
19. Bornert, A.; Doumalin, P.; Dupré, J.-C.; Poilâne, C.; Robert, L.; Toussaint, E.; Wattrisse, B. Short remarks about synthetic image generation in the context of sub-pixel accuracy of Digital Image Correlation. In Proceedings of the ICEM 15–15th International Conference on Experimental Mechanics, Porto, Portugal, 22–27 July 2012.
20. Bornert, M.; Brémand, F.; Doumalin, P.; Dupré, J.-C.; Fazzini, M.; Grédiac, M.; Hild, F.; Mistou, S.; Molimard, J.; Orteu, J.-J.; et al. Assessment of Digital Image Correlation Measurement Errors: Methodology and Results. *Exp. Mech.* **2009**, *49*, 353–370. [CrossRef]
21. Fazzini, M.; Mistou, S.; Dalverny, O. Error assessment in Image Stereo-correlation. In Proceedings of the ICEM 14–14th International Conference on Experimental Mechanics, Poitiers, France, 4–9 July 2010; Volume 6, p. 31009. [CrossRef]
22. Hack, E.; Burguete, R.; Patterson, E.A. Traceability of Optical Techniques for Strain Measurement. *Appl. Mech. Mater.* **2005**, *3–4*, 391–396. [CrossRef]
23. Sutton, M.A.; McNeil, S.R.; Helm, J.D.; Chao, Y.J. Advances in Two-Dimensional and Three-Dimensional Computer Vision. In *Photomechanics*; Rastogi, P.K., Ed.; Springer: Berlin/Heidelberg, Germany, 2000; pp. 323–372.
24. GOM. *ARAMIS User Manual RevA—Software*; GOM: Braunschweig, Germany, 2009.

MDPI
St. Alban-Anlage 66
4052 Basel
Switzerland
Tel. +41 61 683 77 34
Fax +41 61 302 89 18
www.mdpi.com

Applied Sciences Editorial Office
E-mail: applsci@mdpi.com
www.mdpi.com/journal/applsci